Luxury Marketing

Klaus-Peter Wiedmann • Nadine Hennigs (Eds.)

Luxury Marketing

A Challenge for Theory and Practice

Springer Gabler

Editors
Prof. Dr. Klaus-Peter Wiedmann
Director of the Institute of
Marketing and Management
Leibniz University of Hannover, Germany
wiedmann@m2.uni-hannover.de

Dr. Nadine Hennigs
Assistant Professor, Institute of
Marketing and Management
Leibniz University of Hannover, Germany
hennigs@m2.uni-hannover.de

ISBN 978-3-8349-4398-9
ISBN 978-3-8349-4399-6 (eBook)
DOI 10.1007/978-3-8349-4399-6

The Deutsche Nationalbibliothek lists this publication in the Deutsche Nationalbibliografie; detailed bibliographic data are available in the Internet at http://dnb.d-nb.de.

Library of Congress Control Number:

Springer Gabler

Editorial: Barbara Roscher, Jutta Hinrichsen

Printed on acid-free paper

Springer Gabler is a brand of Springer DE. Springer DE is part of Springer Science+Business Media.
www.springer-gabler.de

Foreword

In view of the dynamic growth in the luxury market and the availability of luxury goods to a wider range of consumers, the luxury market has transformed from its traditional conspicuous consumption model to a new experiential luxury sensibility that is marked by a change in how consumers define luxury. In a global context, it is crucial for luxury researchers and marketers to understand why consumers buy luxury, what they believe luxury is, and how their perception of luxury value impacts their buying behavior.

To encourage future research, we are convinced that it is not sufficient to prepare papers for conferences or subsequently for scientific journals. Contributions to the field of luxury marketing also require a platform in the form of a comprehensive book that addresses relevant topics and contributions for a broader range of people from academia and industry. A wider perspective on the field of luxury marketing is useful for both academics and practitioners who wish to better understand consumer behavior; such a perspective might also serve as a basis for successfully creating, marketing and monitoring luxury brands or products in a constantly shifting global economy. This handbook aims to provide a holistic approach to luxury marketing with respect to the characteristics and the key challenges and opportunities of luxury brand management. Therefore, the multifaceted contributions by authors from different parts of the world will offer both a research and management perspective of luxury marketing and deliver a concentrated body of knowledge with contributions from diverse elements. In this way, the book represents a valuable tool for academics in their research and for marketing managers seeking to understand and leverage the potential of luxury brands. Researchers in various disciplines will find the latest research on luxury marketing, which will aid them in their own research endeavors. Additionally, the individual chapters will provide valuable insights to marketing managers working to successfully address the opportunities and challenges in luxury brand management.

To implement the idea of a handbook on luxury marketing, at least two important preconditions are necessary: first, colleagues who are willing and able to contribute, and second, a publisher who is willing to print and distribute such a project. Regarding both prerequisites, we have been extremely fortunate. First, we would like to thank all of the authors of this book for their excellent contributions. We are truly honored to collaborate with such outstanding experts in luxury marketing. Moreover, we wish to thank Barbara Roscher and Jutta Hinrichsen from Springer Gabler Publishing for their support and their great efforts in the publication process. We sincerely thank Christiane Klarmann for her invaluable assistance in all stages of the development of this book. Lastly, we are grateful for the support of Katharina Hotze and Levke Albertsen, who provided their valuable skills and the utmost attention to detail in the final editing of the various chapters.

In closing, given that for us the realization of this handbook was a truly luxurious experience, we hope that all of our readers will enjoy a sense of luxury as well!

Hannover, May 2012 *Klaus-Peter Wiedmann, Nadine Hennigs*

Inhaltsverzeichnis

The Authors

Aaron Ahuvia is a Professor of Marketing at the University of Michigan-Dearborn College of Business. He is widely regarded as the world's leading authority on why consumers actually love certain products; and is also an internationally recognized expert on the relationship between income and happiness.

Gaetano Aiello is a Full Professor of Marketing and the Director of the Management and Business Administration Department at the University of Florence, Italy. He has realized numerous research projects in international retailing, luxury brand management, SME marketing. His published research outputs have appeared in refereed international journals and conference proceedings.

Sascha Alavi works as research and teaching assistant at the Marketing Department of Jan Wieseke at the University of Bochum, Germany. His works are published in the Journal of Service Research and Schmalenbach Business Review.

Glyn Atwal is Associate Professor of Marketing at ESC Dijon-Bourgogne, France. His teaching and research interests include luxury marketing and emerging markets. He is co-editor of The Luxury Market in India: Maharajas to Masses (Palgrave Macmillan). Prior to academia, Glyn worked for Saatchi & Saatchi, Young & Rubicam, and Publicis.

Ilaria Baghi is an assistant professor in marketing at the Department of Communication Sciences and Economics, University of Modena and Reggio Emilia (Italy). Her research interests concern consumer behavior and consumer's decision making. Her papers on these subjects have been published in academic journals and have been presented at international conferences.

Bernardo Balboni is Assistant Professor at the Department of Business Studies, University of Modena and Reggio Emilia. His research interests include business marketing, SMEs' marketing and communication activities, and relational dynamics in consumer and service contexts. On these topics he has published several papers in both international and national journals.

Vincent Bastien is affiliated professor at HEC Paris where he teaches "The Luxury Strategy". He has managed for over 25 years international companies in luxury (Louis Vuitton, YSL, etc.) and in B-to-B (Glass, Print). He graduated from École Polytechnique Paris (MSc), HEC Paris (MBA), and Stanford Business School (SEP).

Stefan Behrens works as scientific assistant at the Institute of Marketing and Management/ Leibniz University of Hannover. His research interest focuses particularly on Luxury and Wine Marketing as well as International Marketing.

Dr Isabella Chaney is a Senior Lecturer in Marketing at Royal Holloway, University of London. Her research interests include luxury marketing, international marketing, wine marketing and in-game advertising. She has presented at a number of international conferences and published in several journals including Asia Pacific Business Review, International Business Review, Marketing Intelligence and Planning, Journal of Interactive Advertising and International Journal of Wine Marketing.

Dr Cheng-Hao Steve Chen is a Senior Lecturer in Marketing, Oxford Brookes University Business School. He earned his PhD and his MSc in Management from Royal Holloway, University of London. Previously he has taught at Royal Holloway (UK), and at Middlesex University (UK). His research interests are in the areas of consumer behaviour and marketing strategies in the online context. His research has been published in the Journal of Targeting, Measurement and Analysis for Marketing and the Journal of Internet Commerce and he has presented at various international conferences.

Dr. Kartik Dave is an Associate Professor in Marketing with BIMTECH, Greater Noida, NCR, India. He has done his doctoral research in the area of marketing of IT products. His research interests lies in services marketing, product and brand management, Luxury, retail management and management education. He has published extensively in these areas in noted journals.

Virginie de Barnier is a Full Chaired Professor of Marketing and the Director of the Marketing Master of Science Programme at the Aix- Marseille University (IAE Aix-en Provence), France. Her research interests are focusing on consumer behaviour, advertising, culture and luxury brand management and have been published in several academic articles and books.

Sara Denize is Deputy Dean in the School of Business at the University of Western Sydney. Her diverse research interests include word of mouth communication and luxury branding in consumer marketing contexts. Dr Denize has also published work in inter-firm contexts where her research has focused on complexity in networks in a number of marketing settings.

Garima Dhamijais Director at Salto Dee Fe – a business enablement firm based in Gurgaon, India. has 15 years of focused experience in the retail space in India across retail strategy, operations, training and process design and implementation. In the corporate, both as an executive and a consultant, she has led teams to deliver retail operations' projects encompassing store launches, VM plan and implementation, merchandise mix strategy, SOPs and training. In the academia , Garima has been associated with various Business schools in India as faculty for their retail programs. She is a post graduate in Economics and Business Management.

Raffaele Donvito (PhD) is Aggregate Professor of Marketing and International Management at the University of Florence, Italy. His research interests include international marketing, brand management, retail marketing, fashion and luxury marketing. His published research outputs have appeared in refereed international journals and conference proceedings.

Wolfgang Fritz is Professor of Marketing and Director of the Institute of Marketing at the Technische Universität Braunschweig, Germany, and Honorary Professor at the University of Vienna, Austria. He is the author of 300 articles and has published 25 books in Europe, USA, and Japan.

Veronica Gabrielli is an assistant professor in marketing at the at the Department of Communication Sciences and Economics, University of Modena and Reggio Emilia (Italy). Her primary research interests are consumer behaviour, branding and marketing communication. She is responsible for a Research Unit on counterfeiting.

Giacomo Gistri is an Assistant Professor of Marketing at the University of Macerata, Italy. His primary research interests are in the field of brand management, consumer behaviour and marketing communications. His papers on these subjects have been published in academic journals such as Marketing Letters and Journal of Brand Management, among others.

Bruno Godey is Associate Professor of Marketing at Rouen Business School. His research interests include consumer behaviour, brand and luxury marketing. He is a member of the Association Française du Marketing (AFM) and the European Marketing Academy (EMAC) and a regular participant at international marketing conferences. Professor Godey is the author of numerous case studies and publishes regular articles on Consumer Behaviour in the Luxury Sector.

Laura Grazzini is an Assistant Researcher at the Department of Management and Business Administration at the University of Florence, Italy.

Wencke Gülow is a research assistant at the Institute of Marketing at the Technische Universität Braunschweig, Germany. She holds a diploma in International Management from the Otto-von-Guericke University in Magdeburg. Her main field of research is Sino-German communication at business level.

Nadine Hennigs is Assistant Professor at the Leibniz University of Hannover, Germany, Institute of Marketing and Management. Her main subjects of research and teaching as well as consulting are: Luxury Marketing, Strategic Management, Brand & Reputation Management, Marketing Research, Consumer Behavior.

Liselot Hudders is a postdoctoral researcher at the department of Communication Sciences at Ghent University, Belgium. Her research focuses on the domains of consumer behavior and marketing communications with a special interest in luxury consumption, the psychology of consumer decision making and consumers' use of persuasion knowledge.

Jutatip Jamsawang is a lecturer of the Department of Business Administration at the University of Vienna, Austria. After obtaining her degree in Economics, she had experience as a market research analyst. Her current research focuses are in the field of consumer behaviour, in-store marketing and pricing.

Jaehee Jung is an associate professor of fashion and apparel studies and a member of International Textile and Apparel Association as well as American Marketing Association. She has done extensive cross-cultural research in consumer behavioral and marketing/branding aspects of fashion and has presented papers through numerous international conferences.

JN Kapferer, Ph.D Kellogg (USA), is a world thought leader on brand management . Co-authoring the reference book "The Luxury Strategy", Professor at HEC Paris, Europe premier luxury research center, he holds the Pernod Ricard Chair on the Management of Prestige Brands and teaches to executives worldwide .

Sabine Kleinsasser graduated in international business administration at the University of Vienna. Besides, being affiliated with the University as research and teaching assistant she finished her doctoral studies on sustainability at the Vienna University for Economics and Business. Currently Ms Kleinsasser is working as senior research specialist for World Vision in Sri Lanka.

Christiane Klarmann is Scientific Research Assistant at the Leibniz University of Hannover, Germany, Institute of Marketing and Management. Her main focus in research and teaching as well as consulting are: Luxury Marketing, Strategic Management, Brand & Reputation Management, Marketing Research, Consumer Behavior.

Koyama Taro is a lecturer in marketing at the College of Business Administration and Information at Chubu University, Japan. His research explores the interconnected issues in the Italian design process and marketing and analyzes brand meanings by applying the Lacanian psychoanalytic theory.

Sascha Langner is Scientific Research Assistant at the Leibniz University of Hannover, Germany, Institute of Marketing and Management. His main subjects of research and teaching as well as consulting are: Neuroeconomics, Word of Mouth Marketing and Consumer Behavior.

Seung-Hee Lee got her B.S. from Ewha Womans University, Seoul, Korea and her M.S. and Ph.D. from the Ohio State University, USA. Her research interests include brand marketing and fashion consumer behaviour. Now she is working as Associate Professor at Dept. of Clothing & Textiles in Ewha Womans University, Seoul, Korea.

Michael Mauer is the owner and CEO of several companies in the luxury segment. He is a professional speaker for luxury marketing and works as a strategic advisor at the Marketing Department of Jan Wieseke at the University of Bochum, Germany.

Kenneth Miller is Professor Emeritus in the UTS Business School at the University of Technology, Sydney, Australia. Professor Miller has had many articles published in leading marketing journals including the Journal of Marketing Research, the Journal of Retailing, the Journal of Consumer Research, Journal of Travel & Tourism Marketing, the Journal of Business and Industrial Marketing, the Journal of Travel Research, Psychology and Marketing, and the Journal of Business Research. He has many years of marketing consulting and teaching experience.

Agnes Neulinger is an Associate Professor of the Institute of Marketing and Media at the Corvinus University of Budapest. She has business experiences related to marketing research and she is author and co-author of several academic publications. Her research interests relate to consumer behavior with a focus on household and family consumption.

Dr. Hyunjoo Oh is the Research Director of the David F. Miller Center for Retailing Education and Research in the Warrington College of Business Administration at the University of Florida, USA. She is specialized in retail consulting. Her research focuses on aesthetic design and its meaning in product/store choice, luxury brand cultural consumption, and interactivity and visual bundling in E-commerce.

Stefano Pace is Assistant Professor at the Department of Marketing of Bocconi University (Milan), where he is the director of MiMeC (Master in Marketing & Communication). His publications include articles in international peer-reviewed journals such as Marketing Letters, International Marketing Review, European Journal of Marketing, Group Decision and Negotiation.

Mario Pandelaere is Associate Professor in Marketing at the University of Ghent, Belgium. His research interests include the relation between materialism, (luxury) consumption and well-being, biases in quantitative decision making, and strategies for interpersonal influence. He has published in JCR, JCP, IJRM, JPPM, JESP, Journal of Advertising, and Journal of Happiness Studies.

Daniele Pederzoli is Full Professor and Chair of the Marketing Department at Rouen Business School. He has many years of experience as a management consultant. He has realized research projects in international retailing and luxury marketing. He has authored books, academic articles in international journals and more than 100 professional articles.

Silvia Ranfagni (PhD) is Aggregate Professor of Marketing at the University of Florence, Italy. Her research interests include innovation, internationalization and brand management with special reference to the fashion and cultural industry. Her published research outputs have appeared in refereed international journals and conference proceedings.

Carmen Rodríguez Santos is a Professor of Marketing at the University of León, and Professor Visitor in different Institutions abroad. Her papers are present in important academic publications. She has a deep experience directing International and National projects. Her background in the consultancy area is also recognized, having worked for different firms. Her main areas of analysis are: Consumer Behaviour, Brand Strategy and Market Research.

Simona Romani is an associate professor in consumer behavior at Luiss Roma (Italy). Her primary research interests are consumer behavior and branding. Her work has been published in several important international academic journals.

Dr. Steffen Schmidt is Research Assistant Professor at the Institute of Marketing & Management/Leibniz University of Hannover with research interest focus on Neuroeconomics in particular.

Silvia Grappi is an assistant professor in marketing at the Department of Business Studies, University of Modena and Reggio Emilia (Italy). Her primary research interests are consumer behaviour and branding. Her papers on these subjects have been published in academic journals and have been presented at international conferences.

Dr Nicole Stegemann is a Lecturer in the School of Business at University of Western Sydney. Concurrently, Nicole consults in brand strategy management and marketing for various industries. Her research area includes consumer behaviour, luxury brand management, and teaching & learning. Nicole has published in international economics, marketing and education journals.

Suzane Strehlau is Marketing professor at Universidade Nove de Julho – UNINOVE, Sao Paulo, Brazil. She obtained her PhD in Management from EAESP-FGV (Brazil) and studied also at HEC (France), WHU (Germany).Her publications are mainly focused on the area of brand management and consumer behaviour.

Doc. Ing. Janka Táborecká-Petrovičová, PhD. is assistant professor and deputy head of the Department of Corporate Economics and Management, Faculty of Economics, Matej Bel University in Banská Bystrica, Slovakia, author of various articles and scientific monographs, former leader of the module Managing Strategic Marketing within MBA study in co-operation with Nottingham Trent University and guarantee of the co-operation between Faculty of Economics and Association of Public Pharmacies. Her scientific interests represent consumer behaviour, strategic marketing and relationship marketing issues.

Pierre Valette-Florence is Full Professor of Marketing and Quantitative Methods at IAE de Grenoble, Grenoble University, France. He has many years of experience as a brand management and marketing research consultant for leading international companies and has published numerous articles in international refereed marketing journals and conferences.

Udo Wagner is chaired Professor of Marketing at the University of Vienna, Austria. He received academic degrees in Management, Operational Research, Mathematics, and Econometrics. He held positions at various national and international universities. His main stream of research concentrates on marketing research and consumer behaviour.

Klaus-Peter Wiedmann is a Full Chaired Professor of Marketing and Management and the Director of the Institute of Marketing and Management at the Leibniz University of Hannover, Germany. He has many years of experience as a management consultant and top management coach and has realized numerous research projects with over 500 academic publications.

Jan Wieseke is a Full Chaired Professor of Marketing and Management and Dean of Research at the Faculty of Business and Economics at the University of Bochum, Germany. His works are published in the leading international marketing journals such as the Journal of Marketing and the Journal of Marketing Research.

Alistair Williams is an Associate Professor at Johnson & Wales University, Charlotte, North Carolina, USA. His research and teaching interests lie in the staging of hospitality marketing experiences and the marketing of hedonistic and luxury hospitality products. Professor Williams has published widely within the field of hospitality marketing.

Meng-Shan Sharon Wu is a doctoral researcher at School of Management, Royal Holloway University (UK). Her research interests are in the areas of luxury marketing and international luxury consumer behaviour. Her work has been presented at various national and international conferences.

Dr. Thomas Wüstefeld is Research Assistant at the Institute of Marketing & Management/Leibniz University of Hannover with research focus on brand management in particular.

Part 1

Luxury Marketing as a Challenge for Marketing Theory and Practice

1 Placing Luxury Marketing on the Research Agenda Not Only for the Sake of Luxury – An Introduction

Klaus-Peter Wiedmann, Nadine Hennigs

1.1 The Reason Why: What is the Relevance of Luxury and Luxury Marketing?

Certainly, luxury might offer a thrilling experience from the perspective of individual researchers, and yes, to a certain extent, we also share this type of motivation for the research topic. However, more serious motives underlie the effort to place luxury and luxury marketing on the research agenda.

1.1.1 The Relevance of Luxury and Luxury Marketing from an Economic Perspective

First, luxury is an enormous and ever-increasing industry. The growth of luxury relates to the societal trend that the *"rich get richer, and the poor get poorer"*. The gaps between the rich and the poor tend to be sharpened, and we increasingly have some type of impasse for the middle class a type of a *"stuck in the middle"*. People in the middle class who are still managing well increasingly attempt to signal that they are among those who are successful. Certainly, we also observe countervailing trends regarding people's general consumption orientation, especially regarding those in the middle class. Several of these aspects will be addressed below. However, we especially must consider that such countervailing trends are much stronger in western industrialized countries than in economically developing societies, especially in Asia. Additionally, regarding consumption in general and luxury consumption in particular, there is in those developing countries a) a high backlog demand and b) growing wealth, due to global societal changes in economic power and prosperity from the old industrial leaders to the new ones. Moreover, luxury consumption is far more supported by culture in those countries. In sum, the societal impacts on empowering the trend to luxury consumption are strong, and the market is globally increasing at a rapid pace.

From an economic perspective, placing luxury marketing on the research agenda seems to be more than feasible. However, one might argue that from an economic point of view, focusing on the challenge of poverty seems to be more important than discussing luxury. Undoubtedly, poverty is much greater and, unfortunately, growing faster than luxury markets not only in third world countries but in formerly leading western industrialized countries, such as the US, Germany (see the German poverty report). Particularly in Europe, the aftermath of the economic crisis will endure for a long period. In such a situation, would it not be even more reasonable for socially responsible researchers and practitioners to address the challenge of poverty appropriately?

1.1.2 Luxury Marketing as an Important Chance for Western Industrialized Countries in the Context of Growing International Competition

Apart from the clear necessity of placing the topic of poverty on the agenda, we deeply believe that especially from a European perspective, it is worth beginning on the high end of possible economic developments. First, this priority relates to international competition and its past, present, and future development. European or, more generally, western countries tend to become caught in the role of victims in the context of an international competitive continuum along which economically less developed countries attempt to compete with those countries that are on a higher level. Countries located at the low end of this continuum have difficulty finding countries with which to compete. If *"Western Industrialized Countries"* (WICs) attempt to react by incorporating the same strategies as the "hunting" countries, e.g., with low prizes, there is certainly not much opportunity to win the race, particularly because, due to their societal conditions, the competing countries satisfy the prerequisites for such strategies better than WICs. To a greater or lesser extent, it is much more promising to move forward to higher levels of quality and innovation. Overall, we might discuss *"innovation-focused high-end strategies"* to characterize the basic direction of strategies that might be successful in the context of international competition. Within such a strategic orientation, different options certainly exist. For instance, based on natural science and technology, one option might be a *"high-tech strategy"* (e.g., the approach of some German companies to leverage the reputation of *"German engineering"*). In addition to engineering and technology, there are also other fields for obtaining high-end strategies, for instance, in *"art and culture"* (high-art-and-culture strategy). Hence, a high-end strategy could also follow ideas, such as high-end service, customer vicinity, producer propinquity (e.g., *"local brands"*), emotional bonding. We might discuss "high-touch strategies" as an umbrella term, and certainly, the different strategic directions might be combined in a *"high-tech/high-touch strategy"*, *"high-tech/high-art-and-culture strategy"* or even a *"high-tech/high-art-and-culture/high-touch strategy"*. Overall, the question is what type of background is used to establish experiences distinguished from *"everyday"* experiences via outstanding, highly innovative, rare qualities.

On the one hand, a *"luxury marketing strategy"* might be regarded as the peak of high-end strategies. On the other hand, such a strategy might be differentiated by special characteristics, such as long-lasting value, superfluity, and highly impressiveness to others. To a great extent, this handbook offers insights for a better understanding of this luxury phenomenon either via its definition from a marketing perspective (e.g., highlighting the *"highest brand prizing power"* or similarly, Bastien & Kapferer) or via the attempt to understand and measure perceived luxury value from a customer perspective (e.g., De Barnier & Valette-Florence). In the latter research context, we might detect that luxury does not always refer to highly prized high-quality products. For instance, for busy managers, having time to sit on a bench for lunch during a busy workday might represent an extreme form of luxury, whereas for job-seeking and homeless people, such activity is part of a sad daily routine and is far from a luxury. In such a situation, being taken seriously and treated kindly by those who fare better will attain luxury status. Thus, luxury is to a great extent a highly subjective phenomenon. However, for luxury marketing, marketers always must address the question of the highest end of quality as perceived by their target group, either by relying on strategies of *"high tech"*, *"high art and culture"* or *"high touch"*.

1.1.3 The Luxury Marketing Strategy as a Success Factor in Different Industries

Discussing different target groups and various luxury perceptions, we might also begin to understand that a luxury marketing strategy is not present only when companies attempt to sell high-end cars, villas or yachts to the upper class. Furthermore, instead of only delivering *"more value for less money"*, as conventional marketing literature demands, it is worthwhile to leverage the idea of luxury when approaching "normal" people attempting to obtain *"more money for value"*. Within our understanding of marketing, customer orientation must be balanced with a beneficial "corporate orientation" following a resource-based view. Indeed, this viewpoint is important not only from a corporate focus but also from a societal perspective. What types of options do we possess to provide jobs for people? We require good workplaces, especially for people who are accustomed to a high standard of living. With regard to the WICs, the only option and possible choice appears to be the *"money for value strategy"*.

In connection with the outlined strategic option, we must consider that the quality of products stemming from countries that are attacking the economic position of the WICs is constantly growing: These countries are increasingly delivering *"good value for small money"*. Companies in the WICs are forced to advance quickly to a true high-end strategy regardless of the type of foundation (engineering, art and culture and/or high-touch). There is no other chance to ensure or in some cases to restore a solid position in international competition, which is the basis of being able to provide good workplaces. From a European perspective, we retain many potential competitive advantages to proceed more strongly to high-end and especially luxury strategies. However, we should not underestimate that competing countries, especially those in Asia, have either dramatically improved their strengths (e.g., in engineering) or learned how to leverage already existing highly developed strengths (e.g., in art and culture). Consequently, the race has begun for the maintenance or restoration and therefore future improvement of knowledge, skills and abilities in all relevant areas (engineering, art and culture, etc.). Moreover, the key ability is to implement marketing concepts that lead to real experiences of *"high end"* in general and luxury in particular.

1.1.4 Luxury Marketing and its Positive Impacts on Societal Developments

On initial examination of the WICs' situation, we might immediately perceive that enormous efforts are required in the areas of education and motivation. On closer examination, the motivation particularly of young people seems to be the most important obstacle. Although there are renowned institutions in education, further improvement is needed. Additionally, there is significant expertise in art and culture, engineering and craftsmanship established in the different industries and especially in small- and medium-sized enterprises (e.g., SMEs, Aiello et al. in this book), which build the foundation of economic prosperity in diverse European countries. The challenge is to a) motivate young people to strive for good education and challenging jobs (also in SMEs) and b) to motivate customers to pay more for good value to ensure the survival of existing companies and the establishment of new companies. In both cases, high-end strategies and luxury marketing strategies provide promising outcomes. With regard to the motivation of young people who strive for a good education and challenging jobs, luxury marketing can foster the perception that, for example, craftsmanship is valued and can lead to a higher quality of life. Certainly, the model of certain types of luxury consumption (e.g., the longing for frivolous luxury products) might not be useful for motivation and guidance from an individual and a societal perspective. Hopefully, the desire for a good life consistent with sustainability and social responsibility is preferable. In this context, it is important to mention that sustainability and luxury are perfectly compatible. Because customers are willing to pay higher prices for higher value, the luxury market is strongly associated with products and brands with a history of heritage and tradition. Therefore, luxury goods are often inherited from generation to generation and yield high prices on the vintage market as well. Additionally, for future challenges, such as the issue of opening markets to e-mobility, the best strategy might be to begin to convince customers at the upper level of affluent society to perceive cars as a luxury. In a subsequent step, customers at lower income levels might also be fascinated by the idea of eco-friendly cars.

The concept of an *"adapted luxury marketing strategy"*, i.e., the adaptation to new issues (e.g., sustainability) and/or new target groups apart from the upper class, illustrates the option of following the basic idea of a luxury marketing strategy in many different markets. In their highly valuable contribution to this handbook, Bastien and Kapferer describe several examples of such cases (e.g., Apple, Mini). Indeed, such examples – without any need for further explanation – illustrate the possibility of implementing basic ideas of luxury to obtain more money for true value. Consequently, better profits can be gained from an individual corporate perspective, whereas from a societal perspective, economic prosperity can lead to more valuable jobs for educated and skilled people. Certainly, the latter development is extremely important for European countries to survive in international competition.

In this context, globally successful luxury brands and products are a valuable basis for developing competitive advantages on a global level. Well-respected luxury products significantly influence the reputation of a nation or country. Such an effect can be further strengthened via a consequent national branding strategy to assure a positive transfer from the outstanding capabilities of globally successful luxury products (incorporated in their design, production, and commercialization) to the culture, typical abilities and skills of the people living in a country. Certainly, those valuable *"country of origin"* (COO) effects are present without specific national branding concepts. Nevertheless, there is enormous potential to selectively strengthen und sustain these positive effects. Undoubtedly, companies that utilize such COO effects in their luxury marketing strategies and effective integrated branding and communication efforts can be globally successful. For instance, companies with strong roots in France or Italy will always be able to build competitive reputational advantages when their businesses engage fashion and design, whereas German companies might have such competitive advantages when engineering is the basic factor of the business.

1.1.5 Combining It All: The Essence of Luxury Marketing

To summarize the societal relevance of luxury and luxury marketing as described above, we highlight the following aspects:

1. As a first step, the luxury market traditionally defined is an enormous and growing market that provides many opportunities, even from a societal perspective, as it offers good jobs to well-educated people and sustaining economic prosperity in the context of increasing international competition.
2. The concept of a luxury marketing strategy is transferable to other markets apart from the traditional core markets of luxury. This idea is a useful basis for gaining and sustaining competitive advantages even in such markets for which we might discuss *"new patterns of luxury"*, including products characterized by outstanding qualities of sustainability.
3. Generally, luxury concepts offer and extend the opportunity to obtain *"money for value"* based on the admiration and appreciation of achievements in the context of designing, producing, and distributing outstanding products. This process may help to mobilize countermovements against the trend toward *"freeconomics"* or the emergence of the *"age of cheap"* (see Fritz & Gülow in this book) and is relevant not only from a corporate but also a societal perspective. Against the background of growing global competition, not many other options can provide future prospects for well-educated and skilled people. For instance, if outstanding craftsmanship is not appreciated and honored in businesses, who would invest significant time and effort to cultivate such skills and abilities?
4. It should be emphasized from a societal viewpoint that effective luxury markets provide the opportunity to motivate young people to invest in education and to use existing and possible new cultural strengths. Additionally, perhaps effective luxury marketing more fundamentally than other strategies has a positive impact on young people's willingness to work hard in their education and subsequently in their professional lives to earn enough money to afford elements of the luxury dream.

Overall, luxury and luxury marketing reveal decisive opportunities not only for business in the context of growing global competition but also for societal development.

Certainly, we are aware that in real life, luxury marketing does not automatically lead to positive effects on business and society, or in the latter case, luxury marketing might not even aim for such positive societal influences or sometimes leads to negative societal effects (e.g., an orientation toward superficial consumption or the feeling of social deprivation). Nevertheless, attempting to unleash the positive effects of luxury and luxury marketing and to identify and stigmatize negative effects is a strong driving force for placing this topic on the research agenda. In this context, a basic framework is required that addresses relevant questions that guide research in the field of luxury marketing, and this framework is presented in the following section.

1.2 The "Know How": Possible Directions for Discussion and Future Research

1.2.1 Developing a Meaningful Typology of Luxury Product/Market Combinations

An important prerequisite for utilizing a luxury marketing strategy striving for corporate success and for appropriate societal development is a better understanding of the specific drivers of success and the challenges related to existing and possible strengths and weaknesses, opportunities and threats. An appropriate basis might be a categorization of the different fields of luxury and luxury markets. In terms of possible issues focusing on relevant aspects of luxury, there are at least four distinct but closely related approaches:

- The *"commodity approach"*: What types of products can be defined as luxury? Are there product characteristics that may help to define a whole product category as luxury (e.g., villas and yachts), or is there a need for specific criteria to distinguish among luxury, premium, and "standard" products within one category (e.g., cars and jewelry)? For instance, what types of household services are luxuries, having a "butler" as the British upper nobility or hiring a housekeeping service once a week? Although several characteristics are useful for differentiating products (e.g., very seldom or extremely expensive), there is still a need for orientation of the customer perspective.
- The *"customer perception approach"*: First and more generally, this approach emphasizes that luxury is a subjective concept. For a young student without financial support from his/her parents, a housekeeping service is indeed a luxury. In contrast, for established couples both working at least from 9 to 5, this service might not at all be related to the idea of luxury. Based on this reasoning, there are several attempts to conceptualize consumer perceptions of luxury value (e.g., based on the differentiation of financial, functional, individual, and social luxury value by Wiedmann, Hennigs, and Siebels, 2009). A profound orientation of luxury value as perceived by different consumers is a useful basis for understanding the meaning of luxury for different people and for identifying different types of luxuries that might lead to new product/market combinations.
- The *"marketing-management approach"*: To identify basic differences between luxury brands and other brands based on such criteria as exclusivity, uniqueness, and price premium power, there are diverse marketing-management guidelines that characterize a luxury strategy. In this context, we can directly refer to the contribution of Bastien and Kapferer in this book.

- The *"functional approach"* or *"instrumental approach"*: This approach is dedicated to identifying the basic drivers that create a luxurious high-end experience. From a marketing-management and a customer perspective, it is useful to highlight aspects that generate a luxury product and more precisely can be regarded as the relevant nucleus of such a high-end experience. As mentioned, we suggest, for example, that high tech, high art and culture and/or high touch are important areas for establishing a high-end perception.

Focusing on the different approaches, we propose a comprehensive framework as a useful basis to better understand different luxury products and markets and to identify challenges in luxury marketing (see **Figure 1.1**).

Figure 1.1 Towards a Comprehensive Framework of Luxury Marketing

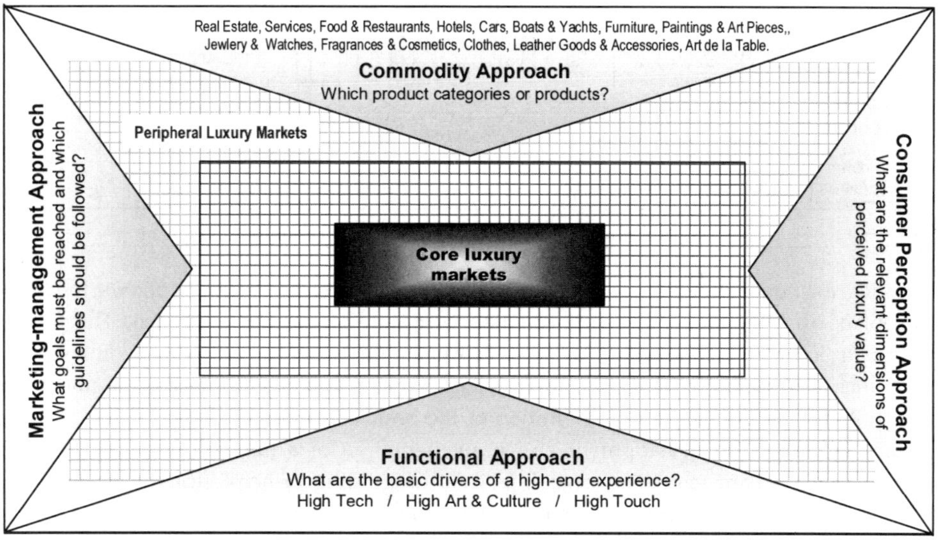

1.2.2 Guiding Future Research along a Comprehensive Contingency Approach

The focus on relevant dimensions that describe luxury products and markets is a prerequisite for incorporating a luxury marketing strategy with specific product/market conditions. Certainly, the focal point should be an elaborated contingency theory to explain and identify under which conditions what types of managerial guidelines, strategies, and operational marketing programs will generate and sustain what types of corporate success with what impacts on business and societal development. Based on this reasoning, the next step requires asking what types of effects on societal, market, and corporate conditions might be attached to different luxury marketing concepts. An initial outline of a research framework that basically addresses relevant research questions in the field of luxury marketing and management in the context of economic success and positive societal development is illustrated in **Figure 1.2**.

Figure 1.2 Basic Research Framework of Luxury Marketing

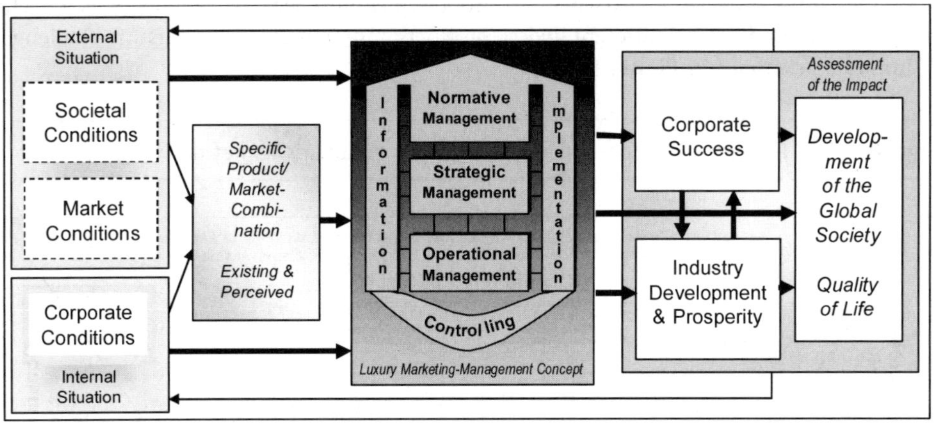

The idea for conceptualizing a *"luxury marketing management concept"* follows the basic concept of an integrated marketing-oriented corporate leadership approach. The pivotal point is a "normative management" that addresses the development of basic managerial guidelines, a corporate philosophy, such as a convincing vision and mission statement, the basic politics and policies, and a definition of the basic corporate goals and values. In particular, regarding the development and implementation of a luxury marketing concept, such normative management is crucial, as it is the guiding principle for the right understanding of luxury and luxury marketing at all company levels.

Although normative management provides the basis for strategically maneuvering a company facing multifaceted internal and external challenges, at the next level, professional strategic management is required. Based on strategic analyses of practical challenges and options (e.g., SWOT analyses), precise strategies and goals must be defined and developed to guide all corporate activities at the operational level. For example, basic strategic concepts include the topics of market positioning, customer targeting, branding, collaboration and the building of networks. Several contributions in this handbook address issues focusing on the planning and realization of these basic marketing strategies and on the appropriate design of marketing mix elements in the luxury industry.

Accompanying the different levels of luxury marketing management, information management and implementation processes are required to efficiently and effectively guide all management processes and structures in a designated direction that is ensured by a professional controlling system.

Relying on the idea of a contingency approach, the design of a comprehensive luxury marketing concept as illustrated here (see **Figure 1.2**) must be adjusted to the specific situation of a company or network of companies that attempt to reach their corporate goals and to positively contribute to society. What type of market positioning, segmentation or branding approach and collaboration strategies best fit a specific luxury product/market-combination under specific internal and external situational conditions? What type of market research is required to identify the desires of target groups, or is consumer research, at least the traditional type, unnecessary in the luxury market? Which corporate culture and especially which type of organizational structure supports the design and realization of a successful luxury strategy? Additionally, what type of corporate culture leads to successful luxury strategies that unleash positive effects on the development of societies?

1.2.3 The "What We Need": A Preliminary Conclusion

Considering the questions exemplified above, the heuristic potency of the research framework outlined in **Figure 1.2** becomes apparent. Based on the elements of this model, various important and interesting research questions emerge related to the design and implementation of luxury marketing. Concurrently, this framework also allows the classification of several directions in luxury marketing research.

- A comprehensive typology of luxury product/market combinations is needed to identify relevant contingency factors and dimensions facing the differing challenges in luxury markets.
- A better understanding of luxury markets requires holistic models of consumer perceptions and behavior with reference to different industries and global settings.
- A specific approach to analyzing relevant challenges, such as globalization, virtualization and counterfeiting, in light of distinct case studies, may provide a basis for the development of effective marketing-management concepts.
- Comprehensive concepts and guiding principles of marketing management in the luxury sector that separate the marketing of luxury brands from conventional marketing principles are required.
- Research insights on the societal impact of luxury might provide valuable results that refer to luxury marketing in a macromarketing perspective.

In sum, although there is significant research that addresses the topic of luxury marketing to date, nonetheless, enormous efforts are necessary to fill existing gaps that can be identified in the context of our research framework. Based on these considerations, the idea for this handbook emerged and was the initial impetus for inviting renowned researchers in luxury marketing to contribute their valuable findings. As a result, based on the submissions we received, primarily two topics can be assumed to be of particular importance currently in the luxury industry: first, the conceptualization, measurement, and analysis of consumer perceptions of luxury and luxury brands, and second, the challenge of counterfeits. In the following section, we will briefly outline the chapters of this handbook with different headings that accentuate important aspects in the field of luxury research.

1.3 The "What We Have": An Overview of the Different Contributions to this Handbook

The primary objective underlying this handbook is to provide a holistic approach to luxury marketing with respect to the characteristics, key challenges and opportunities of luxury brand management. We hope that through a combination of valuable contributions by highly honored colleagues from different parts of the world, this book might be a valuable tool for academics in their research and for marketing managers to understand and leverage the potential of luxury brands. The chapters of this book are divided into eight parts including different chapters:

- Part 1: Luxury Marketing as a Challenge for Marketing Theory and Practice

In the first chapter, in our role as editors, we already attempted to convey in the introductory passage an initial understanding of the challenges of luxury and luxury marketing from a theoretical point of view. However, the main contribution is provided by *Vincent Bastien and Jean-Noël Kapferer*. Against the background of their outstanding knowledge, experiences and contributions to literature in the domain of luxury brands, the authors elaborate on the specifics of luxury marketing from a theoretical and a practical perspective. The authors' chapter *"More on Luxury Anti-Laws of Marketing"* nicely demonstrates that luxury marketing should be differentiated from other marketing approaches and demands the consideration and implementation of different managerial guidelines, rules and values. Although these authors have previously discussed practical strategies and measures of luxury marketing, as stated, we treat their input as an important contribution to what we would call the "normative management" of luxury marketing.

- Part 2: Luxury and Luxury Consumption: A Global Phenomenon or Dependent on Cultural Differences?

Regarding the development of a contingency approach to luxury marketing, the question of to what extent luxury consumption could be treated as a global phenomenon or must be understood as strongly influenced by differing cultures is of great importance. This question arises not only from a theoretical but also from a practical viewpoint, especially because in most cases, luxury marketing must be designed in the context of the internationalization of businesses. Three contributions accentuate this topic from different perspectives.

First, *Virginie De Barnier and Pierre Valette-Florence,* two researchers who have already contributed significantly to the domain of luxury, focus in their chapter, *"Culture and Luxury: An Analysis of Luxury Perceptions across Frontiers,"* on luxury perceptions of consumers from Romantic, Germanic and East Slavonic cultures. Second, in the chapter *"An Intercultural Comparison of the Perception of Luxury by Young Consumers"*, we had the opportunity to collaborate with our appreciated colleagues *Bruno Godey and Daniele Pederzoli* from France and *Gaetano Aiello and Raffaele Donvito* from Italy. The study presented in this book focuses on perceptions of the concept of luxury and ranks numerous adjectives defining luxury among an international sample of young people. Third, based on the results of a collaborative study with colleagues from ten countries, in the chapter, *"Consumer Value Perception of Luxury Goods: A Cross-Cultural and Cross-Industry Comparison,"* we investigate the question of possible differences and/or similarities in the luxury value perceptions of consumers in distinct parts of the world.

- Part 3: Luxury and Luxury Consumption from a Theoretical Perspective

Among the papers to be included in a "consumer perception" or more generally a "consumer behavior" approach, three papers particularly focus on a more basic theoretical perspective.

For instance, *Taro Koyama* describes in the chapter, *"Luxury Brands from a Psychoanalytic Perspective,"* the value of an interdisciplinary perspective to better understand luxury consumers. Additionally, in their contribution, *"Indulging the Self: Positive Consequences of Luxury Consumption," Liselot Hudders and Mario Pandelaere* investigate the impact of obtaining luxury brands on an individual's self-esteem level. Furthermore, in the third chapter of this section, *Hyunjoo Oh* examines the question of *"How do Exclusivity Perceptions of Independent and Interdependent Consumers Influence Their Desires for Luxury Products?"*

- Part 4: Different Approaches to Understanding Luxury and Luxury Brands from the Customer's Perspective within Specific Cultural Contexts

Following the idea of a contingency approach to luxury marketing research, two highly valuable papers explicitly elaborate the specific context of different cultures.

The first chapter, *"Luxury Buying Behaviour and the Role of Culture: An Indian Context,"* by *Kartik Dave and Garima Dhamija,* focuses on the increasing demand for luxury brands in Asian countries with special focus on India, which can be considered the next significant opportunity in the luxury market place. In the second contribution to this part of our handbook, *Nicole Stegemann, Sara Denize and Kenneth E. Miller* examine how consumer perceptions drive attitudes toward luxury brands in Australia in *"Perceptions, Attitudes and Luxury Brands: An Australian Study of Consumers' Perceptions of and Attitudes to Luxury Brands".*

- Part 5: Luxury and Luxury Consumption in the Context of Societal Change

Another avenue that encourages more attention to important contingencies is to accentuate and analyze specific challenges involved with societal change. Among the multifaceted trends that definitely impact the design and implementation of luxury marketing, the topics of globalization and virtualization must be especially highlighted. Whereas the issue of globalization is already included in Part 2 and Part 4, this part addresses the question of virtualization and the conflicting priorities of exclusivity versus ubiquity.

In their contribution, "Luxury Marketing in the Age of Cheap: An Exploratory Survey on Consumer Behavior in Online Shopping Clubs," Wolfgang Fritz and Wencke Gülow examine the prospects of luxury marketing in light of the unfolding megatrend known as the

"Age of Cheap" in the current economy and society. Focusing on the question of how trust can be established in the online marketplace for luxury branded products, Meng-Shan Wu, Cheng-Hao Chen and Isabella Chaney concentrate on "Luxury Brands in the Digital Age – the Trust Factor".

- Part 6: Counterfeits Challenging the Luxury Industry, Consumers and Society

As stated, our call for chapters has led to valuable contributions dealing with the phenomenon of counterfeits. This problem is indeed a crucial challenge not only for the afflicted companies but also for the countries in which the companies are located. Among their contributions, the authors also perfectly illustrate the societal relevance of luxury brands and luxury consumption.

In the first chapter, titled *"Luxury goods vs. counterfeits: An intercultural study," Udo Wagner, Seung-Hee Lee, Sabine Kleinsasser and Jutatip Jamsawang* present an intercultural project on consumer attitudes toward false luxury goods, ethical issues surrounding counterfeits, and the resulting purchase intentions. Following this, *Suzane Strehlau* focuses on the Brazilian market in *"Brazil: Luxury and counterfeits"*. To identify groups of consumers who differ in their values and risk perceptions related to the trade-off between genuine and counterfeit luxury goods, in our chapter, *"Luxury Longing and Counterfeit Complicity: A Consumer Typology based on the Perception of Luxury Value and Counterfeit Risk,"* we examine consumers' motives for counterfeit consumption. Arguing that research on counterfeits has significantly challenged the conventional view, the fourth contribution by *Aaron Ahuvia, Giacomo Gistri, Simona Roman, and Stefano Pace* is dedicated to the question, *"What is the Harm in Fake Luxury Brands? Moving Beyond the Conventional Wisdom."* Subsequently, in their structured investigation of the effects of the counterfeiting phenomenon on luxury brands, *Silvia Grappi, Ilaria Baghi, Bernardo Balboni and Veronica Gabrielli* analyze all aspects of brand equity that might be affected by counterfeiting in *"Counterfeiting of Luxury Brands: Opportunity beyond the Threat"*.

- Part 7: The Management of Luxury and Luxury Brands

Additionally, our call for chapters has generated papers in the field of managing luxury brands. Certainly, this aspect has been addressed in the realm of analyzing consumer perceptions and behaviors.

With special focus on strategic luxury marketing, an important contribution is provided by *Gaetano Aiello, Raffaele Donvito, Silvia Ranfagni and Laura Grazzini* in their chapter, titled *"Luxury SMEs Networks,"* which demonstrates the importance of the country and territory of origin as a firm's intangible resource and specific creative knowledge in networks of collaborative interaction. Furthermore, focusing on the value of brand luxury, our chapter, *"Managing the Value of Luxury: The Effect of Brand Luxury Perception on Brand Strength,"* presents antecedents and outcomes of consumers' perceptions of a luxury brand and effects resulting in attitudinal brand strength. Finally, *Jan Wieseke, Michael Mauer and Sascha Alavi* examine in their manuscript, *"Personal Selling for Luxury Brands: The Effect of Cognitive and Affective Influence Strategies on Customer Value Perceptions,"* the effect of affective and cognitive sales strategies on customer value perceptions of luxury brands.

- Part 8: Luxury and Luxury Marketing in the Wine Industry

One important access to a contingency approach in luxury marketing is to establish research in specific industries that, on the one hand, are traditional luxury industries in which luxury strategies are successfully implemented or, on the other hand, might be interesting fields of "new" luxury. The two chapters included in this handbook address luxury marketing in the wine industry. For many reasons, this field is notably interesting for attempting to implement a luxury strategy to receive money for value.

The first chapter, *"The Hedonistic Consumption of Luxury and Iconic Wines"* by *Alistair Williams and Glyn Atwal*, explores the ways in which luxury wine products and brands can develop effective strategies focusing on experiential symbolic dimensions of consumption. Second, we present *"Wine as Luxury Experience: A Taxonomy of Consumers Based on Best-Worst Scaling"* based on the methodology of best-worst (BW) scaling and a segmentation of wine consumers according to their choice criteria for wine.

1.4 Acknowledgments and Outlook

Luxury marketing is a research topic of great importance, not only from a corporate perspective but also from a societal point of view. This conviction has guided us deeper into this field of research, beginning with the question of how we can measure perceived value first from a consumer perspective. Fortunately, we quickly gained access to many colleagues who shared our opinion about the relevance of luxury and luxury marketing.

Against the backdrop of such a fascinating range of articles included in this handbook, we would like to thank all of our contributors. We are deeply honored and grateful for being able to collaborate with such wonderful colleagues; the experience was truly luxurious.

Moreover, to encourage further studies on luxury marketing, we welcome readers' feedback on the numerous issues raised in the different chapters and are keen to discuss forthcoming ideas with researchers and managers in the domain of luxury. Please email any correspondence to us at *luxury@m2.uni-hannover.de.*

2 More on Luxury Anti-Laws of Marketing

Vincent Bastien, Jean-Noël Kapferer

2.1 Introduction

In 1989 Ford bought the Jaguar brand, symbol of British luxury worldwide, endowed with heritage, status, glamour, prestige, almost a cult brand with iconic models, for 2.2 billion dollars. Nine years later, on March 26th 2008, after having spent 6 billion dollars [15], Ford sold it to the Indian conglomerate Tata, along with another mythical British brand – Land Rover – for 2.3 billion dollars. As Land Rover brand was estimated at 2.5 billion dollars, it means that Jaguar brand was just given for free: meanwhile it had lost its pricing power, its glamour and was still not profitable. Surely there are many causes of such a failure: a brand turn over is a difficult endeavor. Re-instilling rapidly a culture of high quality in a British company that had lost it would have been a feat. But the main reason that led Ford to this disaster was the implementation of classical management and marketing methods for a luxury brand. What worked very well for Ford automobiles destroyed the value of the brand Jaguar. At first, cost controllers were hired to drastically reduce the cost of manufacturing whenever possible: instead of building value by promoting uniqueness, their objective was to build margins by sharing as many parts as possible with Ford cars themselves… and not only invisible ones. For instance, instead of proposing to the clients a unique Jaguar stick shift – sensual sign of a really different species of car, making it stand apart from the mass cars however good they are – cost analysts slowly destroyed the perception of privilege attached to Jaguar. Then, using classical market research, they soon measured that there was a latent demand for a more accessible Jaguar able to provide big sales to the brand, hence help amortize part of the fixed costs. New models were introduced: the six cylinders S Type (a revival of the MK2 model), and then the four cylinders X type. This down-market model boosted the sales, but at the expense of distinctiveness [14]. Marketers had answered point by point to the demands of the average consumer, as expressed by clinics and surveys, but had failed to reinforce the dream of the brand – that intangible magic lever that drives passion and justifies a high price, far above what functional utility and performance would command.

This is no surprise. Analyzing the long term behavior of those luxury brands that have earned the highest brand value measured by either Interbrand or Millward Brown methods (Louis Vuitton, Hermès, BMW, Bulgari, Gucci, Ferrari, Porsche, etc.) we have unveiled (in "The Luxury Strategy" / Kogan Page [10]) that they had enacted the same very specific strategy, named "the luxury strategy" in this book, based on very strict principles, and turning most of common marketing principles upside down.

In fact, we coined the term *anti-law of marketing* to designate the counterintuitive managerial principles, empirically carved through time by the founders and most often owners of these brands, which made these brand command their incredible pricing power and margins [16]. This power drove them safely through the 2008-2010 economic crisis without the need to reduce prices – more: they even kept on growing by increasing their prices. This "Luxury Strategy" calls for a total integration upstream and downstream as a sure way both to control quality and uniqueness, and deliver the highest consumer experience at retail in one to one relationships. It is the objective of this article to briefly remind of these anti-laws but above all to analyze in depth some of the most provocative yet powerful anti-laws of marketing followed by the luxury brands, at least those who are consistently profitable through time.

2.2 A source of managerial confusion: the six meanings of luxury

One of the reasons why the specificity of luxury marketing has not been enough recognized so far is due to the extended and loose usage of the word luxury nowadays. Since the word is fashionable, everyone likes to use it: Coach says it is an accessible luxury, Abercrombie and Fitch defines itself as a casual luxury, Silverstein [13] promoted the word 'new luxury' to put under the same umbrella realities as diverse as mass prestige brands such as L'Oréal Paris skin care and make up, and downward extensions of high end prestige brands. Today the word luxury is always used with a qualifier: new, old, accessible, über, hyper, modern, classic. Would the term 'luxury' have lost its intrinsic meaning?

In fact before one can even discuss the specificities of luxury marketing, one needs to make clear essential distinctions between six meanings of luxury.

a. *"Luxury"* as an absolute concept systematically evokes out of reach products, services and lives. Typical worldwide luxury surveys [8] reveal that when hearing the word luxury, consumers spontaneously think of yachts, private jets, private islands in the Caribbean Sea, a world of privilege [4].

b. *"My luxury"* in contrast evokes something very intimate, most often not a product but a moment or a rare experience. For some, it is having a break with one's family and spending time together in some original place. For some other it is going fishing alone in North Canada. These are dreamed experiences not really out of reach but yet quite rare or hardly implementable.

c. *"A luxury"* is a very subjective concept: underlying this concept are dimensions of non necessity yet high desirability, of a high price far beyond what performance and utility should command, of hedonistic experience, of high quality, of personal relationships and finally of prestige and status. Luxury is tied to some kind of social hierarchy. Many studies have been made, factor analyzing consumers' perception of luxury [9], [17], [7], [18]. Most recently De Barnier et al. [6] factor analyzed the perceptual items used in these different studies. Two major dimensions emerge from this work: elitism and hedonism. This reminds that luxury has two indissociable facets: luxury for self and luxury for others – one of the fundamental bases of the "Luxury Strategy", as explained in our book.

d. The "luxury sector" refers to the brands and companies that are considered by their peers as manufacturing and marketing luxury products or services. Whether they are perceived as luxury by consumers or not is not relevant here, as it is far too subjective: Armani is luxury for some, not for others. To define the members of this economic sector Professional Syndicates have been created – Comité Colbert in France considers that all its 75 members are luxury brands. The same holds true in Italy for the Fundazione Altagamma. Note however that the Italians do not use the word luxury even though they talk about Bulgari, Armani, Zegna, or Prada. Altagamma means "high end".

e. The "luxury market" is the estimation of the sales past, present and future of the above-mentioned actors. Bain and co., the consultancy company specialized on luxury, produces every year the estimates of the size of this luxury market. It was 77 billion Euros in 1995, and 170 Euros in 2007, just before the economic crisis. Note that the Bain estimates do not comprise the sales of cars, neither those of yachts, hotels, services etc. They measure the sales from the classic sectors of apparel, leather, jewelry, watches, fragrance and skin care, wine and spirits. Bain in fact defines luxury brands as those "selling premium products at premium prices in premium environments" (Bain & co, 2011): here again, the question is not whether they are perceived as such. After all such consumer perceptions can vary through time and space, thus making the compilation of sales data too unpredictable. In China, Lacoste is perceived at now as a luxury brand by the Chinese (Time Design supplement), whereas it is seen as a casual premium product in Europe. For the sake of building sales forecasts and backdata on the "luxury market", one cannot rely on consumer perceptions.

f. Finally one should also talk of *"luxury strategy"*. Our own analysis of the behavior of so-called luxury brands – included Bain & co's data – reveals three business models that it is fundamental to distinguish: the luxury strategy, the fashion strategy and the premium strategy [10]. They are lumped together when one talks of the "luxury consumer " or "the luxury market". To make it short the "luxury strategy" strictly defined was invented by such brands as Louis Vuitton, Chanel, Hermès, as well as Tiffany, BMW, Ferrari, or Ritz Carlton.

2.3 Distinguishing luxury, fashion and premium strategies

There are many consequences of the recognition that luxury is a precise strategy too, not to be confounded with a fashion, or a premium strategy. The difference between these three strategies is huge. It does not change much in most of basic consumers' eye, at least at short term. But, when one has to manage a brand, the difference is pivotal.

The luxury strategy aims at creating the highest brand pricing power by leveraging all intangible elements of singularity (time, heritage, country of origin, craftsmanship, man made, small series, prestigious clients). By mobilizing all these intangibles, the brand is to be made non comparable to any other. The luxury strategy commands a full vertical integration upstream and downstream – to control the manufacturing and quality on one side, and consumer retail experience on the other. The core tenet of the luxury strategy is one to one direct relationships with the clients: hence the importance of DOS (directly operated own stores). The goal is to create long selling products and not best selling products. As a matter of fact Porsche 911 was born in 1964 and Chanel N°5 in 1921. In the luxury strategy there are no sales promotion, no licensed products.

"Fashion strategy" is a totally different business model: here heritage, time, are not important: fashion sells being fashionable that is to say a very perishable value. After a few weeks clothes lose their appeal and need to be sold with rebates, once the season gone. This is why fashion brands have no problem delocalizing production in low wages countries (China, Vietnam, Indonesia.) to reduce their manufacturing costs. This is also to alleviate the burden of loss of profitability when 'super sales' start in order to get rid of the inventory before the new collection comes. Also, a fashion item is not bought to last. Its quality needs not be at the highest level. In fact the buttons of any Burberry cloth must be reinforced after the purchase. Because fashion is now expressed through accessories more than apparel, fashion brands have developed extensive licensing policies. Burberry is a typical case.

Finally, the "premium strategy" can be summarized as "pay more get more". Here the goal is to prove – through comparisons and benchmarking – that this is the best value within its category. Quality/Price ratio is the motto. This strategy is by essence comparative, and relies on the tangible dimensions of value – those who can be objectively measured, whereas in a luxury strategy one will rely on the intangibles, those who create a structural non-comparability. The Korean brands engaged in trading up are typically using this strategy: Samsung has now achieved its goal of being the leader of the Smartphone market, ahead of Apple… but in unit sold, not in turn-over nor in profit. The same holds true for Hyundai cars as it used to be at the early times of Audi and Lexus.

The "luxury strategy" has been originally developed for the "luxury market", broadly defined, and it is there that the more often you can find it – in fact, it is the most efficient strategy on this market; it is seldom met on other markets, even though it can be very successful there, as Apple, or to a lesser extent Nespresso, have demonstrated. In fact Apple is structurally following a luxury strategy. Even though few consumers would perceive Apple as luxury, its strategy is a luxury strategy: more and more directly operated stores, exclusive channels of distribution of iPhone, limited range, high price far beyond the mere utilitarian function, high hedonism, latent social stratification between 'the have' and 'the have not', etc. However, on one critical aspect Apple does not follow the luxury strategy: the manufacturing of iPhone and iPad is not only outsourced, but relocated in Asia and not in California – with the slogan "designed in California". And the price paid for this infringement of one of the major laws of the luxury strategy is high – severe downgrading of the social image (workers suicides at Foxconn), counterfeits, copycats, look-alikes and patent fights with Samsung. "Anti-laws" are a consistent corpus; you must apply all of them or none to be really successful long term.

To summarize at this point, the notion of "luxury market" or of generic "luxury customer" are loose concepts, when the "luxury strategy" is not. Now for practical purposes it makes some sense to use a broad or even loose definitions of "luxury" to get the sales figures of the global "luxury market" or to make surveys on "the luxury consumer": not all of the so-called luxury brands are following a luxury strategy. On a long period, several high status brands move their strategy between luxury, fashion and premium ones while staying on the same market: Gucci went from luxury to fashion strategy in the 90's with Tom Ford at the helm of the brand and is going back to luxury now. Dior is trying to move from a fashion strategy to a luxury one. Louis Vuitton itself, by hiring Marc Jacobs as creative director and making "Haute Couture défilés" put some fashion aspects in its luxury strategy. Mercedes went from luxury to premium strategy in the 90's (Class A launching) and is now trying to come back to a luxury strategy. Revival of the Maybach brand in 2002 for the top of the range – Mercedes brand having lost its aura, sale of Chrysler in 2007 – Audi and Lexus are moving from a premium strategy to a luxury one today.

On the other hand, knowing that a luxury strategy is a precise and exacting one, brand extension and brand stretching are not the same.

2.4 Do not confuse brand extension and brand stretching

In fact, a big part of the luxury market to day is made up of luxury brands that have grown through moving out of their original core business. Fendi is a former furrier. Former dealers in fine leather moved into footwear and then into ready-to-wear (Gucci, Ferragamo); a great name in the jeweller's craft (Cartier) now puts its name on watches, pens, and fine leather goods. The creator of glass and crystal objects, Lalique, now sells jewellery. Some brands have moved classically from haute couture to ready-to-wear and then into accessories (Chanel, Dior, etc); others have followed the opposite trajectory: Hermès, Louis Vuitton.

All this seems at first very confusing, but the origin of this confusion should now be very clear. It comes from the multiplicity of senses of the word luxury to day. All along this article, we make a clear difference between the two meanings of the word "luxury": in "the luxury strategy" and in "the luxury market" (or when one talks of "the luxury consumer"). In "luxury brand stretching", the same confusion occurs. Anyone knowing the business makes immediately the difference between

- Brand "extension": in this case, the luxury brand implements its luxury strategy in a new territory – like Cartier the jeweler in watches, Louis Vuitton the luggage maker in leather goods or Hermès the leather goods maker in silk ware
- Brand "stretching": in this case, the luxury brand does not apply the luxury strategy in a new territory, but a fashion or a premium one – like Cartier for perfume, or Louis Vuitton for ready-to-wear.

In the first case, the luxury brand becomes a full luxury player of the new market by controlling progressively the whole process – as did Chanel with the jasmine fields of Grasse for n°5 or Hermès with silk ware when they took control of their suppliers from Lyon. This is of course the best way to do it, but it is very demanding in money and time. In the second case, you are closer to basic licensing – which has nothing to do with the luxury strategy. You just hire a designer and look for suppliers.

2.5 The anti-laws of marketing

We have exposed elsewhere the strict rules followed by the leading luxury brands, with high pricing power. We just remind them briefly before analyzing three anti-laws which require an in depth treatment.

1. Forget about 'positioning', luxury is not comparative
2. Does your product have enough flaws to give it soul ?
3. Don't pander to your customers' wishes
4. Keep non-enthusiasts out
5. Don't respond to rising demand
6. Dominate the client
7. Make it difficult for clients to buy
8. Protect clients from non-clients, the big from the small
9. The role of advertising is not to sell
10. Communicate to those whom you are not targeting
11. The presumed price should always seem higher than the actual price
12. Luxury sets the price, price does not set luxury
13. Raise your prices as time goes on in order to increase demand
14. Keep raising the average price of the product range
15. Do not sell
16. Keep stars out of your advertising
17. Cultivate closeness to the arts for initiate
18. Don't relocate your factories

2.6 Why luxury brands should not pander to their customers' wishes?

One of the major anti-laws of marketing, respected by those brands following the luxury strategy, looks like heresy to classical marketing theory. After all isn't the credo of all well managed companies to be customer oriented? Shouldn't any marketing plan start by summarizing the voice of the consumer? Yes for all brands but not those who follow the luxury strategy. This does not mean being deaf of course but the function of luxury brands is to create dreams, not to answer to problems and needs. As once said by Tom Ford, then COO of Gucci Group: "I look at research then throw it away and forget it for this is about yesterday. I must create dreams for tomorrow".

Luxury is a non-necessity made desirable: it sells promotion emotions (self elevation, pleasure, recognition) not prevention emotions (risk reduction, absence of problem and discomfort) [5]:

- Promotion emotions lead to thrill, excitement and delight
- Prevention emotions lead to security, confidence, and satisfaction

Luxury is a non-necessity made desirable on the basis of emotional values (surprise, beauty, elevation of self through hedonism and elitism): asking people what they want is by definition a contradiction in this respect. They will answer something that generally destroys the dream – that is to say the lever of pricing power.

Now there are brands belonging to the "luxury sector" which remarkably thrive by extensively listening to the market through market studies. Coach the leather bag seller claims to spend 2.5 million dollars worldwide in tests and market research asking consumers what they expect from bags and what they like. Coach soon adopts its clients' recommendations, has all its bags made in China to reduce the cost of goods, and launches many collections per year. Today Coach claims to dominate the 'accessible luxury leather bag market' it created. Now it should be reminded that most of the Coach bags range sells below 800 Euros, whereas the Hermes iconic Kelly bag starts at 3,000 Euros. Coach is following a premium/fashion strategy not a luxury strategy. The closer to mass, the more traditional marketing re-becomes relevant.

Fast fashion low cost retail brands such as Zara or H&M do also show remarkable profitability: the luxury strategy is indeed not the only one able to make a company profitable. It is the one getting the most efficient results when one stops thinking mass. Both Zara and H&M have remarkable feedback information systems telling them almost immediately what sells very well in their thousands stores worldwide and what does not. They immediately react and adapt the production and creation to this feedback.

H&M brand value is estimated by Millward Brown (2011) at 13 Billion Euros, higher than Hermès at 12 Billion dollars. Yet global sales of H&M are 14 billion Euros in 2010 and those of Hermès 2.4 billion Euros. As a result H&M is worth less than its sales, Hermès 5 times more. The price earnings ratio of H&M is 6.5 and that of Hermès 28. Hermès does follow a strict luxury strategy.

When one asks consumers what luxury brands should do to please them, they generally do not understand the long-term interest of these brands, which is to preserve their attractivity based on prestige, creativity, surprise and high pricing power. During the economic crisis, The Luxury Institute, a survey company based in New York, kept on asking so-called "luxury consumers" – recruited on the basis of past purchases – what they expected from luxury brands. What answer came first, second and third? To reduce the prices, then to offer discounts, and at last to make more products accessible on Internet discount sites. It looks like if consumers have no idea of their own psychology! In fact, they have a clear understanding of it: they would like all these amenities reserved to them, not to the other customers... All studies show that when luxury brands reduce their prices they reduce their prestige. In Asia, prestige is strictly correlated with price [3]: the brand would better accept a loss of volume but never compromise on price.

Buying a luxury product from a brand adopting the luxury strategy is a long-term investment (unlike a fashion product or a premium one, both subject to obsolescence). In times of crisis, consumers need to know that there remain sure values, long term. Hermes, one of the few brands which never reduced its prices during the 2009 crisis, showed a superior resilience – notably in its core leather handbag business which represented 43% of sales last year. The upscale strict price positioning of Hermes made it immune to the financial crisis, at least in handbags, which account for more than 60% of Hermes profits. As a result, Hermes was – with Louis Vuitton and for the same reason (no price reduction) – among the very few luxury good players posting positive results in 2009.

In addition, making products suddenly accessible destroys the ethics of your pricing and endangers trust: why would a Porsche Panamera 4S usually sold new at 107,000 Euros be suddenly available at 90,000 Euros, then go back to 100,000 Euros?

This does not mean that overall price reduction is not an issue for those brands who cannot implement anymore a luxury strategy. Mauboussin, a traditional jeweller (founded in 1898) located Place Vendôme – the jeweller holy place in Paris – is still endowed with heritage among the many. But in 1998, the jeweler lost its main customer and started to suffer years of huge financial losses. It could not compete anymore in terms of high quality, creativity, cultural advance and social stratification. In 2002 it was sold to an entrepreneur businessman, who repositioned the whole product range in the very accessible price segment (below 500 Euros). The brand went further: it communicated heavily on its low price. It was a success – the brand was turned around and net result became positive. But maybe a short-term success – the dream value of the brand is very low now. It is a good example of trade off between net result and brand value – one could say "milking the cow" – by giving up the luxury strategy and going mass, while trying as long as possible to be perceived as a Prestige brand. But what else could be done for Mauboussin? So many luxury brands prefer dying that loosing their aura… At least, Mauboussin is still alive, even with little prestige left and no more pricing power. Pierre Cardin is still perceived as a "luxury brand" by the masses in France as well as in China –the surveys show. But its pricing power is low: he does not follow a luxury strategy anymore, and has not for a long time. Instead he manages the brand with the most licenses in the world, a typical milking strategy of what is left of dream among the masses. For long he is no more the dream of the cultural, creative and economic elites.

There is another structural reason why traditional marketing is prohibited in the luxury strategy: it leads to a 'regression to the mean'. Recall that luxury brands are cultural forces. Luxury is about taste education. This is why it flirts so much with art, 'Avant Garde' art. Luxury brands do not aim at being popular (that is to say liked by everybody today), but at setting the long lasting standards of taste for tomorrow. By definition market studies like big numbers: they will tell P&G which Hugo Boss future fragrance (a license) is most appreciated today by the target (called the ambitious social urbans, name of a segment in an European consumer segmentation study). This is fine to launch the Hugo Boss fragrance of the year, followed by another one next year and so on, in mass prestige distribution (Sephora, Douglas, Marionnaud, Department Stores, Tax Free Shops, etc.). But this is not the way to launch a long lasting success, a tomorrow's classics. Capitalizing on what the majority of average present targeted consumers declare they like today is not the route to build the future Chanel N°5, almost a century old World best seller. P&G does not follow a luxury strategy but a premium strategy where indeed marketing methods are needed. P&G fragrance licenses managed by its Prestige Division leverage P&G unique know how at repetitively launching new products on a world scale basis in mass prestige channels. Time, risk, art, imagination, standing apart – are not core values in the P&G corporate culture. This is why they never launched any premium brand in the FMCG markets.

Using market studies to listen to the consumer provide paradoxical results: why the „best Saab ever» (Saab 9-3) according to the automotive press did not sell and did not save Saab? It was even one the best cars ever made (if you rate it objectively on a multi attribute clinic test, according to JD Powers). Yet General Motors had to sell Saab in January 2010. They applied classical marketing and listened too much to the consumer instead of following a luxury strategy for Saab. But is GM corporate culture –premium oriented – able to host a luxury strategy?

GM managed with a premium strategy, based on classic marketing not on luxury rules (heritage, authenticity, exclusive value principle.). Authenticity is not a masquerade, like keep some iconic details (superficial features) of the brand.

As the goal was to expand the volume of sales, and not to pursue a value strategy like Hermès, GM asked non-buyers what would make them buy a Saab. Accordingly GM....

- Abandoned the stylistic audace of the brand design: as a result it looked like more an Audi A4 or a BMW series 3
- Let prospective clients who read the press know the engines were now made by GM and no more by Saab. In a luxury strategy, a key rule to sustain the dream is never to delocalize the production of the essential parts (anti-law n°18). A rule well respected by BMW now managing MINI: they are made in the Oxford factory (UK).
- Decided to abandon the becquet on the rear hood (for surveys revealed that "people did not like it" plus it added a cost)
- Softened the driving and acceleration experience, which makes the Saab drivers fans of their brand. For sure soft turbos rate better in mass-market questionnaires.

BMW understands what is the luxury strategy. Their managers know that the BMW frame of mind, brand personality, values appeal to only 20 % of the segment that can afford a BMW. Now what should BMW do? Expand these 20 % by diluting all the elements that make BMW stand apart and above (as GM did with Saab) or instead look worldwide after these 20 % in each new high growth country, and raise the prices?

This is just what BMW is doing with MINI. Although very few people in the world would declare that MINI is luxury, BMW actually enacts a luxury strategy to manage it, with a price far above what rational performance and utility values would command – at the great surprise of BMW German engineers themselves. In the car business, price is a function of size: despite a size comparable to that a Renault Twingo, MINI sells at a double price (22,000 Euros), notwithstanding the important need for uniqueness which leads many clients to pay up to 38,000 € with all options that personalize their car, transforming the production waiting time as an opportunity.

In a luxury strategy one sells emotionally loaded cars, reassured by quality to let their clients indulge and find the high price very normal. MINI sells the Swinging London heritage, also an iconically British and forever classic car. MINI has kept defaults that are continuously mentioned in all consumer surveys: too small coffer, to low, too small. Why not take them into account? Because they are the price to be paid for:

- Keeping the soul of the product, an intangible plus that makes it stand apart in an era of banalization, standardization creating boredom
- Keeping the Go-Kart driving experience ("bone-shaker")
- Keeping the iconic design
- And, last but not least, children love to sit in it

What luxury sells is excitement, not security, not problem reduction.

2.7 Why always raise the average price of the brand?

One of the key words of the luxury sector is democratization. For sure the sector gained 100 billion Euros between 1997 and 2008 (see Bain & Co.) because it now sells to non-riches, mostly the growing middle class. This is why more accessible products were launched in the stores, not to talk about accessories. Yet one of the key principles of the luxury strategy is to keep raising the average price of the brand.

This does not mean having one or two alibi products, exceptionally high priced, and created just to launch the buzz: this is a classic PR game. Not a real luxury strategy. Why then raise the prices continuously?

The essence of luxury is to sell a dream of exception. Luxury sells "the ordinary of extraordinary people, and the extraordinary of ordinary people". Dreams have a dimension of non-accessibility yet remaining at a close distance. People do not dream of 300 million Euros yachts: this is too far from them. However they look at 100,000 Euros automobiles. To remain the dream brands must remain in the radar of the extraordinary people, not just recent celebrities or soccer champions recruited for that purpose.

Recall also that the global wealth is growing. Moscow is now the town with the most billionaires on earth. There are more than one thousand $ millionaires in China today (Forbes, 2011). This means that the luxury status of the brand is earned on this target. Going back to our introductory Jaguar – when managed by Ford – example: the brand kept on launching new models downward, thus getting closer to the middle class. Per se, launching accessible lines is not the problem as long as two essential rules are respected:

- The brand regularly launches new expensive models upwardly. Jaguar never launched its S Class – this was a major signal of weakness: the brand admits it lacks the creative power, the technological know how, and the class distinction to do it.
- The price of the lowest priced models should be above the luxury price threshold, that is the price above which people feel it is luxury [11]. Instead the Jaguar X and S Types were not priced above their competition.

Inspired by sociologist Pierre Bourdieu's theory of the dynamics of social distinction [2], Ameldoss et al [1] showed that the luxury market is divided in two groups: the "snobs" and the "conformists" (we regret their use of such negatively value laden words to describe consumer behavior). Now the first ones value brands as long as they feel not too many people carry them: this is why their demand goes when the price goes up. Conformists are just the opposite: the number of people wearing the brand in the streets reassures them. This number grows when price is more accessible. The talent needed in the luxury strategy is to manage both targets.

2.8 How to always keep raising the price point

Systematically increasing price point is a real challenge, as it goes against all the legitimate habits of company management. As price/volume elasticity is usually negative, reducing costs and price increases the volume and the market share. It improves competitive position, which is linked to market share according to Boston Consulting Group analysis, and has proven to be true on most markets.

More: luxury customers are educated customers. They are ready to pay more... but for getting much more. So, just increasing price without adding significant value leads to disaster – as 'luxury brands' relying purely on Veblen effect have quickly discovered at their expense.

Creating value is the motto in the luxury strategy. But being creative is not enough to sustain a systematic price increase. You need lots of creativity in the low-cost industry to invent new business models, sell at a significant lower price than competition, and be profitable – but it is the job of the CEO. You need lots of creativity in the fashion industry to keep on selling at the same price point – but it is the job of the designer. In luxury, you must install the whole the company in the creating value process: luxury value creation does not rely only on the talent of a creator, but on each employee of the company

- Production people: lots of new ideas originate on the workshop floor. This is the reason why a luxury company must make its products and not relocate – creation teams live in symbiosis with artisans.
- Sales staff: new ideas come from customers – not by pandering to their wishes, but by understanding their dream. This is why you must have your own sales staff – they are

fully part of the company – and why they must be local – the customer must be able to talk in his language with sales people sharing his or her culture

- And of course top management.

Louis Vuitton huge success in luxury leather goods is a good example of this management: every worker in the workshop, every sales person in the shops is focused, in addition to do his job as well as possible, to find ways of improving the product or the service in such a way that new products are always more relevant to the dream of the customer – existing or new. Of course, productivity is searched, but always discarded if it is, even slightly, detrimental to the Louis Vuitton dream. New products are not introduced to replace an exiting one – they keep their standing – but to add value to the whole range. They are sold at a higher price, but this higher price is always explained and justified in the shop by the sales people. And never by saying "this new product is better or more fashionable than this old one", but by saying "we have added this new product to our line to bring a new idea".

In fact, the increasing price point is not due to the price increase of existing products – which price stays the same – but to the introduction of new complementary lines.

We are talking here of standard – or 'ready-to-wear' – products. In addition, most of luxury brands have a service of customized products. This service has been part of the luxury universe since it exists. A customized product delivered at home was the original luxury business model of artisanship, long before the concept of brand was born (see our book 'the luxury strategy' [10]). However, if those customized products raise both the average price point of the offer and the prestige of the brand, as quantities sold are small, they do not in fact raise the average selling price – except for very small luxury brands.

2.9 Why beware of celebrities?

This third anti law has been received with surprise by many professionals and executives of the luxury sector. Intuitively also, when one browses through the pages of glossy magazines, celebrities are everywhere in the advertisements and in the pages relating who attended what select event sponsored by a luxury brand.

It is important to remind the reader that all along this article we do not use the word luxury in a broad or loose way (like when one talks for sake of convenience of the luxury sector, the luxury market, or the luxury consumer, hastily mixing luxury, fashion and premium brands). Celebrities are to be used with caution in the luxury strategy, if the brand wants to build its pricing power, distinction, style and sustained appeal. They are not used as selling agents, for new customers to buy the product through an imitation model ("I want to buy the bag because this celebrity has it") – this is the fashion business model. They must be used, when used, as a testimonial ("this famous person is also using my bag, staying in the hotel I went last year") for existing customer – it comforts the status of "extraordinary product for a usual customers, which is also an ordinary product of extraordinary people".

However, although brand ambassadors or "égéries" cannot be used fully in luxury, they can be used locally – as a nation have one ambassador in a specific country. The goal, especially when the brand is not well known in a specific country or if its dream seems too exotic, is to give a relevant incarnation to the dream.

For instance, when Louis Vuitton decided to penetrate Japanese market in the 80's by opening its own brand shops, specific Japanese celebrities in Arts and Culture were used to explain that LV was not just a foreign brand for tourists, but also products nice to have for highflying Japanese people who appreciated the French culture – like Seiji Osawa. But, here again, we are more in the register of testimonials.

For the same reason, in China, Louis Vuitton is using Chinese celebrities in its communication – but always several ones, not a unique ambassador or "égérie".

2.10 Implementing the luxury strategy beyond the luxury market

When you consider those anti-laws, you see that they define a whole consistent and original marketing strategy – what we called "the luxury strategy". You also see that this strategy is not limited to the existing "luxury market", although it is the place where it is the most relevant – so its name. Apart of the anti-law n°12 ("luxury sets the price, price does not set luxury"), the word "luxury" does not appear in their expression. This means that the marketing strategy defined by those laws can be implemented beyond the luxury market.

For instance, Apple, which is not a luxury brand, has been immensely successful in applying the luxury strategy in the computer and phone market, where the pure "luxury market" is just a niche, even if it can be a beautiful one as Vertu has demonstrated for the mobile phones. As soon as you have a unique product and service, this strategy is worth to be considered. Same when you are facing scarcity of human talent – Mc Kinsey is a very good example in the consulting business. Or when you are facing scarcity of resources – think of the problems of sustainable development and the example of Tesla strategy in full electric cars. Think also of the issues of agriculture and bio-food. In fact the luxury strategy is often the best business model to make sustainable products or services profitable at the launch phase.

2.11 The limits of the luxury strategy

Whatever powerful this strategy is, it has limits. Firstly it is very demanding in day-to-day management as most of the usual, well-known and very powerful marketing techniques – and among them market surveys and benchmarking – do not apply. Secondly, and worse, growth has a limit for a luxury brand wanting to stick to the luxury strategy. Take Louis

Vuitton: to-day, it is the most valuable luxury brand in the word, and has been able to keep on growing fast owing to the development of huge new markets like China, and India to-morrow. But it cannot go too far in market penetration without losing its aura. And moving to the fashion model to keep on growing is a strong temptation – fashion has no volume limit. Hiring Marc Jacobs as creation director was a clever move for Louis Vuitton, as he brought a more fashionable image without harming Louis Vuitton luxury standing: traditional Monogram bags and luggage is still the core of the brand. But it is also a big risk for the brand if Marc Jacobs' influence goes further than just creation management: the brand will become a usual fashion brand, a pure logo LV put on miscellaneous products, losing its luxury image… and its pricing strength.

References

[1] Ameldoss, W./Jain, S. (2005): Pricing of conspicuous goods, in: Journal of Marketing Research, Vol. 42, No. 1, pp. 30-42.
[2] Bourdieu, P./Nice, R. (1987): Distinction: a social critique of the judgement of taste. Harvard University Press.
[3] Chadha, R./Husband, P. (2007): The cult of the luxury brand. Nicholas Brealey Ed.
[4] Chevallier, M./Mazzalovo, G. (2008): Luxury Brand management, Wiley.
[5] Chitturi, R./Raghunathan, R./Mahajan V. (2008): Delight by Design: The Role of Hedonic Versus Utilitarian Benefits, in: Journal of Marketing, Vol. 72, No. 3, pp. 48-63.
[6] De Barnier, V./Falcy, S./Valette-Florence, P. (2012): Do consumers perceive three levels of luxury? A comparison of accessible, intermediate and inaccessible luxury brands, in: Journal of Brand Management, February, pp. 1-14.
[7] Dubois, B./Laurent, G./Czellar, S. (2001): Consumer Rapport to Luxury, Consumer Research Working Paper, No. 736, HEC , Jouy en Josas.
[8] Ipsos (2011): World Luxury Tracking, HEC Conference, Paris, May 2.
[9] Kapferer, J-N. (1998): Why are we seduced by luxury brands, in: Journal of Brand Management, Vol. 6, No. 1, pp. 44-49.
[10] Kapferer, J-N./Bastien, V. (2009): The Luxury Strategy: Break the Rules of Marketing to build Luxury Brands, Kogan-Page, London.
[11] Kapferer, J-N./Laurent, G. (2012): Is there a minimum price for luxury products ?: a multisector international empirical study, HEC Research Paper, HEC Paris.
[12] Karpik, L./Scott, N. (2010): Valuing the unique. Princeton University Press.
[13] Silverstein, M./Fiske, N. J. (2003): Luxury for the masses, in: Harvard Business Review, Vol. 81, No. 4, pp. 48-57.
[14] Stones, J. (2004): Crushed under the weight of car giants, in: Marketing Week, 23 September, p. 24.
[15] Strach, P./Everett, A.M. (2006): Brand corrosion, in: Journal of Product & Brand Management, Volume 15, No. 2, 2006, pp. 106–120.
[16] Tabatoni, O./Kapferer, J-N. (2010): Is Luxury Really a Financial Dream, HEC Research Reports, CR 935/2010, HEC Paris.
[17] Vigneron, F./Johnson, L.W. (1999): A review and conceptual framework of prestige seeking behavior, in: Academy of Marketing Science Review, Vol. 3, No. 1.
[18] Wiedmann, K.P./Hennigs, N./Siebels, A. (2007): Measuring Consumer Luxury value perception: a cross cultural analysis, in: Academy of Marketing Science Review, Vol X, No. 7.

Part 2

Luxury and Luxury Consumption: A Global Phenomenon or Dependent on Cultural Differences?

3 Culture and Luxury: An Analysis of Luxury Perceptions across Frontiers

Virginie De Barnier, Pierre Valette-Florence

3.1 Introduction

According to geographic proximity, linguistic similarities, population migration and historic developments, European cultures can be grouped into three main cultures Germanic, Romance and Slavonic. Research shows that the origin of this cultural typology is not quiet clear because of its ancient roots. Some researchers refer it to ancient Romans [4], [9], others to St Bede, a British writer of 735, one of the first history writers [1]. Germanic cultures include ethnic groups of Germanic origin or the linguistic, cultural and racial descendants of the old Germanic tribes, such as modern UK, Germany and the Scandinavian countries [1]. Romance cultures are descendants of Italic origin, ethnic groups that spoke Vulgar Latin dialects and were part of the Roman Empire, such as modern Italy, France, Spain and Portugal [1]. Slavonic cultures include ethnic groups of Slavic peoples speaking the Slavic languages, such as modern East and Central Europe, as well as the Balkans [39].

Taking into account the growing importance of Europe luxury consumption, the current chapter focuses on luxury perceptions of consumers from Romantic, Germanic and East Slavonic cultures. Among those cultures, three countries have been selected France, the United Kingdom and Russia. France has been selected as a representative culture of the Romantic origin for its status of a reigning fashion capital and for its high concentration of aging population. French luxury brands are the heritage of France. French people are often considered as the most style and fashion-conscious culture in Europe. The UK has been chosen as a representative to develop luxury perceptions of Germanic cultures because of the highest concentration of millionaires (47,000 people) in Western Europe with tastes for luxury. Russia was lastly selected as a representative of East Slavonic cultures because it is considered one of the most perspective markets for luxury goods consumption. Together with India and China, it is part of what is seen as a "golden triangle" of the newly wealthy who are "thirsty" for luxury goods.

This chapter is structured around three main parts. First, main issues and results stemming from the literature related to luxury are presented. Then, a second part will focus on the results obtained from a qualitative survey undertaken within the three aforementioned countries. Finally, we will conclude on the main contributions of this research and on the future research that could be undertaken.

3.2 Luxury concept: different definitions according to different perspectives

Academic literature on luxury concept reveals three main interrelated perspectives that were used to study and to define luxury brands: an economic perspective, a psychological view and a marketing standpoint.

3.2.1 Luxury: an economic perspective

Initially economic theories focused mostly on distinctions between luxury and necessity. Veblen was one of the first who analysed luxury goods in the social economic context in his celebrated "The theory of the leisure class" [36]. Nowadays the primary focus of economic theories is the influence of pricing strategies on exclusivity of luxury goods. They strongly associate luxury with a "high" or "exclusive" pricing. Based on the connection of price and exclusivity, Groth and Mc Daniel (1993) [13] developed the Exclusive Value Principle as a framework for "marketing strategies to achieve brand exclusivity". According to the authors, the Market Price for a product is a sum of the Pure Utilitarian Value of the product and the Exclusive Value Premium. For luxury goods, sources of utility include product quality, aesthetic design, excellence of service, etc. Exclusive Value Premium incorporates external factors that motivate luxury goods consumption behaviour, such as advertising and promotion campaigns.

3.2.2 Luxury: a psychological view

Social and behavioural psychology define luxury from luxury goods consumption motivations based on "interpersonal" or "external" factors, such as "opinions, influences, approval and suggestions of or interaction with others" [13] and "personal" or "internal" factors, such as feelings and emotions that stimulate luxury brands consumption [37]. Luxury goods are consumed either for social recognition, status and positive impression management purposes [35]; [36]; [26]; [29]; [3] or for hedonic and pleasure-seeking ones [12]; [35]. Recent studies demonstrate that luxury brands possess "emotional values in access of their functional utility" and "are likely to provide subjective intangible benefits" [36]. Indeed, Dubois and Laurent (1996) [7] already pointed out the emotional value as a vital characteristic of luxury products. They stated that *"a vast majority subscribes to the hedonic motive (...). One buys luxury goods primary for one's pleasure"*.

3.2.3 Luxury: a marketing standpoint

Marketing theories on luxury brands are grounded on economic and psychology theories. They deal primarily with differentiations of luxury goods from non-luxury ones, as well as definition of salient product features that could constitute luxury. Vickers and Renand (2003) [35] developed a three dimensional model differentiating luxury brands from non-luxury ones based on symbolic meanings of luxury brands in terms of functionalism, experientialism and symbolic interaction. Functionalism is defined as product features that could 'solve a current problem' or 'prevent a potential one', such as 'superior quality and strength, durability, confidence of items replacement'. Experientialism incorporates features that could stimulate sensory pleasure and hedonic consumption, such as 'traditional and exclusive designs', 'special richness and tone of decoration', 'elegance of days gone by'. Symbolic interaction implies product components related to status, self-enhancement, and 'group membership, such as 'prestigious name' or recognizable designer style. The most recent and important studies to define luxury perceptions on a cross-cultural level have been carried out by Dubois, Laurent and Czellar (2005) [9]. They performed qualitative and quantitative cross-cultural consumer-based studies in Western Europe, USA and Asia Pacific and gave a general definition of luxury as a combination of the following 6 dimensions: Price, Quality, Uniqueness, Aesthetics, Personal History and Superfluousness (see **Table 3.1** for the six main facets of luxury).

However, this literature review reveals that, despite a rapid growth of the luxury goods market and rich accumulated knowledge on the subject, there is no clear consensus of what constitutes a luxury brand [37]; [35]; [6] although the synthesis proposed by Kapferer and Bastien (2012) [22] is considered to date as the most accurate one.

Table 3.1 The six main facets of luxury

Facet	Description
Excellent quality	Exceptional ingredients, components, delicacy and expertise, craftsmanship
Very high price	Expensive, elite and premium pricing
Scarcity and Uniqueness	Restricted distribution, limited number, tailor-made
Aesthetics and Poly-sensuality	Piece of art, beauty, dream
Ancestral heritage and personal history	Long history, tradition, pass-on to generations
Superfluousness	Uselessness, non-functional

Source: Dubois, Laurent and Czellar, 2005[9]

It is noteworthy that no further cross-cultural research has been undertaken since then to verify whether the existing 6 luxury facets can stimulate consumers of different cultures to purchase luxury goods or whether consumers changed their luxury perceptions and search for something else in luxury. Hence, the objective of this research is to find out what are the main luxury perceptions among the consumers of three selected countries (France, Russia and the United Kingdom) and to verify to what extent the six luxury dimensions mentioned by Dubois, Laurent and Czellar (2005) [9] have evolved over the last years.

3.3 Luxury concept: different facets according to different countries

This section presents first the research design chosen. Then results common to the three countries are first presented, followed by a presentation of the luxury dimensions specific to each country.

3.3.1 Research design: a qualitative perspective

In order to assess cultural effects on consumer perceptions of luxury and to elicit the facets of luxury most relevant to Romantic, Germanic/Anglo-Saxon and Slavonic cultures and due to the very exploratory nature of our study, qualitative research methods have been undertaken. According to Miles and Huberman (1994) [28], *"qualitative samples tend to be purposive, rather than random"*. Therefore, a total of 30 people were interviewed, 10 per each culture, English, French and Russian. The sample contained active users of luxury brands of different gender, age and occupation. To avoid narrow sampling and to gain a better understanding of the studied phenomenon, a "peripheral sampling" was also chosen as recommended by Miles and Huberman (1994) [28]. It included respondents *"who are not central to the phenomenon but are neighbours to it"* [28]. They are not active users of luxury goods but have bought at least one luxury product during the past year. In total 15 additional people were interviewed, 5 representing each culture, English, French and Russian.

The interview guide was based on "a progressive approach" [27] starting from a description of the latest luxury goods shopping experience and luxury products acquisitions to general perceptions of luxury. The six facets of luxury defined by Dubois, Laurent and Czellar (2005) [9] were included in the interview guide as probes in order to test their importance and relevance to respondents. The interview type was an unstructured, narrative one aiming at getting information about luxury meanings relevant to consumers [27]. On average, each interview lasted about one hour and a half. All interviews were recorded and transcribed. A content analysis of each interview was performed to identify the major themes or clusters appearing in the text [28].

3.3.2 Main categories of luxury: analysis for the three countries

The in-depth interviews with French, British and Russian interviewees revealed five facets common to the three countries: Aesthetics, Premium Quality, Personal History, Self-Pleasure and Expensiveness. A new dimension of Self-Pleasure common for all the three cultures, as well as new facets of Product Conspicuousness and Aspiration for France, Functionality and Luxury Atmospherics for the UK and Functionality for Russia emerged during the interviews.

Figure 3.1 displayed below shows the frequencies of enunciation of these facets as well as the facets specific only to one or two countries. Appendix at the end of the chapter presents the main themes and sub-themes frequencies for the three countries.

One of the interesting research findings is that for all respondents the luxury dimension of Superfluousness is irrelevant. For the French and the British respondents the facet of Uniqueness is also unimportant (see **Table 3.2** for excerpts). They are referred to as "very vague", "intangible" and "hard to assess". The respondents never think about Superfluousness and Uniqueness when they purchase luxury goods. The in-depth interviews reveal that consumers search for quality benefits in luxury brands associated with comfort, security or other special characteristics that could make life easier. Purchase of luxury goods is considered as an investment. For financial reasons consumers can't afford useless things and can't put up with the idea that luxury goods are superfluous.

Figure 3.1 Luxury perceptions: country comparison

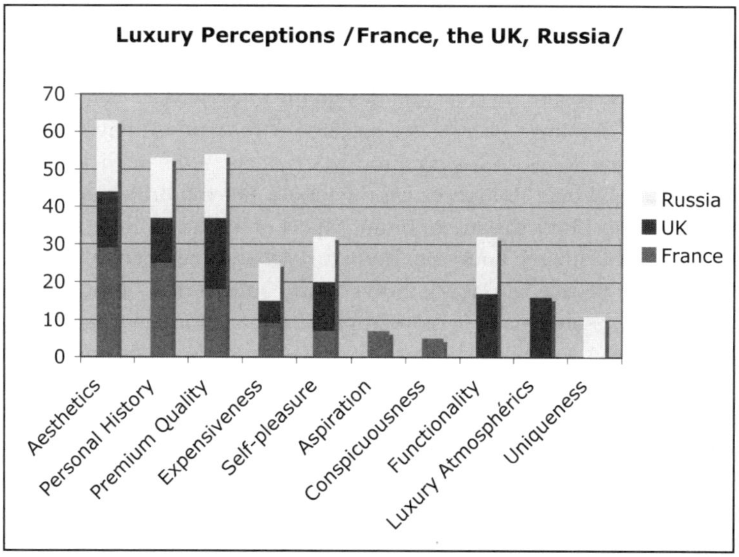

Source: De Barnier, Valette-Florence

Table 3.2 Irrelevance of Uniqueness and Superfluousness
Excerpts from question: "Why should others buy this product?"

France	UK	Russia
"The products of this brand are created for people and not only for models. The brand offers extremely practical things that could last forever and enjoy the feel of a good quality."	"People should buy it because of quality, originality and usefulness. It could last for a long time and will make several generations of your family happy. Uniqueness and Superfluousness are unimportant for me because I do not understand how it can be measured and what is meant by uniqueness".	"People should buy it for usefulness and practicality. I can't understand and accept superfluousness. How can I assess it? I prefer to have useful luxury things around but not the ones that take up space in my house. I would say that mass produced goods are superfluous because they last for a short time and you have to invest money in purchasing them again."

Source: De Barnier, Valette-Florence

3.3.2.1 Aesthetics

The aesthetics dimension is composed of design, colour and style that create beauty. They are of primary importance. When people see a luxury product they are attracted in the first place by originality of its design, colour and style. They motivate consumers to touch and to feel the product. Colour and its combination with materials are considered the most important facets of design and style because they bring out beauty and originality of style and design. On the one hand, design and style create a beautiful overall image of a luxury consumer emphasising his or her best features. On the other hand, design and style give a practical value to a luxury product allowing consumers to use it for different occasions from every day to special ones. In style and design the interviewees value a combination of fashion and tradition. For them fashion is associated with trends, innovation and modernity. Tradition stands for codes of luxury products that were invented years ago, such as straps on Chanel bags, Louis Vuitton or Dior symbols. For the British and the Russian respondents luxury atmospherics is an important aspect of luxury aesthetics. It is achieved by the interior design of a luxury boutique, by luxury goods presentation in boutiques and adverts, as well as by people who work in the luxury sphere. They inject life into intrinsic luxury goods aesthetics and make it more appealing to customers (refer to **Table 3.3** for excerpts). For France the Aesthetics dimension earned the highest position, 29%, in comparison to Russia, 19% and the UK, 15%.

Table 3.3 Aesthetics
Excerpts from question: "What does luxury mean for you?"

France	UK	Russia
"For me, luxury is beauty, elegance, refinement, as well as quality, which is a selection of materials, fabrics and a special cut and design. Beauty of the product is extremely important because it motivates me to touch it and to feel it."	"It is style and design that incorporates fashion and tradition. When I say fashion – I mean modern trends when I say tradition it is the adaptation of the brand codes to modernity. This combination makes a product sophisticated and desirable."	"Luxury should combine modern and traditional that was invented a long time ago but still attracts attention, like straps on Chanel bags or Dior symbols."

Source: De Barnier, Valette-Florence

3.3.2.2 Premium quality

Premium quality of luxury is relevant for the respondents of all the 3 countries as much as Aesthetics. In fact, this is the main criteria leading to a luxury purchase. For the interviewees it is associated with superior materials and colours that insure product longevity and durability. Materials should not only perform an aesthetic role but should also be of practical value to the customer. For the respondents the superior quality should ensure longevity when the product is passed on from one generation to another. It should never lose its appeal but should become better with time. From the respondents' perspective "luxury is an expensive investment". Therefore, a high price has to be justified by an outstanding quality and aesthetics that mass-produced goods can't guarantee. Taking into account expensiveness of luxury goods, aesthetics and exceptional quality should be interconnected (refer to **Table 3.4** for excerpts). For the UK the Aesthetics dimension earned the highest position, 19%, in comparison to France, 18% and Russia 17%.

Table 3.4 Premium quality
Excerpts from question: "Why should others buy this product?"

France	UK	Russia
"Luxury is a quality. It should be so high that you can pass it on from one generation to another. For me quality of color and materials should be very high and resistant to anything. They should bring out beauty of the product."	"For me luxury is quality in the first place. Beauty is not the main luxury criteria for me. It has to be combined with quality because of a high price ones pays for luxury. Beauty and design should be backed up by an outstanding quality to last forever and to become better with time."	"I also loved the quality because it is very strong and beautiful. From practical purposes, it is very easy to wear this thing and look after it. The quality of wool is so high that you can wear the thing in any bad weather conditions and look after it without any problems after."

Source: De Barnier, Valette-Florence

3.3.2.3 Personal history

One of the most interesting research findings lies in the fact that French and Russian respondents associate product personal history with famous brand names. They regard it as one of the most crucial stimuli to purchase luxury because brand name guarantees quality and security that the product will last for a long time. The respondents prefer to buy luxury products from luxury brands that are considered experts in their fields such as for instance, bags from Hermes or Chanel, accessories from Gucci, jewellery from Cartier etc. Surprisingly, with comparison to the French and the Russian, the British do not associate luxury with famous brand names. They define Product Personal History as a good reputation that a luxury product achieved through favourable consumer references. Personal history is created by consumers' word of mouth rather than advertising and promotional campaigns (consult **Table 3.5** for excerpts). For France the Personal History dimension obtained the highest position, 25%, in comparison to Russia, 16% and the UK, 12%.

Table 3.5 Personal history
Excerpts from question: "What effected your final decision?"

France	UK	Russia
"Image of the brand as being innovative, traditional and fashionable. A design that is fashionable. Quality of combining the design with materials. Brand name as a security of a costly purchase."	"It is the quality of the product, in the first place, associated with longevity. Luxury is quality that you can pass on from generation to generation and it will never lose its attraction. I would trust a reference of my friends about a product and its good reputation."	"Brand name with long tradition, quality and design effected my final decision."

Source: De Barnier, Valette-Florence

3.3.2.4 Self-pleasure

Respondents mentioning an association between luxury and self–pleasure explained that luxury makes them feel comfortable, fills their life with enjoyment and makes it more interesting spiritually and intellectually. Above all, luxury gives pleasure derived from outstanding quality that creates comfort, security and harmony backed up by original aesthetics. Self-pleasure is also associated with self-cultivation or the intellectual aspect of luxury. It teaches people how to be sophisticated and refines their internal and external world (refer to **Table 3.6** for excerpts). For the UK the Self-pleasure dimension reached the highest position 13%, in comparison to Russia 12% and France 7%.

Table 3.6 Self pleasure
Excerpts from question: "Why should people buy luxury goods?"

France	UK	Russia
"People should buy luxury brands for pleasure and for cultivating a good taste and elegance".	"People should buy luxury brands to enjoy in the first place and to be different from others. "	"People should buy them for pleasure. Life is very boring without them. I think that everyone should afford some amount of luxury. "

Source: De Barnier, Valette-Florence

3.3.2.5 Expensiveness

Respondents consider Expensiveness as a legitimate characteristic of luxury. One has to pay a high price for expensive materials and craftsmanship to ensure quality and longevity. It takes a lot of time and craftsmanship to create something original and durable. Durable materials guarantying longevity are costly. Therefore, the effort should be paid off by a high price (consult **Table 3.7** for excerpts). For Russia the Expensiveness dimension earned the highest position, 10%, in comparison to France, 9% and the UK, 6%.

Table 3.7 Expensiveness
Excerpts from question: "What is the first thing that comes into your mind when I say luxury?"

France	UK	Russia
"The expensiveness of the materials was vivid that created a positive impression of quality".	"It is a costly investment and a high price should guarantee quality that could last forever. It is necessary to pay a high price for a great effort, quality and craftsmanship. "	"Expensiveness because one has to pay a high price for quality, craftsmanship and originality."

Source: De Barnier, Valette-Florence

3.3.3 Main categories of luxury: analysis for countries' specificities

Despite similarities, the three studied countries show some differences in luxury perceptions. For instance, for France only Aspiration and Conspicuousness dimensions are relevant. Luxury Atmospherics is important for the British only, Product Uniqueness is mentioned only by the Russian and Product Functionality is a stimulus for the British and the Russian. Such differences can be explained by cultural differences derived from climate, language and habitual practices relevant for certain cultures [14]; [17]; [33]; [24].

France is representative of Romance cultures that Lewis (2003) defines in general as "aesthetic", "conscious of beauty values", "affective", "aware of the value of feelings", "romantic", and "expressive". Hall and Hall (1990) [14], Hofstede (1991) [17] and Trompenaars (1993) [33] characterize Romance cultures as extremely creative, artistic, perfectionist in style, elegance and manners, people-centered and humanistic. Therefore, the emergence of Dream and Aspiration dimensions can be understood as specificities of Romance cultures.

The UK represents Germanic cultures that are characterized as neutral with "opaque emotional state" when people are "subtle in verbal and non-verbal expressions" [16]. The formation of Germanic cultures has been influenced to a large extent by severe cold, snowy or rainy climatic conditions that affected communication and expression models, as well as daily practices [24]. Climatic conditions affect selection of clothes and colors [24]. Brits, as well as Nordics and Germans, *"are interested in the protective aspects of clothes (quality of materials, warmth and durability) for many months of the year"* rather than appearance and style [24]. In connection with that, Luxury Atmospherics and Functionality dimensions can be explained by characteristics specific to Germanic cultures.

Russia is representative of East Slavonic cultures. Its creation was profoundly influenced by Vikings, Scandinavian, Germanic and Baltic tribes in the north and by Khazar, Avar and Tartar tribes in the south, as well as Byzantine (X – XIV centuries), France, Germany and Italy (XVIII – beginning of XX century) that impacted their value and attitude toward wealth and luxury ([31]; [32]). Such contrasting influences of the North and the South, as well as changeable climatic conditions, partly explain the ambivalence of East Slavonic cultures. In their expressiveness, they are described as "loquacious, emotionally self-indulgent and reserved" [32]. During 70 years of communism Russian people were deprived of luxury. It impacted their perception of luxury as something unique, accessible to the elite [31]. Therefore, the dimensions of Uniqueness and Functionality can be understood as specific to East Slavonic cultures.

3.3.3.1 Aspiration for France

The Aspiration dimension scored 7% among the French interviewees. For the respondents, luxury is an aspiration or a dream to achieve perfection, sophistication, self-refinement and class through the ownership of luxury products that are in their nature perfect and sophisticated. They enhance to a large extent external image and internal world of the consumer, his character and manners. Luxury is the world of beauty (refer to **Table 3.8** for excerpts).

Table 3.8 France excerpts for aspiration

France
"Luxury is a dream, something that people should aspire to. It is a dream because nowadays only luxurious things provide comfort and easy use. Through luxury products one learns to develop his tastes and achieves self-refinement."
"Luxury is a dream of what somebody should become in terms of refinement and sophistication."

Source: De Barnier, Valette-Florence

3.3.3.2 Conspicuousness for France

The Conspicuousness dimension scored 5% among the French interviewees It is noteworthy that the respondents do not associate conspicuousness with personal ambitions to show off status or wealth. It is rather an intrinsic quality of luxury goods. They are conspicuous by their nature because of their aesthetics and superior quality that makes them visible among other products. The respondents envision conspicuousness in a positive light because it is one of the stimuli to purchase luxury products. They admit that conspicuous products help differentiate themselves from others but it is not the key stimuli. They purchase conspicuousness for pleasure to own a piece of art (consult **Table 3.9** for excerpts).

Table 3.9 France excerpts for conspicuousness

France
"Its design and quality were so wonderful that the product was like a piece of art. It attracted attention by its beauty and a selection of high calibre materials. It was showing off its sophistication and class. To possess such a product meant refinement and elitism. It was conspicuous and could differentiate the owner from others."
"It was the product attractiveness and conspicuousness. It stood out of the others by its beauty and quality. To possess such a product is to be different from others, to be conspicuous. But you do not show off yourself. It is the product that creates attractiveness and conspicuousness."

Source: De Barnier, Valette-Florence

3.3.3.3 Luxury atmospherics for the UK

The Luxury Atmospherics dimension scored 16% among the British interviewees. According to them, luxury is a philosophical notion. A non-luxury object can become a luxury one depending on atmospherics. It is associated, in the first place, with the extremely high standards of services that are linked to the selling skills and image of vendors. The primary goal of vendors is to create an appropriate luxury shopping experience for the consumers. Boutique interior design is also a crucial component of luxury atmospherics and experience. The British respondents prefer boutiques with cosy atmosphere of elegance and class. They do not have to be dominating and loud, but sophisticated, simple and original, bringing out the essence of luxury products. Finally, merchandise presentation is considered as an important aspect of the boutique atmospherics and luxury experience, because for British respondents the consumers are attracted, not by the products, but by their presentation (see **Table 3.10** for excerpts).

Table 3.10 UK excerpts for luxury atmospherics

UK
"For me luxury is a certain unforgettable experience created by people who sell luxury products. There are a lot of extremely beautiful things, but they become more appealing if somebody creates a luxury atmosphere around them."
"We were attracted by the shop atmospherics which was very quiet with subdued lights. The shop assistant was not over smartly dressed but looked in tune with the shop atmosphere. In my view, luxury starts with impeccable services."

Source: De Barnier, Valette-Florence

3.3.3.4 Uniqueness for Russia

The Uniqueness dimension scored 11% among the Russian interviewees. It is associated with a limited number of luxury products that allow their owners to distinguish themselves from others and to create a special original and sophisticated self-image. Due to their art status, pieces of art can't be mass-produced. Therefore, luxury goods which the respondents regard as pieces of art can't be mass manufactured either. Uniqueness also serves as an indication of quality and expensiveness (refer to **Table 3.11** for excerpts).

Table 3.11 Russia excerpts for uniqueness

Russia
"Uniqueness is an important aspect of luxury. You possess a piece of art created by a famous artist, in this case the company designer that no one can copy. Unique original things also have interesting association for history, people and events."
"...the uniqueness of the collection. The owner could be different from others because it was not a bag for everyone because of a high price and a limited number. I liked the design and the color which reflected the color of horse skin."

Source: De Barnier, Valette-Florence

3.3.3.5 Functionality for the UK and Russia

The functionality dimension is an important research finding which shows that the British and the Russian search for usefulness in luxury. They look for intrinsic product qualities that provide them with functional benefits of outstanding comfort, easy operation and security that mass-produced goods lack of. Nowadays, people prefer to invest it in something that should serve their needs to the best. For the British and the Russian it is the mission of luxury products to ensure the premium satisfaction of human needs through superior quality and functionality. They invest money into multi-purpose luxury that could be utilised several times in different combinations, but not once in a lifetime (see **Table 3.12** for excerpts). For the UK the Functionality dimension scored 17% and for Russia it scored 15%.

Table 3.12 UK and Russia excerpts for functionality

UK	Russia
"Quality and practicality of the jacket that I could combine with different things and wear on daily bases, as well as for special occasions. It is an extremely functional product because of its practicality and easy care, as well as resistance to unfavorable weather conditions."	"It is the original design and expensive materials, used for the jockey themes, as well as an impeccable quality of leather. It is an extremely practical thing for Russia because it closes on top. It is wise to buy bags that could be closed on top for rainy and snowy Russian weather."

Source: De Barnier, Valette-Florence

3.4 Conclusion

Although exploratory in its very nature, this research brings out several interesting results, either from a conceptual stand point or a more operational perspective. This chapter analyses consumer perceptions of luxury of Romantic-French, Germanic-British, and East Slavonic-Russian, cultures choosing the cross-cultural approach proposed by Dubois, Laurent and Czellar (2005) [9]. Results enrich the previous studies and contribute to the understanding of the luxury phenomena in the following ways.

First of all, this research confirmed the relevance of defined luxury dimensions of Aesthetics, Premium Quality, Personal History and Expensiveness as pointed out by Dubois, Laurent and Czellar (2005) [9]. However, one of the most interesting research finding is that it revealed new trends in luxury goods consumption with the emergence of Self-pleasure dimension and the relative lack of importance of Uniqueness and Superfluousness facets.

The emergence of a new Self-pleasure dimension can be linked to previous findings of social and behavioural psychology showing that goods may be consumed for experientialism and hedonic interaction [18], [35]. This dimension was found relevant for all three cultures probably because of luxury internationalisation through global advertising and promotion campaigns emphasising hedonic consumption of luxury, on the one hand. On the other hand, people travel a lot, exchange ideas, develop certain international tastes for luxury and adopt new common trends for luxury consumption [35].

It is noteworthy that the discovered Self-pleasure dimension challenges the traditional positivist perception of consumer behaviour. This traditional perspective assumes that *"product purchase is the end result of an objective, rational, thoughtful process by the buyer and that because of this rational process, it is possible to "predict" consumer behaviour"* [19]. This dimension is in line with the interpretive approach of consumer behaviour stating that *"emotions, feelings, and subjective contextual meaning play an important role in product purchase. Each consumption experience is unique and findings are not generalizable"* [2].

The Self-pleasure dimension of luxury consumption confirms and enriches previous studies on experiential marketing carried out by Holbrook and Hirschman (1982) [18], Bourgeon and Filser (1995) [2], as well as Vickers and Renand (2003) [35]. It demonstrates that luxury goods possess *"emotional values in excess of their functional utility"* and *"are likely to provide subjective intangible benefits"* [35]. This finding points out the emotional value as a vital characteristic of luxury products [7]. It shows the importance of hedonic aspects in luxury goods consumption stemming from a personal experiential interaction with a luxury product. Hedonic interaction is linked to sensory pleasure derived from an outstanding product aesthetics backed up by a premium quality and personal history [13]; [9]; [35]; [37]. It would be interesting to compare these findings with those stemming from the recently new luxury scale proposed by Wiedmann, Hennigs and Siebels (2009) [38]. The scale is structured around four factors impacting the consumer's luxury value perception and consumption at an international level: individual, social, financial and functional dimensions. Consumer's hedonistic approach to luxury may be close to the individual facet proposed by Wiedmann, Hennings and Siebels (2009) [38], this facet being composed of a materialistic value, a self-identity value and a hedonic value. In addition, the verbatim could also be analyzed with regards to other and perhaps more accurate scales such as Kapferer [21] or Vigneron and Johnson [37].

This chapter shows that it is necessary to take into account consumer perceptions of luxury. A company can stimulate consumer purchase behaviour of its luxury products through careful management of its marketing communication mix by addressing specific factors and dimensions relevant for modern consumers on an international level. The cultural hierarchies of luxury perceptions can be used in a boutique environment as guidelines for salesmen to sell successfully products to customers of different nationalities. They can indicate the key selling points of luxury products relevant for consumers.

In addition, this research can help managers to understand that luxury goods possess certain dimensions and attributes of high relevance to consumers that differentiate them from standard or non-luxury goods. The studies reveal that despite some differences in luxury associations French, Russian and UK consumers have a hedonic approach for luxury derived from aesthetics, quality and personal history. Therefore, any marketing communication mix should emphasise these aspects. Also, the research pointed out that beauty, quality, personal history and expensiveness differentiate luxury goods from standard ones. By examining these differences managers can create a more valued image of luxury products that could enhance their competitive positions versus non-luxury ones.

Finally, as consumers expect to have a unique piece of art when purchasing a luxury good, it is recommended that luxury companies get organised to fight the spreading of fake luxury products on the market. Such practices harm luxury industry through low quality level, low prices and rendering common place of luxury products.

This research reveals a new trend in luxury goods consumption and enriched the existing definition of luxury with the Self-pleasure dimension. Consequently, our results are closely in line with the most up to date and accurate definition of luxury given by Kapferer and Bastien (2012) [22]: *"Luxury designates objects or services which are needlessly expensive: non necessary–one can live without it–no functional argument can ever justify their price, only the feeling of privilege made of rare quality, hedonistic experience, symbolic elevation and conspicuousness"*. Our research also demonstrates that Superfluousness and Uniqueness facets are no longer relevant to present day consumers due to the lack of tangible characteristics that could help them to assess and to appreciate these dimensions. Consumers also look after utilitarian values in luxury goods associated with aesthetics and quality.

Moreover respondents of the three cultures have expressed a hedonistic approach to luxury. They purchase luxury goods for self-pleasure. This self-pleasure is linked to aesthetics, quality and personal history. Therefore, it would be of prime importance to explore to a greater extent which sub-dimensions could constitute the self-pleasure orientation, on a broader international and multicultural level. Also, the research gave surprisingly few results on symbolic consumption of luxury. Finally, further research is needed to understand the way in which self-pleasure is assessed by consumers and how this dimension is linked to symbolic interactions while consuming a luxury product.

References

[1] Bede/Webb, J.F./Farmer, D.H. (1996): The Age of Bede: Revised Edition, in: Penguin classics (eds.).

[2] Bourgeon, D./Filser, M. (1995): Les apports du modèle expérientiel à l'analyse du comportement dans le domaine culturel : une exploration conceptuelle et méthodologique, in: Recherches et Applications en Marketing, Vol. 10, No. 4, pp. 5-25.

[3] Brinberg, D./Plimpton, L. (1986): Self-Monitoring and Product Conspicuousness on Reference Group Influence, in: Advances in Consumer Research, Lutz, R.J., Provo, Utah (eds.): Association for Consumer Research, Vol. 13, pp. 297-800.

[4] Carlton, C.M. (1965): Studies in Romance Lexicology, in: Oxford University Press (eds.), 102 p.

[5] Dubois, B./Czellar, S. (2001): Prestige Brands or Luxury Brands? An Exploratory Inquiry on Consumer Perceptions, in: European Marketing Academy 31st Conference Proceedings, Cd-Rom.

[6] Dubois, B./Duquesne, P. (1993): The Market for Luxury Goods: Income versus Culture, in: European Journal of Marketing, Vol. 27, No. 1, pp. 35-44.

[7] Dubois, B./Laurent, G. (1996): The Functions of Luxury: A situational Approach to Excursionism, in: Advances in Consumer Research, Corfman, K. P./ Lynch, J. G., Provo, Utah (eds.) : Association for Consumer Research, Vol. 23, pp. 470-477.

[8] Dubois, B./ Paternault, C. (1995): Observations: Understanding the World of International Luxury Brands: The Dream Formula, in: Journal of Advertising Research, Vol. 35, No. 4, pp. 69-76.

[9] Dubois, B./Laurent, G./Czellar, S. (2005): Consumer Segments Based on Attitudes Toward Luxury: Empirical Evidence from Twenty Countries, Marketing Letters, Vol. 16, No. 2, pp. 115-128.

[10] Edcock, W.D. (1960): The Romance Languages, in: Oxford University Press (eds.).

[11] Erickson, G.M./Johansson, J.K. (1985): The Role of Price in Multi-Attribute Product Evaluations, in: Journal of Consumer Research, Vol. 12, No. 2, pp. 195-199.

[12] Fenigstein, A./Scheier, M. F./Buss, A. H. (1975): Public and private self-consciousness: Assessment and theory, in: Journal of Consulting and Clinical Psychology, Vol. 36, No. 1, pp. 241-250.

[13] Groth, J.C./McDaniel, W. (1993): The Exclusive Value Principle, in: Journal of Consumer Marketing, Vol. 10, No. 1, pp. 10-16.

[14] Hall, E.T./Hall, M.R. (1990): Understanding Cultural Differences, in: Intercultural Press (eds.), USA.

[15] Hirschman, E. C./Holbrook, M. B. (1982): Hedonic Consumption: Emerging Concepts, Methods and Propositions, in: Journal of Marketing, Vol. 46, No. 2, pp. 92-101.

[16] Hoecklin, L. (1995): Managing Cultural Differences, Strategies for Competitive Advantage, in: Addison-Wesley Publishers Ltd and the Economist Intelligence Unit (eds.), 2nd edition, UK.

[17] Hofstede, G. (1991): Cultures and Organisations, in: McGraw-Hill International (eds.), Great Britain.

[18] Holbrook, M. B./Hirschman, E. C. (1982): The Experiential Aspects of Consumption: Consumer Fantasies, Feelings, and Fun, in: Journal of Consumer Research, Vol. 9, No. 2, pp. 32-40.

[19] Huang, M-H. (2001): The Theory of Emotions in Marketing, in: Journal of Business and Psychology, Vol. 16, No. 2, pp. 239-47.

[20] Jaffe, E.D./Nebenzahl, I.D./Usunier J.C. (2003): Personifying Country of Origin Research, in: Management International Review, Vol. 43, No. 4, pp. 383-406.

[21] Kapferer, J-N. (1998): Why are we seduced by luxury brands?, in: Journal of Brand Management, Vol. 6 , No 1, pp. 44 – 49 .

[22] Kapferer, J-N/ Bastien, V (2012): The Luxury Strategy: Break the Rules of Marketing to Build Luxury Brands, 2nd edition, Kogan Page Ltd.

[23] Kemp, S. (1998): Perceiving Luxury and Necessity, in: Journal of Economic Psychology, Vol. 19, No. 5, pp. 591-606.

[24] Lewis, R. (2003): The Cultural Imperative, in: Nicholas Brealey Publishing (eds.), London.

[25] Lipovetsky, G./ Roux, E. (2003) : Le luxe éternel, in: Gallimard (eds.), Paris.

[26] Mason, R. (1992): Modelling the Demand for Status Goods, in: Provo, Utah (eds.): Association for Consumer Research, special volume, pp. 88-95.

[27] McCracken, G. (1988): The Long Interview, Newbury Park, Sage (eds.), UK.

[28] Miles, M. B./Huberman, A. M. (1994): Qualitative Data Analysis: An Expanded Sourcebook, in: Sage (eds.), 2nd edition, UK.

[29] Novak, T.P./MacEvoy, B. (1990): Segmentation Schemes: The List of Values and Lifestyles, in: Journal of Consumer Research, Vol. 17, No. 1, pp. 105-109.

[30] Nueno, J.L./Quelch, J.A. (1998): The Mass Marketing of Luxury, in Business Horizons, Vol. 41, No. 6, pp. 61-69.

[31] Skrinnikov, I.M. (1997): Russians and Luxury, in: Exmo Press (eds.), Moscow.

[32] Suslina, E. (2003): Daily Life of Russian Fashionistas from 11th to 21st centuries, in: Molodaya Guardiya (eds.), Moscow.

[33] Trompenaars, F. (1993): Riding the Waves of Culture, in: Nicholas Brealey Publishing (eds.), London.

[34] Usunier, J.C. (2000): Marketing Across Cultures, in: Prentice-Hall (eds.), 3rd edition.

[35] Vickers, S.J./Renand, F. (2003): The Marketing of Luxury Goods: An Exploratory Study – Three Conceptual Dimensions, in: The Marketing Review, Vol. 3, No. 4, pp. 459-478.

[36] Vigneron, F./Johnson, L. (1999): A Review and a Conceptual Framework of Prestige-Seeking Consumer Behavior, in: Academy of Marketing Science Review, Vol. 1999, No. 1, pp. 1-17.

[37] Vigneron, F./Johnson, L. (2004): Measuring Perceptions of Brand Luxury, in: Journal of Brand Management, Vol. 11, No. 6, pp. 484-506.

[38] Wiedmann, K-P./Hennigs, N./Siebels, A. (2009): Value-Based Segmentation of Luxury Consumption Behavior, in Psychology and Marketing, Vol. 26, No. 7, pp.625-651.

[39] Zelinsky, F. M. (1965): Typology of Slavonic Cultures, Exmo Press (eds.), Moscow.

Appendix: Themes and subthemes for France, UK and Russia

Table 3.13 Themes and sub-themes for France

Themes	Sub-themes	Word frequencies		Main themes frequencies %
		Number 517	%	
Aesthetics	Design and style	32	22%	**29%**
	Beauty	29	20%	
	Simplicity and chic	23	15%	
	Elegance	21	14%	
	Colour	18	12%	
	Colour and Materials	15	10%	
	Tradition and Fashion	11	07%	
	Total	**149**		
Personal History	Famous brand name	30	23%	**25%**
	Security guarantee	25	19%	
	Reflect consumer values	22	17%	
	Know-how	20	15%	
	Craftsmanship	20	15%	
	Cultivates consumer tastes and image	14	11%	
	Total	**131**		
Premium Quality	Superior materials and colours	28	31%	**18%**
	Longevity / Durability	27	29%	
	Pass on to generations	27	29%	
	Become better with time	10	11%	
	Total	**92**		
Expensiveness	High price	20	41%	**9%**
	Exclusive	18	37%	
	Elite	11	22%	
	Total	**49**		
Self-pleasure	Comfort	11	32%	**7%**
	Pleasure	11	32%	
	Enjoyment	8	22%	
	Interesting life	5	14%	
	Total	**35**		
Aspiration	Dream	11	31%	**7%**
	Sophistication	8	24%	
	Class	8	24%	
	Self-refinement	7	21%	
	Total	**34**		
Conspicuousness	Show-off	10	37%	**5%**
	Different from others	10	37%	
	Association with the trendy	7	26%	
	Total	**27**		

Source: De Barnier, Valette-Florence

Table 3.14 Themes and sub-themes for the United Kingdom

Themes	Sub-themes	Word frequencies		Main themes frequencies %
		Number 560	%	
Premium Quality	Longevity	31	28%	**19%**
	Tradition	31	28%	
	Pass on to generations	28	26%	
	Becomes better with time	20	18%	
	Total	**110**		
Functionality	Usefulness	34	36%	**17%**
	Comfort	33	35%	
	Protection and security	27	29%	
	Total	**94**		
Luxury Atmospherics	Impeccable service	25	27%	**16%**
	Soft selling versus hard selling	20	22%	
	Interior design	18	19%	
	Merchandise presentation	16	17%	
	Classy	14	15%	
	Total	**93**		
Aesthetics	Boutique beauty and quality	23	26%	**15%**
	Presentation	20	23%	
	Selling skills of vendors	20	23%	
	Beauty	15	17%	
	Design	10	11%	
	Total	**88**		
Self-pleasure	Make life more interesting	27	37%	**13%**
	Enjoyment	20	28%	
	Pleasure	15	21%	
	Self-cultivation	10	14%	
	Total	**72**		
Personal history	Know-how and tradition of manufacturing	25	37%	**12%**
	Good reputation	23	33%	
	Favourable references	21	30%	
	Total	**69**		
Expensiveness	High price	34	100%	**6%**
	Total	**34**		

Source: De Barnier, Valette-Florence

Table 3.15 Themes and sub-themes for Russia

Themes	Sub-themes	Word frequencies		Main themes frequencies %
		Number 852	%	
Aesthetics	Beauty and Originality	32	21%	**19%**
	Design	30	19%	
	Materials and Colours	30	19%	
	Magic atmospherics	25	17%	
	Never out of fashion	22	13%	
	Piece of Art	20	11%	
	Total	**159**		
Premium Quality	Premium Quality	36	26%	**17%**
	Longevity	33	23%	
	Expensive materials	28	20%	
	Pass on to generations	24	17%	
	Resistance to unfavourable conditions	20	14%	
	Total	**141**		
Personal History	Famous brand name	35	26%	**16%**
	Famous designer	30	22%	
	Traditional know-how	28	20%	
	Long story and interesting associations	25	18%	
	Quality guarantee	20	14%	
	Total	**138**		
Functionality	Usefulness	38	29%	**15%**
	Multi-purpose	32	25%	
	Comfort	25	19%	
	Easy to operate	20	15%	
	Security	15	12%	
	Total	**130**		
Self-Pleasure	Enjoyment	28	29%	**12%**
	Intellectual growth	25	26%	
	Sensorial satisfaction	24	23%	
	Admiration from others	20	22%	
	Total	**97**		
Uniqueness	Limited number	35	36%	**11%**
	Different from others	31	33%	
	Original image	29	31%	
	Total	**95**		
Expensiveness	Premium price	37	40%	**10%**
	Quality guarantee	30	33%	
	Craftsmanship guarantee	25	27%	
	Total	**92**		

Source: De Barnier, Valette-Florence

4 An Intercultural Comparison of the Perception of Luxury by Young Consumers

Bruno Godey, Daniele Pederzoli, Gaetano Aiello, Raffaele Donvito, Klaus-Peter Wiedmann, Nadine Hennigs

4.1 Introduction

Recently, has there been a change in the perception of luxury: from an elitist concept to a consideration of luxury as a many-faceted issue [3], [10]. In the last few years, it has become clear that luxury consumption is open to different kinds of consumers, and that it therefore contributes to defining differentiated identities.

This article analyses the perception of the concept of luxury and ranks some adjectives defining luxury among an international sample of young people.

In order to test the research questions empirically, a statistically non-representative sample of 233 undergraduate management students from European, Asian and American Universities was set up. These students worked in teams to respond to a two steps research process:

- In the first step, students were asked to define luxury in their native language and in English;
- In a second phase, respondents were asked to rank 10 adjectives proposed by researchers and defining luxury depending on the relative weight of each adjective to explain the concept of luxury.

Our results show that the perception of luxury is multi-faceted and includes elements both conspicuous and status oriented elements and others that are more oriented towards the hedonistic and emotional dimensions. Also We found strong cross cultural differences when evaluating the different dimensions of luxury through the proposed adjectives; these differences, however, are not defined by the level of maturity of the luxury marketing in each country, as we had supposed at the beginning of our research, but they also present variations inside the most developed markets.

The first part of our article is devoted to presenting the concept of luxury and its evolution in the literature concerning luxury perception and luxury evaluation

In the second part, we develop the research questions and we explain the methodology adopted for this research.

The third part presents the main empirical results concerning the young people's perceptions of luxury, the evaluation of the different adjectives proposed and the cross cultural differences discovered.

Finally, we present a general discussion and the conclusion of this research.

4.2 The Concept of Luxury

Following the recent difficulties experienced in 2008-2009 and despite the return to growth in 2010-2011 (See for example forecasts from the Bain & Company's study "Luxury Goods Worldwide Market", spring 2011), businesses need to make smart strategic choices in international marketing. This research is part of the managerial perspective creating a better understanding of the cultural differences in perception and purchasing behavior linked to luxury brands.

There is general agreement about the lack of "substantial" definitions of luxury. That is to say, based on general intrinsic characteristics of products, techniques employed or the techno-economic characteristics of the industry [10]. However, a certain consensus is emerging about the major characteristics desired by the consumer of luxury brands. The common denominators are beauty, rarity, quality and price but also the existence of an inspirational brand endorsing the product. In this context, brands compete on the basis of their ability to evoke exclusivity, identity, awareness and perceived quality for the consumer.

Researchers in consumer behavior have taken two theoretical directions to explain the consumption of luxury goods. They were initially inspired by the work of social psychology to focus almost exclusively on socially oriented motivations [8], [6]. More recently, this trend has been complemented by theoretical explanations based on a personally oriented vision [37].

Motivations behind the acquisition of luxury brands have traditionally been reduced to the notion of "conspicuous purchase". This idea tends to be still more or less the strategic foundation for the management of luxury brands [5], [6], [37], [27]. From this perspective, which has its origins in sociology and social-psychology, via the Theory of Impression Management, consumers strongly orient their behavior towards the creation of a favorable social image that they can build through their purchases [8][6]. Brands are then used as vectors to implement two distinct consumption strategies. On the one hand, they are the visible symbols of consumer tastes (i.e.: "social salience") and, secondly, they are regarded as icons representing certain social groups, and thus help consumers to strengthen their membership of these groups (i.e.: "social identification").

A number of researchers have enriched the traditional vision of luxury consumption [36], [32], [34], [33], [37]. In this revised paradigm, two types of luxury consumption orientation (social and personal) must be considered in the management of luxury brands. Wong and Ahuvia [37] were the first to show that the personal orientation towards luxury brands was more important for some consumers than others. When these consumers choose a luxury brand, there are usually utilitarian, emotional and symbolic dimensions that underlie their personal orientation.

Regardless of the perspective used, previous research in the field of luxury showed the particular importance played by the brand as the vector of strategies used by consumers in their decision making process [14], [33][34].

On the basis of the empirical and theoretical contributions outlined so far, it can thus be noted that the concept of luxury and the attitude towards luxury consumption among the young from different countries display similarities as well as differences. On the grounds of this consideration, we believe it is useful and interesting to endeavour to formulate a more detailed cross-cultural analysis of the concept of luxury amongst the young.

4.3 Brand Functions and Brand Relational Dimension

The current crisis is encouraging companies to look more deeply into the links between consumers and luxury brands. Consumption of luxury products is based on two main reasons: the purchase for one's self, for pleasure and purchase as a symbol of success. For Kapferer [16], the future of luxury brands is based on the search for balance between these two orientations of consumers, according to geographical areas. In addition, luxury is often consumed for gift-giving. Nevertheless, whatever the perspective chosen, the brand remains the main vehicle for connecting with the consumer.

A brand may influence customers' perceptions and attitudes in several ways. The dominant components of the brand influence upon customers' purchasing intentions may be analysed through the investigation of two complementary key issues: a functional dimension and a relational dimension.

4.3.1 The functional brand

It is commonly accepted that the brand usually represents "the memory" of a firm, which encompasses all of the investments, research activities, and process technologies or innovations the firm carries out over time [37]. Nevertheless, customers may use brands as a vehicle or mode of expression of attitudes, individualism, and needs [18]. According to Keller and Lehmann [22], brands can simplify choices, be synonymous with a particular level of quality, reduce risk and generate trust.

The development of the concept of brand equity has resulted in a significant evolution of the brand concept itself. The model of brand equity proposed by Keller [19] became dominant, providing the link between its two dimensions: brand awareness and image.

The image of a brand is characterized in the consumer's mind by a set of associations [20], [22]. In this perspective, the brand meaning can be formed through functional elements of performance or more abstract elements related to the imaginary [21], [19]. Brand performance is linked to its intrinsic properties and to how the consumer perceives the fit between the brand and his functional needs (features, quality of product, services related to the brand, style and design, price). Brand meaning also involves extrinsic properties and how the brand meets the psychological or social needs of consumers (user profiles, situation, personality and values, heritage and experiences).

The examination of specific functions the brand performs further illuminates the analysis of brand influence over consumers' perceptions, and purchasing decisions. Kapferer & Thoenig [17], in addition to Lambin [25], classify a series of utility functions that can be attributed to the brand. These functions can be useful both for customers (placement, guarantee, personalization, practicality, pleasure functions), and for manufacturers (protection, positioning, capitalisation).

According to Keller [19],[21] the brand is structured upon the basis of three fundamental components: firstly, the identity component (signs of recognition); secondly, the perceptual component (cognitive associations and perceptions) [28]; and lastly, the trust component (confirmation of expectations).

4.3.2 The brand relational dimension.

In continuation of work on brand equity and on the process of creating abstract associations, current research aims to consider the brand as a symbolic entity with which the consumer maintains an interpersonal relationship [1], [37]. Thus, some researchers have extended and enriched work on possessions [2], [37], [23] to apply them to the brand.

From this perspective, consumers search for emotional elements, which are sometimes anchored in socio-cultural trends towards which customers feel a sense of belonging [12], [11]. Customers search for emotional elements when they are tangible, and objective elements play a secondary role. Therefore, firms attempt to create a "symbolic" universe surrounding their products as a way to reinforce consumers' brand loyalty.

This research stream has led to interest in the long-term emotional bond that is developed by a consumer with a specific brand through the concept of brand attachment [24].

4.4 Research Questions and Methodology

4.4.1.1 Research questions

Our research aims to analyse perception of the concept of luxury and luxury brands in an international sample of young people. In particular, the following research questions are developed:

Question 1a: Is the perception of the concept of luxury among the young consistent with the recent theoretical interpretations developed by scholars?

Question 1b: What are the elements that characterise the perception of luxury among the young today?

Question 2: Are the concepts of luxury and luxury brands perceived differently among young people of different nationalities?

Question 3: What are the preferences of young people in a list of 10 well known luxury brands proposed by the researchers?

Question 4: Are there differences between brands perception according to the country of the respondent?

4.4.1.2 Research methodology

In order to test the research questions empirically, a statistically non-representative sample of 233 undergraduate management students from European, Asian and American Universities was set up. Since we are more interested in basic psychological processes than generalisations, the sample size is considered appropriate [13], [26]. The students of the sample were subdivided into six geographic units (Italy, France, Germany, China, Japan, and USA); subsequently, the respondents (average age 22.6) from each of these geographic units spontaneously group themselves into sub-groups composed of 3-5 members each. Thus an "overall", total of 53 groups was obtained. For each geographic unit, respondents were asked to produce the following outputs:

1. Written definition of the concept of luxury, in their native language and in English;
2. Ranking of 10 adjectives (proposed by the researchers) that illustrate the concept of luxury;
3. Ranking of 10 luxury brands proposed by the researchers according to the preferences of respondents;
4. Profile of the members of the group, containing general and demographic information.

Table 4.1 Composition of the sample analyzed

Country	Groups[1]		Young people		Male		Female		Age (average)
	%	n.	%	n.	%	n.	%	n.	
Italy	32.08%	17	25.32%	59	40.68%	24	59.32%	35	22.2
France	9.43%	5	6.87%	16	37.50%	6	62.50%	10	22.8
Germany	22.64%	12	25.75%	60	23.33%	14	76.67%	46	24.2
China	7.55%	4	9.44%	22	36.36%	8	63.64%	14	21.7
Japan	13.21%	7	15.45%	36	63.89%	23	36.11%	13	21.1
USA	15.09%	8	17.17%	40	45.00%	18	55.00%	22	22.5
Total	**100.00%**	**53**	**100.00%**	**233**	**39.91%**	**93**	**60.09%**	**140**	**22.6**

The research steps are defined on the basis of a specific methodological perspective (of a predominantly qualitative nature) followed in this research. Firstly, "qualitative text content analysis" is used [31], which aims to measure the recurrence of specific units of analysis within a text. In the second place, we devised a semiotic representation of the signifiers (lexical units illustrative of a concept) contained in the definitions of the concept of luxury as expressed by the young people in the survey [9].

[1] To reflect the fact that the number of groups per country is unequal, weighted means were used for the rest of the article for the "overall".

4.5 General Empirical Results of Young People's Perceptions of Luxury: A Semiotic Analysis of the Concept of Luxury and the Ranking of Luxury Brands

In this part of our paper an aggregate analysis of respondent's perceptions of the concept of luxury and luxury brands is presented.

Table 4.2 contains the results of the analysis of the lexical units contained in the definitions expressed by 53 different groups from 6 countries. The percentages of definitions in which a given lexical unit is used to describe luxury are calculated with the aim of giving an estimate of the relative importance of the main lexical units that emerged from the spontaneous definition of luxury. In order to avoid redundancy in our analysis, lexical units that have the same meaning or are synonymous are grouped in the same category.

The data in **Table 4.2** shows that the first concept associated with luxury is "expensive", followed by "quality", while in the third place concepts correlated with "exclusiveness" can be observed. Furthermore, respondents quote also "unique-uniqueness" (luxury as a dimension that cannot be rivaled), "desire" and "dream" as key elements to define luxury.

Less important for respondents appear some other sets of definitions characterized by the value of the product ("price" and "cost") or by the ostentatious side of luxury goods ("excess", "sophisticated", "extravagant").

Table 4.2 Analysis of the lexical units in the spontaneous definitions of luxury

Lexical units	Occurrence (%)
[Expensive]	36
[Quality]	29
[Exclusive-exclusivity-exclusiveness]	24
[Desire-desirable]	22
[Dream] – [Unique-uniqueness] – [Symbol-symbolic-symbolize] – [Material]	20
[Excess] – [Satisfy-satisfaction] – [Emotion- emotional]	18
[Price] – [Unnecessary]	13
[Immaterial] – [Prestige-prestigious] – [Rare-rarity] – [Sophisticated]	11
[Cost] – [Superior-superiority] – [Elite-elitist] – [Not essential] – [Comfort]	9
[Elegance] – [Aesthetic] – [Useless] – [Extravagant] – [Superfluous] – [Pleasure] – [Exceptional]	7
[Democratic] – [Conspicuous] – [Snob]	2

The results suggest that young people's perceptions of luxury are consistent with both the theories of ostentatious consumption behavior ("hetero-referring luxury") and with hedonistic consumption theories ("self-referring luxury"). Among the different definitions of luxury there are concepts explaining luxury as self-referred ("desire", "dream", "emotional") and others that much more hetero-referred ("expensive", "exclusiveness", "price", "quality", "extravagant").

Through an interpretation drawing on the semiotic square [11], we endeavor to establish a link between the different lexical units emerging from the survey and the theoretical perspectives explaining luxury and its consumption **Figure 4.1**. According to our data analysis, two "poles" can be observed; both exhibit a fairly marked power to attract these lexical units. Thus on the one hand we see the pole of "conspicuous and status-related luxury"; while on the other we find "emotional-hedonistic luxury". As is made clear in the Figure 1, while some lexical units are more firmly located close to one of the two poles, others fall within an intermediate area which is not fully identified with either extreme.

Figure 4.1 Semiotic representation of the lexical units contained in the definitions of the concept of luxury

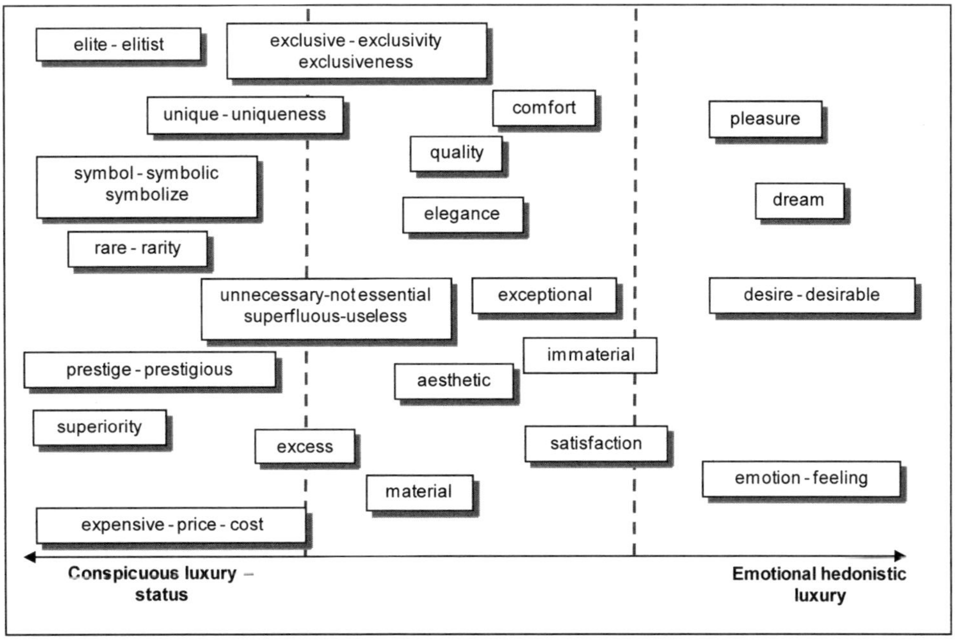

The conspicuous or status-related pole is characterized by the interpretation of luxury based on a clear cut hetero-referring perspective, typified by concerning "uniqueness", "superiority", "prestige", "symbolism", "rarity", and "price". The emotional-hedonistic pole is characterized by the lexical units relating to "desire", "dream", "emotion", "feeling", and "pleasure" which in turn take on a self-referring nuance associated with emotions and feelings. The self-referring dimension can also be seen in some definitions (16.7%) that highlight the relativity of the concept of luxury, suggesting that it is influenced by individual conditions. Finally, the area covering the intermediate area between the two poles may be represented as a "hybrid semiotic space" within which the lexical units undergo a twofold and simultaneous "attraction" from the two opposite poles. In our view, the interpretation of this specific space would need further investigation which, for reasons of space, cannot be performed on this occasion.

Groups of respondents were also asked to evaluate the extent to which ten adjectives proposed by the researchers from the literature [15], [7], [34], [32] were consistent with their concept of luxury. The results are shown in **Table 4.3**, which presents the mean for each adjective, both for all the groups from the 6 countries and for each country analyzed. From a methodological point of view, the respondents attributed a score (1 to 10) to each adjective, with the maximum consistency corresponding to 10 and the minimum to 1.

For each country, we also introduce in **Table 4.3** the ranking of the first four adjectives to facilitate understanding of the table and comparison with the overall results.

Table 4.3 Evaluation of the consistency of ten adjectives to explain luxury (mean, ranking, gap (+/-) compared to Overall)

Adjectives	Overall	Italy			France			Germany			China			Japan			USA		
Exclusive	8.37	9.50	1	+	7.20	4	-	9.00	1	+	3.75		-	7.29	3	-	9.00	1	+
Prestigious	7.64	7.38	3	-	8.60	1	+	6.58	4	-	8.50	1	+	8.14	2	+	8.29	2	+
Expensive	6.77	6.19		-	8.40	2	+	6.67	3	-	7.25	3	+	9.29	1	+	4.71		-
Elitist	6.46	7.44	2	+	7.60	3	+	5.92		-	4.25		-	7.00	4	+	5.14		-
Desirable	6.23	6.56	4	+	4.40		-	7.33	2	+	6.75		+	4.14		-	6.57	3	+
Sophisticated	4.91	4.63		-	4.80		-	4.17		-	3.50		-	7.00	4	+	5.57		+
Conspicuous	4.09	3.50		-	3.60		-	3.50		-	7.00	4	+	5.57		+	3.14		-
Extravagant	4.07	2.75		-	2.60		-	5.33		+	8.00	2	+	1.57		-	6.14	4	+
Emotional	3.66	4.31		+	3.60		-	3.67		=	2.75		-	1.71		-	4.43		-
Snobbish	2.77	2.75		=	2.80		=	2.83		+	3.25		+	3.29		+	2.00		-

As can be noted, the overall evaluation highlights "exclusiveness" as the most adequate adjective for explaining luxury, followed at a certain distance by two other definitions, "prestigious" and "expensive". The hetero-referred adjectives seem to predominate in this suggested ranking, while it is interesting also to underline that the adjectives expressing a sort of negative perception of luxury ("conspicuous", "extravagant", and "snobbish") are ranked in the lower end of the table.

The cross cultural analysis appears very interesting:

- **Italian** respondents put "exclusivity" in the first place, followed far behind by "elitism" and "prestige", but we can also see a self-referred item, "desirable", not far from them. Italians seem to have a well stratified image of luxury.

- **French** respondents have a traditional concept of luxury, predominantly hetero-referring and based on "prestige", "expensiveness" and "elitism".
- **German** interviewees put in the first place, far above every other adjective, "exclusive", but at the second place in the ranking is "desirable", followed by "expensive" and "prestigious" almost with the same score. So, for German respondents, we can draw similar conclusions as for Italians.
- **Chinese** respondents express a concept of luxury predominantly linked to "prestige" and "extravagance". The general image of luxury as something unusual is confirmed for Chinese respondents by the 4th position occupied by "conspicuous". For both "extravagant" and "conspicuous", the Chinese interviewees' score is the highest out of the six countries.
- For **Japanese** respondents, "expensive" is clearly the most appropriate adjective for luxury, followed by a group of four other adjectives linked with "sophistication", "prestige" and "exclusiveness/elitism".
- **US** respondents express a concept of luxury grounded on the adjective "exclusive" and "prestigious", but their vision of luxury appears well stratified too, because of the ranking of "desirable" in third place and "extravagant" just below.

Going on clarify whether there are differences between countries, we work on the differences between these criteria by each area of consumption. For this we use a variance analysis (one-way Anova). This method makes it possible to use a univariate analysis of variance on a quantitative dependent variable by one factor (independent variable). Analysis of variance is used to test the hypothesis of equality of means.

The Fisher's F-test provides the opportunity to test the equality of two variances by the ratio of two variances and verifying that this ratio does not exceed a certain theoretical value. Thanks to this test, it is possible to say that the greatest differences are observed for the variables "extravagant", "exclusive" and "expensive".

Wishing to then determine those countries for which there are significant differences between group means, we conducted Duncan's post-hoc tests.

If we analyze the results amongst groups of countries (Case ① on the **Figure 4.2**), we can see that for only two adjectives it is possible to find three different well differentiated groups, but with strong similarities concerning the countries inside each group; this situations concerns:

- "Exclusive"; we can form a first group composed by Italy, USA and Germany, a second group composed by Japan and France, when China is isolated and far from the other countries;
- "Extravagant"; for this adjective, China is isolated at the top of the ranking, USA and Germany follow in a second group, Italy, France and Germany are grouped at the bottom of the ranking.

Figure 4.2 The perception of luxury (Mean and Duncan's post-hoc tests) [1]

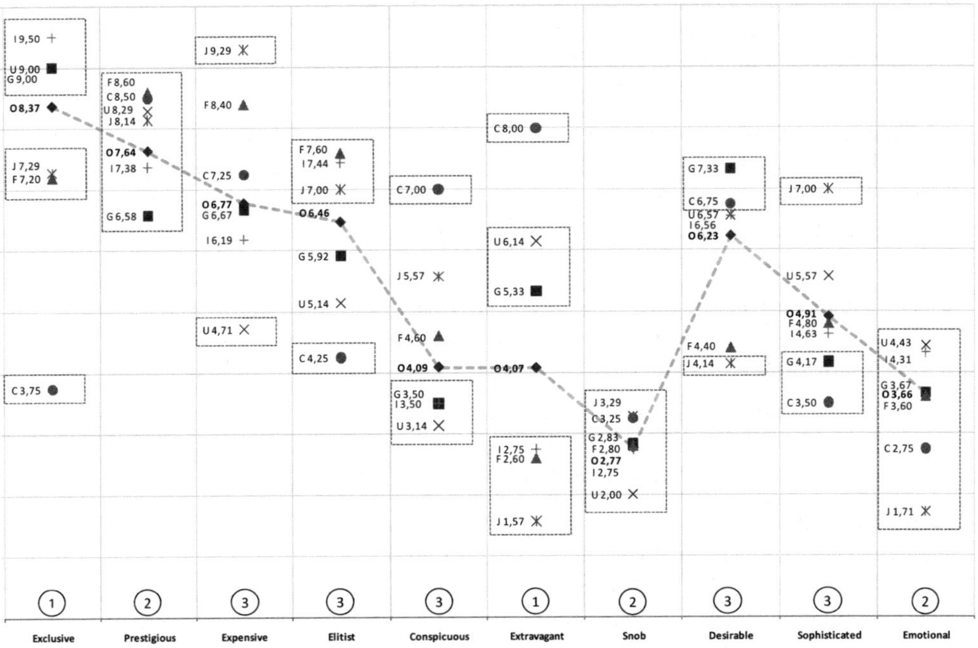

For three other adjectives ("prestigious", "snob", and "emotional" – Case ② on the **Figure 4.2**), analysis of similarities between countries result in the formation of one statistically significant group composed of all the countries.

Concerning the five other adjectives ("expensive", "elitist", "conspicuous", "desirable", and "sophisticated" – Case ③ on the **Figure 4.2**), there is no clearly delineated grouping, adjective by adjective. It is therefore necessary to examine whether stable groups of countries could be identified.

When we started our research, intuitively we supposed that we would have found differences between countries where luxury consumption is long established and other countries that started luxury consumption more recently. Our findings are more complicated than our first idea:

1 [dashed box] Groups of countries from the Duncan's post-hoc tests - O = Overall; I = Italy; F = France; G = Germany; C = China; J = Japan; U = USA

- USA and Germany can be grouped together based on the consumers' evaluation of "exclusivity" and "conspicuousness" respectively at the top and bottom of the ranking. These results are confirmed by some spontaneous definitions of luxury by young respondents:

  "Luxury is what a person considers as rare, prestigious, exclusive and therefore hard to reach. The definition is not only to be understood in a material way, it can be also be in an ideal way. (Germany group Spring Break)

  "Our concept of luxury can be a way of living that is above the standard and is only accessible for few people. Luxury from this perspective contains mostly monetary goods like money, wealth in general, brands, status, etc. (Germany group Deluxe).

- A second group emerges from our analysis and it is composed by France and Japan; these countries share opposite evaluations for "elitism", where they are at the top of the ranking, and "extravagance", the least valuable adjective to define luxury for the respondents of the two countries. To confirm this situation we can quote some spontaneous definitions of luxury written by respondents in the first part of our research:

  "Luxury is a product or a nice aesthetic service which is very qualitative, exceptional, expensive, and rare; it brings a positive and flattering image of the consumer. Consuming luxury goes beyond primary need satisfaction" (France group Q3)

  "Luxury means that it is not necessarily gorgeous or new but rich and expensive. It is what can't get without much money. And it is something clean, sophisticated, aesthetic, and shine. Therefore, it refers to something like commodities, foods and places..." (Japan group Zeta)

- The situation concerning Italy is more multi-faceted because Italy appears in the same group as the USA and Germany for "exclusive", but also for the less evaluated "conspicuous"; at the same time, we can find Italians sharing the same group with France and Japan for "elitist" and again for "extravagant", the last adjective for the three countries. This analysis confirms the well stratified image of luxury for Italian young respondents that we have underlined above. Some definitions of luxury written by Italian respondents confirm these findings:

  "Luxury is about exclusivity, something that only "happy-few" can reach. Something which is rare, bound to decade and disappears is a luxury for those who can achieve it. Luxury is beauty, aesthetic and elegance…." (Italy group Mirror)

  "... includes both material goods and feelings connected with an exclusive situation, exceeding individual necessity. Luxury is always expensive, on the one hand it concerns spending money and on the other hand it can be related with opportunity costs. (Italy, group Marylin Monroe)

- China presents a very differentiated position for most of the adjectives proposed; young respondents from this countries occupy the first place between the six countries analyzed for two adjectives, "conspicuous" and "extravagant", when they rank at the last place for "exclusive" and "elitist". The image of luxury for young Chinese consumers are clearly oriented more towards "status" than "emotion" in the luxury sector, but at the same time they tend to prefer the more self-directed elements of luxury. Some spontaneous definitions confirm these points:

"Something related to better quality and expensive. It is used to differentiate oneself to the others in order to satisfy one's vanity and happiness. Luxury is an unlimited desire. It is something neither practical nor essential for everyone" (China group 2).

Luxury is related to three concepts. First, luxury items are value-added and made of expensive materials. For example, diamond, lace, silk and fur. Luxury is related to western countries and privilege of western people. (China group 3).

4.6 The ranking of luxury brands

In this part of the research, we asked groups of young respondents to rank 10 luxury brands according to their preferences. The ten brands were proposed by the researchers after a two step selection process: firstly, each research unit proposed a list of ten luxury brands considered to be the most representative in the field of fashion luxury; then, the brands appearing in all the list, were retained and an exchange was organised among the researchers to choose the brands that could complete the list. All the brands in the list were finally chosen, with the agreement of all the research units.

The brands have been ranked by young people respondents using a scale from 1 (the preferred one) to 10 (the least evaluated).

The results of this last part of the research are presented in **Table 4.4**.

Table 4.4 Ranking of the ten luxury brands proposed by researchers – Overall

Brand	Rank
Cartier	1
Chanel	2
Hermès	3
Gucci	4
Louis Vuitton	4
Bulgari	6
Dior	7
Ferragamo	8
Prada	8
Armani	8

As it is possible to see by analyzing the results presented in **Table 4.4**, the first positions in the ranking are occupied by brands with a long history and a strong reputation built over time, whereas younger and less "historical' brands are ranked at the bottom of the list.

We do not know exactly when these brands first entered in each of the foreign market we analyzed, but the presence and history of each brand in the different countries of our research can help to explain this general ranking.

The overall ranking, however, hides important differences amongst the preferences expressed in each country of our research.

Table 4.5 Ranking of the ten luxury brands proposed by researchers – by country

	Overall	Italy		France		Germany		China		Japan		USA
Cartier	1	1		4	-3	1		2		2		1
Chanel	2	6	-4	1		2		3		4		4
Hermès	3	7	-4	2		8	-5	1		1		2
Gucci	4	3		3		4		7	-3	7	-3	5
Louis Vuitton	4	2		6		9	-5	6		6		3
Bulgari	6	5		8		6		4		3	+3	8
Dior	7	9		5		7		7		8		6
Fer-ragamo	8	8		10		10		5	+3	5	+3	9
Prada	8	10		7		3	+5	9		9		7
Armani	8	4	+4	9		5	+3	10		10		10

In **Table 4.5** we indicated the rank for each brand in every country and the main differences between the overall ranking and the position in the national one.

It is possible to notice that even same historical and well known international brands do not occupy the same place everywhere in our survey, but their perception from young consumers shows strong differences. It is the case for Cartier that occupies one of the top two positions in the overall ranking and for 5 countries out of the 6 analyzed. For French young respondents, however, Cartier ranks at the 4th position between the luxury brands proposed.

In the case of Chanel, it is the Italian consumers who classify the brand well below the overall ranking. Hermès is perceived well below the overall ranking by German and Italian respondents. German young respondents again rank Louis Vuitton at the bottom of their national ranking. Gucci holds a strong position in all the countries except Japan and China.

When we look at the second half of the overall rankings, we can underline that some brands appear to have a stronger perception in some specific markets; these perceptions are well above the overall for Bulgari in Japan, for Ferragamo in Germany and Japan, for Prada again in Germany and for Armani in Italy and Germany.

Again for this part of the research, our initial hypotheses, that differences can be found between countries where luxury brands are well established for several decades and other countries that are newer in the field of luxury consumption is not confirmed; reality is more multifaceted and the relative position of each brand in each country is certainly explained by the marketing strategy adopted, the strength of the store network and the communication adopted. The analysis of these elements, however, was not an objective of our current research, but can open many perspectives for future research concerning brand perception.

All the brands analyzed are from Italy or France, so it can be interesting to study if there are differences in the perception between European and non European young respondents. The results of this analysis are presented in **Table 4.6**

Table 4.6 Ranking of the ten luxury brands proposed by researchers between European and non European young respondents

	Cartier	Chanel	Hermès	Gucci	Louis Vuitton	Bulgari	Dior	Ferragamo	Prada	Armani
European ranking	1	3	5	2	4	7	8	10	9	6
Non European Ranking	2	3	1	6	4	4	8	6	9	10
Diff European vs non European ranking	-1	0	+4	-4	0	+3	0	+4	0	-4

Four brands out of ten obtain the same ranking between European and non European young respondents, but only Chanel and Louis Vuitton are in the "top four" for both. Cartier does not present a strong difference between the two groups of respondents and the brand shows a very strong perception both in Europe and outside Europe.

For the other five luxury brands proposed, differences in the ranking are very strong, with three brands much better perceived outside Europe (Hermès, Bulgari and Ferragamo) and, on the contrary, two brands having a stronger position in Europe, namely Gucci and Armani.

If we consider the strong investments made by some brands in the USA, in Japan and in China, these differences can be explained by the strategy adopted by some brands that have been targeting mainly non European markets for many years now.

For other brands, the better position in Europe can be explained by their strategy that has been more oriented towards the construction of a strong European base before moving to other continents.

Finally, some brands appear to have a very strong reputation everywhere; they can be considered very "institutional" in the field of luxury and they are well known and appreciated in Europe and outside Europe.

It is interesting to underline that Louis Vuitton occupies the fourth place in the two cases analyzed; the brand is well ranked, but below other brands that occupy the "top three ranking" in Europe (Cartier, Gucci and Chanel) and outside Europe (Hermès, Cartier, Chanel).

If we consider the ranking of Dior, in eighth position for both the regions analyzed, we can suppose that the "democratization" of historical brands can reduce their luxury image. According to our results, Louis Vuitton and Dior are not perceived as being so exclusive and elitist as some other brands with a similarly long history. If we remember that the first adjectives used by our respondents to define luxury are "expensive", "quality" and "exclusiveness" the brand extensions made by Louis Vuitton and Dior have probably reduced the brand equity, at least for young consumers.

4.7 Discussion

The strongest point of our research is the cross-country analysis made possible by the cooperation of six research groups from three continents.

Our findings confirm the multi-faceted concept and evaluation of luxury that emerged from previous studies, but at the same time show clearly that strong cross-cultural differences exist in the evaluation and meanings of luxury.

Although this study was only exploratory in nature, a number of comments can be made to highlight the congruence between the concept of luxury for young people and recent academic literature.

Our results suggest that young people's perceptions of luxury are linked with both the theories of ostentatious consumption behavior and with hedonistic consumption theories but some differences emerge in the cross-country analysis.

The characteristics of the "democratization" of luxury do not appear to be supported by our findings (Research Question 1a). Concerning the main elements that characterize luxury, respondents give a special importance to "exclusivity"; "prestige" and "elite", whereas the first self-referred item, "desirable", appears only in the fifth position of the suggested list ranking (Research Question 1b).

Some relevant differences can be observed in the cross-country analysis (Research Question 2), and it is possible to find these differences amongst three different groups of countries, at least for some of the adjectives proposed. The distinction is not between developed and emerging countries, but also inside the developed countries of our research.

Managerial implications of our research indicate that international luxury companies should take into consideration the multi-faceted concept of luxury in general, but also the main differences between countries in the continuum between the "status" and the "emotional" dimensions of luxury. According to our research, luxury companies should adopt a multinational strategy when addressing the six countries analyzed, taking into consideration the specificities of each country in the perception of luxury.

There are however also some elements common to all the countries analyzed: the most important appear to be the concepts of exclusivity and prestige. This point should be considered very carefully by luxury companies because the strategy of luxury democratization that started some years ago could dilute the value of luxury brands. Some extension of luxury brands into categories not clearly linked with their core business should be analyzed because the search for increased turnover and market penetration can represent a threat to the exclusivity perceived by customers.

Luxury companies have an important job to do to create emotional relations with their customers; according to our results, young consumers relations with luxury brands is based much more on rational concepts than on emotional ones. It appears that the efforts of the brands so far to build this emotional relation have been not so successful, at least for young consumers. This result can be explained by two factors: on the hand, young respondents are not so familiar with luxury brands because of their age; on the other hand, they have generally a limited purchasing power, so luxury goods are seen as desirable and a sign of achievement.

Clearly, if luxury brands are interested in establishing and consolidating a relationship with young consumers, they have to reconsider the tools required to achieve this objective.

Another managerial implication from our study appears to be the "western style" image for luxury goods. This is clearly stated by a team of Chinese respondents, but we can underline also that for Japanese consumers the first adjective defining luxury is "expensive" and it is well known that western luxury products are very expensive in the Japanese market and the positioning of some western brands in Japan is clearly more upmarket than in their home country because of this expectations of Japanese consumers concerning western brands and products.

Concerning the perception of luxury brands, two main points can be relevant for managers:

- In the six countries surveyed, only very few brands are perceived at the same level and strong differences exist in the evaluation of some other brands ;
- When we compared European and non European evaluations of the brands, some important differences emerged. These differences can be explained by the strategy followed by the brands in different regions of the world, but for some brands the general evaluation can also be justified by the extension of the brands in categories not so clearly linked with the concept of luxury.

The limits of our research are firstly in the use of a convenience sample not composed of the main target for luxury goods. It remains, however, that the students in question may be luxury brands customers from time to time and may in the long term become more regular consumers. Our research is qualitative and exploratory, so we cannot draw conclusions concerning the behaviour and attitude of customers in the countries we have studied.

The possibilities for future research are numerous; they can include the possibility of a quantitative research on the perception of luxury addressed to real customers; the possibility of studying the antecedents of brand perception for luxury goods; the comparison between developed countries and emerging markets with regard to the perception of luxury goods.

References

[1] Aaker, J.L. (1997): Dimensions of Brand Personality, in: Journal of Marketing Research, Vol. 34, August, pp. 347–357.

[2] Belk, R.W. (1988): Possessions and the Extended Self, in: Journal of Consumer Research, Vol. 15, No. 2, pp. 139–168.

[3] Berry, C.J. (1994): The Idea of Luxury. A Conceptual and Historical Investigation, Cambridge University Press, New York, NY.

[4] Bomsel, O. (1995): L'industrie du luxe ou comment associer objets et représentations, [Luxury industry or how to combine objects and their representations], in: Annales des Mines – Réalités Industrielles, July-August, pp. 14-20. (in French)

[5] Corneo, G./Jeanne, O. (1997): Conspicuous consumption, snobbism and conformism, in: Journal of Public Economics, Vol. 66, pp. 55-71.

[6] Dittmar, H. (1994): Material possessions as stereotypes: Material images of different socio-economic groups, in: Journal of Economic Psychology, Vol. 15, pp. 561-585.

[7] Dubois, B./Czellar, S./Laurent, G. (2005): Consumer segments based on attitudes toward luxury: Empirical evidence from twenty countries, in: Marketing letters, Vol. 16, No. 2, pp. 115-128.

[8] Eagly, A.H./Chaiken, S. (1993): The Psychology of Attitudes, Harcourt Brace Jovanovich, New-York, NY.

[9] Eco, U. (1976): A Theory of Semiotics, Palgrave Macmillan, London.

[10] Evrard, Y. and Roux, E. (2005): Culture vs. luxury: the paradoxes of democratization, in: working paper 710, Centre d'Etudes et de Recherche sur les Organisations et la Gestion, IAE Aix-en-Provence, April.

[11] Floch, J.M. (2001): Semiotics, Marketing and Communication. Beneath the Signs, the Strategies, Palgrave Macmillan, London.

[12] Fournier, S. (1998): Consumers and Their Brands: Developing Relationship Theory in Consumer Research, in: Journal of Consumer Research, Vol. 24, March, pp. 343–373.

[13] Grewal, R./Metha, R./Kardes, F. (2000): The role of the social-identity function of attitudes in consumer innovativeness and opinion leadership, in: Journal of Economic Psychology, Vol. 21, pp. 233-252.

[14] Kapferer, J.N. (1997): Managing luxury brands, in: Journal of Brand Management, Vol. 4, No. 4, pp. 251-260.

[15] Kapferer, J.N. (1998): Why are we seduced by luxury brands?, in: Journal of Brand Management, Vol. 6, No. 1, pp. 44-49.

[16] Kapferer, J.N. (2009): The specificity of luxury management: Turning marketing upside down, in: Journal of Brand Management, Vol. 16, No. 5/6, pp. 311-322.

[17] Kapferer, J.N./Thoenig, J.C. (1989): La Marque : moteur de la compétitivité des entreprises et de la croissance de l'économie [Brand : Driver of business competitiveness and economic growth] McGraw-Hill, Paris (in French).

[18] Keegan, W./Moriarty, S./Duncan, T. (1992): Marketing, Englewood Cliffs, Prentice Hall.

[19] Keller, K.L. (1993): Conceptualizing, Measuring, Managing Customer-Based Brand Equity, in: Journal of Marketing, Vol. 57, No. 1, pp. 1–22.

[20] Keller, K.L. (2008): Strategic Brand Management, 3rd ed. Upper Saddle River: Prentice Hall.

[21] Keller, K.L. (2001): Building customer-based brand equity: A blueprint for creating strong brands, in: Marketing Management, Vol. 10, July/August, pp. 15–19.
[22] Keller, K.L./Lehmann D.R. (2006): Brands and Branding: Research Findings and Future Priorities, in: Marketing Science, Vol. 25, No. 6, pp. 740–759.
[23] Kleine, S.S./Kleine, R.E./Allen, C.T. (1995): How Is a Possession "Me" or "Not Me"? Characterizing Types and an Antecedent of Material Possession Attachment, in: Journal of Consumer Research, Vol. 22, No. 3, pp. 327–343.
[24] Lacoeuilhe J. (2000): L'attachement à la marque: proposition d'une échelle de mesure, [brand attachment: Proposition of a measurement scale], in: Recherche et Applications en Marketing, Vol. 15, No. 4, pp. 61-77. (in French)
[25] Lambin, J.J. (1991): Marketing, 1st ed., McGraw-Hill, Milano.
[26] O'Cass, A. (2000): An assessment of consumers' product, purchase decision, advertising and consumption involvement in fashion clothing, in: Journal of Economic Psychology, Vol. 21, pp. 547-576.
[27] O'Cass, A./Frost, H. (2002): Status brands: examining the effects of non-product-related brand associations on status and conspicuous consumption, in: Journal of Product & Brand Management, Vol. 11, No. 2, pp. 67-88.
[28] Peter, P./Olson, J. (1987): Consumer Behaviour. Marketing Strategy Perspectives, Richard Irwin, Inc.
[29] Phau, I. /Prendergast, G. (2000): Consuming luxury brands: The relevance of the "rarity principle", in: Journal of Brand Management, Vol. 8, pp. 122-138.
[30] Rego, L.L./Billett, M.T./Morgan N.A. (2009): Consumer-Based Brand Equity and Fim Risk, in: Journal of Marketing, Vol. 73, November, pp. 47–60.
[31] Schmidt, M. (2010): Quantification of transcripts from depth interviews, open-ended responses and focus groups. Challenges, accomplishments, new applications and perspectives for market research, in: International Journal of Market Research, Vol. 52, No. 4, pp. 483-509.
[32] Tsai, S. (2005): Impact of personal orientation on luxury-brand purchase value: An international investigation, in: International Journal of Market Research, Vol. 47, No. 4, pp. 429-454.
[33] Vigneron, F./Johnson, L.W. (1999): A review and a conceptual framework of prestige-seeking consumer behavior, in: Academy of Marketing Science Review, No. 1, available at: http://www.amsreview.org/articles/vigneron01-1999.pdf.
[34] Vigneron, F./Johnson, L.W. (2004): Measuring perceptions of brand luxury, in: The Journal of Brand Management, Vol. 11, No. 6, pp. 484-506.
[35] Wallendorf, M./Arnould, E.J. (1988): My Favorite Things. A Cross-Cultural Inquiry into Object Attachment, Possessiveness, and Social Linkage, in: Journal of Consumer Research, Vol. 14, No. 4, pp. 531–547.
[36] Wiedmann, K.P./Hennigs, N./Siebels A. (2009): Value-based segmentation of luxury consumption behavior, in: Psychology and Marketing, Vol. 26, No. 7, pp. 625-651.
[37] Wong, N.Y./Ahuvia, A.C. (1998): Personal taste and family face: Luxury consumption in Confucian and Western societies, in: Psychology & Marketing, Vol. 15, pp. 423-432.

5 Consumer Value Perception of Luxury Goods: A Cross-Cultural and Cross-Industry Comparison

Nadine Hennigs, Klaus-Peter Wiedmann, Christiane Klarmann (Germany)

In collaboration with: Suzane Strehlau (Brazil), Bruno Godey, Daniele Pederzoli (France), Agnes Neulinger (Hungary), Kartik Dave (India), Gaetano Aiello, Raffaele Donvito (Italy), Koyama Taro (Japan), Janka Táborecká-Petrovičová (Slovakia), Carmen Rodríguez Santos (Spain), Jaehee Jung, Hyunjoo Oh (USA)

5.1 Introduction

The moderate growth in US and European markets coupled with an increasing demand for luxury goods in emerging economies such as Brazil, Russia, India, and China lead to a substantial growth in the global market for luxury goods. Reaching consumers all over the world across national and cultural borders, the luxury sector is according to Bain & Company expected to grow by 10 percent in 2013. Against this backdrop, international luxury businesses are challenged by the identification and satisfaction of the common needs and desires of global market segments; thus, related research methods and business models require a global perspective [20].

Reasoning that *"classical marketing is the surest way to fail in the luxury business"* [3], p. 2, in an international setting, it is a key challenge for luxury brand managers to understand why consumers buy luxury goods, what they believe luxury is and how their perceptions of luxury value affect their purchase behavior. The objective of the present study is to investigate the question of possible differences and/or similarities in the luxury value perception of consumers in distinct parts of the world. Our paper is structured as follows: first, the concept of luxury and luxury product categories are introduced; second, we focus on the key dimensions of luxury value as conceptualized by Wiedmann, Hennigs, and Siebels [43], [44] as the theoretical framework of this study; and third, we present the methodology and results of an empirical study in collaboration with American, European, and Asian researchers. The results reveal that even if significant differences in the perceived importance of financial, functional, individual, and social value aspects of luxury products exist between countries and with reference to diverse industries, the basic value components of luxury consumption are stable on a global level. Finally the results are discussed in light of implications for further research and managerial practice.

5.2 The Market for Luxury Goods

5.2.1 The Concept of Luxury

"Luxury is a necessity that begins where necessity ends."
Coco Chanel, 1883–1971

Given that the understanding of luxury may differ between individuals, is situational contingent and depends on the experience and individual needs of the consumer [43], the concept of luxury is difficult to define [9]. Even if the term "luxury" is routinely used in our everyday life and is a key component in marketing management, there is little consensus on the meaning of luxury. **Table 5.1** provides a chronological overview of selected definitory approaches:

Table 5.1 Defining the Luxury Concept

Author(s), year	Definition of Luxury
Hume 1752, 1965, p. 48 [21]	*"Luxury is a word of an uncertain signification, and may be taken in a good as well as in a bad sense. In general it means great refinement in the gratification of the senses; and any degree of it may be innocent or blamable, according to the age, or country, or condition of the person. The bounds between the virtue and the vice cannot here be exactly fixed, more than in other moral subjects."*
Rae 1834, 1965, p. 272 [33]	*"The degree of pleasure thus experienced is different in different individuals, and it is scarcely possible to ascertain what its exact amount is in any one; hence the difficulty in most cases, of determining what is, or is not, luxury."*
Kambli 1890, Preface, cit. in Valtin 2008, p. 248 [39]	*"No other moral or social issue is as unclarified as that of luxury, and what behaviour toward it can be considered to be well benefit."*
Davidson 1898 [11]	*"The definitions of luxury have been so various that the perplexity of the public is almost excusable."*
Sekora 1977, p. 23 [35]	*"anything unneeded"*
Grossman and Shapiro 1988, p. 82 [19]	*"Those goods for which the mere use or display of a particular branded product confers prestige on their owners, apart from any utility deriving from their function."*
Oxford Latin Dictionary 1992 [30]	*"soft or extravagant living, (over)-indulgence"* *"sumptuousness, luxuriousness, opulence"*
Dubois and Duquesne 1993, p. 43 [15]	*"Motivated by a desire to impress others, with the ability to pay particularly high prices, this form of consumption is primarily concerned with the ostentatious display of wealth."*
Dubois and Paternault 1995, p. 71 [16]	*"More than other products, luxury items are bought for what they mean, beyond what they are."*
Kapferer 1997, p. 253 [22]	*"Luxury defines beauty; it is art applied to functional items. Like light, luxury is enlightening...Luxury items provide extra pleasure and flatter all senses at once...Luxury is the appendage of the ruling classes."*
Kemp 1998, p. 593 [25]	*"The status of a good as a luxury is partially determined by its natural desirability, and not simply by whether it is an object for conspicuous consumption."*
Nueno and Quelch 1998, p. 62 [29]	*"Luxury brands are those whose ratio of functionality to price is low while the ratio of intangible and situational utility to price is high."*
Bernstein 1999, p. 48 [4]	*"one person's functionality may be another person's luxury"*
Phau and Prendergast, 2000, p. 123, 124 [31]	*"Luxury brands evoke exclusivity, have a well-known brand identity, enjoy high brand awareness and perceived quality, and retain sales levels and customer loyalty."* *"the extreme end of the prestige brand category...where brand identity, awareness, perceived quality and loyalty are important components"*

Author(s), year	Definition of Luxury
Webster's Third New International Dictionary 2002 [42]	*"non-essential items or services that contribute to luxurious living; an indulgence or convenience beyond the indispensable minimum"*
Vickers and Renand 2003, p. 459 [40]	*"symbols of personal and social identity"*
Silverstein and Fiske 2003, p. 3 [36]	*"New luxury – products and services that possess higher levels of quality, taste, and aspiration than other goods in the category but are not so expensive as to be out of reach."*
Merriam-Webster's Collegiate Dictionary 2004 [27]	*"a condition of abundance or great ease and comfort"* *"something adding to pleasure or comfort but not absolutely necessary"*
Vigneron & Johnson 2004, p. 486 [41]	*"Goods for which the simple use or display of a particular branded product brings esteem on the owner, apart from any functional utility."*
Danziger 2005, p. 56 [10]	*"Luxuries are the extras in life that make it more fulfilling, more rewarding, more comfortable, more enjoyable."*
Reich 2005, p. 38 [34]	*"Luxury brands embody an image of products or services which is deep-seated in the mind of the consumer. Based on a specifically designed offering system, it expresses itself in specific associations (i.e. excellent quality, premium price, and exclusivity), resulting in a dominance of the ideal brand function."*
Bernstein 2007 [5]	*"Luxury depends on what the person believes others believe about the product or event."*
Truong et. al. 2008, p. 190 [38]	*"While the rich and affluent may consume luxury goods to assert status and membership to the elite class, the modest may consume the same goods to gain status but with a purely conspicuous intention."*
Chevalier and Mazzalovo 2008, p. viii [7]	*"A luxury brand is one that is selective and exclusive, and which has an additional creative and emotional value for the consumer."*
Atwal and Williams 2009, p. 339 [2]	*"Luxury has traditionally been associated with exclusivity, status and quality."*
Berthon et al. 2009, p. 47 [6]	*"Luxury has an intensely individual component as well: what might be luxury to one person will be commonplace, or perhaps even irrelevant and valueless, to another."*
Kapferer and Bastien 2009a, p. 210 [23]	*"In luxury you communicate in order to create the dream and to recharge the brand's value, not in order to sell."*
Kapferer and Bastien 2009b, p. 321 [24]	*"Luxury is an appreciation of fine works, fine craftsmanship, creativity and the making of a legend."*
Wiedmann et al. 2009, p. 627 [44]	*"Luxury is a subjective and multidimensional construct, a definition of the concept should follow an integrative understanding."*
Shukla 2011, p. 243 [37]	*„Luxury goods are conducive to pleasure and comfort, are difficult to obtain, and bring the owner esteem, apart from functional utility."*

Against the backdrop of the rich facets of the luxury concept, a definition of luxury should not follow a narrow but rather an integrative understanding of the luxury concept. Given that luxury is a subjective and multidimensional construct, as described in the following section, there are numerous product categories that are competing in the market for luxury goods.

5.2.2 Luxury Product Categories

Reasoning that the meaning of luxury varies based on subjective evaluations, there are luxury brands in diverse industry sectors that try to seduce the consumer with the desire for luxury: *"Luxury refers to a specific tier of offer in almost any product or service category"* ([14], p. 115).

Apart from traditional luxury houses mainly from France and Italy with a rich history and long heritage of craftsmanship in product categories such as fashion, leather goods and accessories (e.g., Louis Vuitton and Gucci), brands in other categories such as cars (e.g., Ferrari), jewelry (e.g., Cartier) and watches (e.g., Rolex), consumer electronics (e.g., Bang & Olufsen), and home furnishing (e.g., arts and antiques) are part of the luxury market. Besides, the luxury category encompasses experiential luxuries such as dining in an exclusive restaurant or travelling to luxurious destinations and hotels (e.g., The Ritz Hotel).

In an attempt to identify measures that can be used across industries for purposes of assessing consumers' response to luxury products in different countries, in our empirical study, we focus on the following product categories:

- the purchase of a car,
- the purchase of fashion and accessories,
- the purchase of jewelry and watches,
- dining in an exclusive restaurant,
- the purchase of consumer electronics,
- the purchase of home furnishing.

As a basis to successfully market and monitor luxury brands or products in a cross-cultural context, we present in the following section the four-dimensional model of luxury value perception as conceptualized by Wiedmann, Hennigs, & Siebels [43], [44].

5.3 Conceptualization: Luxury Value Perception

From a consumer's point of view, perceptions of and motives for purchasing luxury products can be explained by four dimensions: the financial, functional, individual and social evaluation of the luxury brand [43], [44]. In this context, the concept of customer perceived value, understood as the *"consumer's overall assessment of the utility of a product (or service) based on perceptions of what is received and what is given"* ([45], p. 14), is conceptualized as a key component that drives consumer attitude and behavior. Based on an integral perceived value concept, **Figure 5.1** illustrates the components of consumer value perception in the context of luxury products.

Figure 5.1 Dimensions of Luxury Value Perception

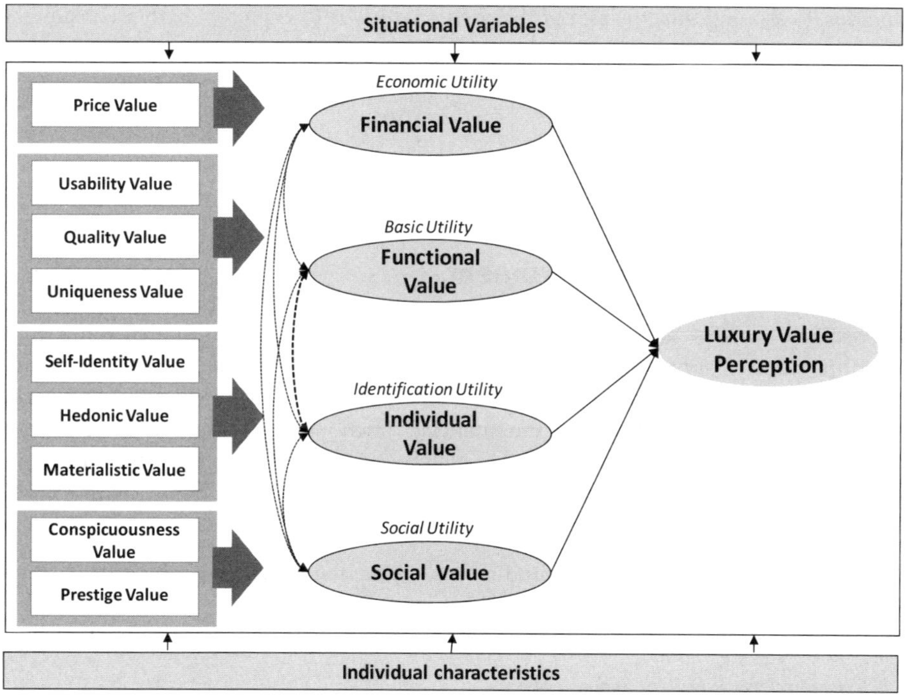

As discussed by Wiedmann, Hennigs, & Siebels [43], [44], the financial dimension of luxury value perception addresses direct monetary aspects, such as price, resale cost, discount, and investment. Referring to core product benefits and basic utilities, the functional dimension focuses on product aspects such as quality, uniqueness, usability, reliability, and durability. Based on customer's personal orientation toward luxury consumption, the individual dimension of luxury value addresses personal issues, such as materialism, hedonism, and self-identity. Finally, the social dimension refers to aspects such as conspicuousness and prestige and focuses on the perceived utility that individuals acquire from products or services that are recognized within social groups.

Along with the increasing global demand for luxury products, the question arises of what are the possible differences and/or similarities in the luxury value perception of consumers in distinct parts of the world. Wiedmann, Hennigs & Siebels [43] suggest that in a cross-cultural context, the key dimensions of luxury value are perceived differently by different sets of consumers, even if the overall luxury level of a brand may be perceived equally. They state that *"...the needs of luxury consumer segments cross national borders and common structures in luxury value perception exist cross-culturally – even if the relative importance of the decision determinants may vary"* ([43], p. 1).

Consequently, the impact of each of the four dimensions on the overall luxury value perception is supposed to differ significantly across cultures and with reference to diverse industries. Reasoning this, the aim of the present study is to explore whether consumers in different parts of the world vary in their perception of the financial, functional, individual, and social dimension of luxury value with respect to diverse product categories.

5.4 Methodology

5.4.1 Measurement Instrument

To investigate whether and to what extent consumer perceptions of luxury value dimensions differ across countries and industries, we focused on the four luxury value dimensions of Wiedmann, Hennigs and Siebels [43], [44] as described in the preceding section. Specifically, after having introduced the financial, functional, individual, and social aspects of luxury consumption to the respondents, they were asked to rate the importance of the specific value dimensions with regard to personal, experiential, and home, luxuries. Referring to the different product contexts, the question read as follows:

There are different types of luxury product categories; the purchase of these product categories is influenced by different aspects. Please imagine...

a. *the purchase of a car*
b. *the purchase of fashion and accessories*
c. *the purchase of jewelry and watches*
d. *dining in an exclusive restaurant*
e. *the purchase of consumer electronics*
f. *the purchase of home furnishing*

...how important are the following aspects?

1. *Financial Aspects (e.g. price, resale price)*
2. *Functional Aspects (e.g. quality, uniqueness, usability)*
3. *Individual Aspects (e.g. self-identity value, hedonic value)*
4. *Social Aspects (e.g. status, prestige)*

All items were rated on a five-point Likert scale (1=absolutely unimportant, 5=absolutely important). The first version of the questionnaire was face-validated twice using exploratory and expert interviews and pre-tested with 40 respondents.

5.4.2 Sample and Data Collection

The data were collected in collaboration with 11 partners from ten countries: Brazil, France, Germany, Hungary, India, Italy, Japan, Slovakia, Spain, and the USA (Northern and Southern regions). To enhance the homogeneity of the respondents across cultures [13], [17], [18] and to reduce minor random errors [8], we decided to draw comparable samples from identifiable subgroups of the population [26], [28], [32]. In accordance with Dawar and Parker [12], we chose student respondents, as they show similar levels of age, education, professional aspirations, general income and other demographic characteristics across countries [12], [1]. For the data collection, we used a purposeful sampling method in which the sample choices were arranged to be comparable across countries: we aimed for a sample target of 100 university student participants (with an equal number of males and females) in each region. The respondents were contacted directly by the researchers at each university to secure the interest of the students in the domain of luxury brands/goods and the willingness of these students to participate in the research project. In March 2011, a total of 1275 valid and completed questionnaires were obtained. **Table 5.2** describes the key characteristics of the sample.

Table 5.2 Sample Characteristics

Countries/ Nationalities	n	%	Female (%)	Mean Age (years)	Income Level/ Family Status
Brazil	106	8.3	54.7	21.4	Middle to high income
France	47	3.7	72.3	23.5	Middle income
Germany	108	8.5	50.5	23.7	Middle income
Hungary	116	9.1	58.6	20.6	Middle income
India	159	12.5	54.7	25.6	Middle income
Italy	64	5.0	53.1	23.5	Middle income
Japan	101	7.9	50.5	20.3	Middle income
Slovakia	101	7.9	65.3	23.4	Middle income
Spain	125	9.8	51.2	20.8	Middle income
USA (Northern states)	131	10.3	51.1	20.6	Middle income
USA (Southern states)	143	11.2	41.3	26.1	Middle income
Other Nationalities	74	5.8			
Total	1275	100.0	53.5	22.7	Middle income

As **Table 5.2** illustrates, 1275 usable questionnaires were received. A total of 53.5% of the respondents in the final sample are female with an average age of 22.7 years and a self-reported middle income level. Given that this study requires a certain level of experience in the domain of luxury goods, all of the respondents who were included in the final sample reported that they are highly interested in luxury, that they are well informed about the luxury world, and that they purchase luxury brands on a regular basis. Even if our sample is not representative with respect to the given exploratory research focus, the sample offers a balanced set of data from each country.

5.5 Results and Discussion

To examine possible differences and/or similarities across countries and industries in the four dimensions of customer luxury value perceptions, we used analysis of variance (ANOVA). As illustrated in **Figure 5.2**, the results reveal significant differences in the perception of the luxury value dimensions across countries and with reference to different industries. In the following paragraph, the results are described in detail.

Figure 5.2 Overview: Cross-Cultural and Cross-Industry Comparison

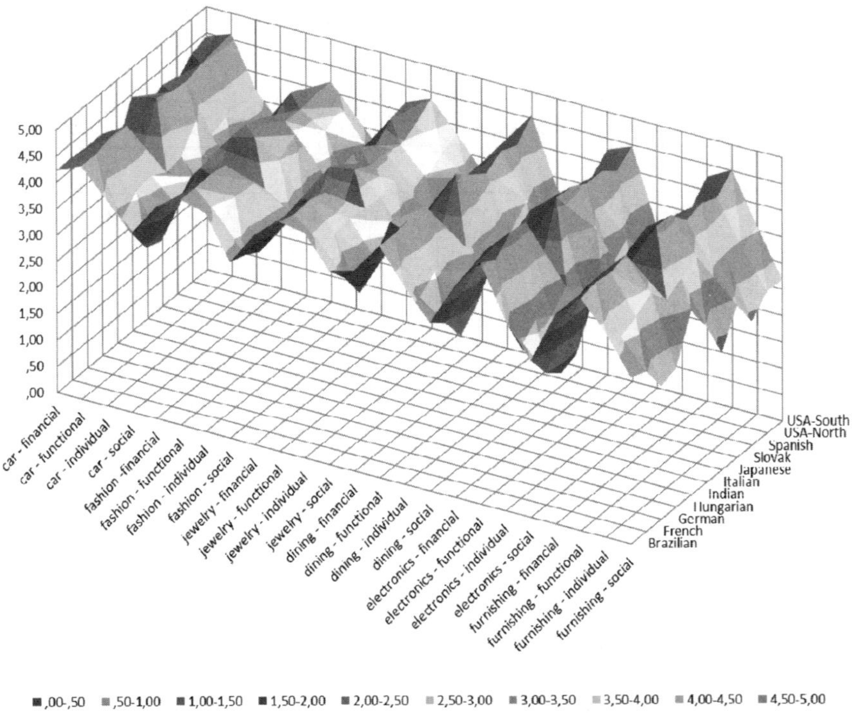

In the context of cars, as shown in **Figure 5.3**, we found significant differences across countries in all luxury value aspects (see **Table 5.3**). Referring to the importance of financial and functional aspects, country mean scores were at high levels (>4.0) for all countries, with the exception of India (3.73; 3.71). The individual and social aspects were perceived to be less important, values ranged between 3.32 (France) and 4.04 (Japan) for the individual value dimension, and between 2.69 (Japan) and 3.54 (India) for the importance of social aspects.

Table 5.3 ANOVA Results: Cars

	Financial Aspects	Functional Aspects	Individual Aspects	Social Aspects
Brazil	4.22	4.42	3.91	3.40
France	4.23	4.43	**3.32**	2.89
Germany	4.19	4.54	3.61	2.79
Hungary	4.22	4.50	3.67	3.23
India	**3.73**	**3.71**	3.49	**3.54**
Italy	**4.45**	4.56	3.56	3.02
Japan	4.01	4.45	**4.04**	**2.69**
Slovakia	4.21	4.41	3.52	3.03
Spain	4.05	4.50	3.62	2.70
USA-North	4.34	4.56	3.64	3.40
USA-South	4.44	**4.61**	3.66	3.14
F	**7.024**	**17.228**	**3.334**	**8.682**
Sig	.000	.000	.000	.000

Figure 5.3 Country Comparison: Cars

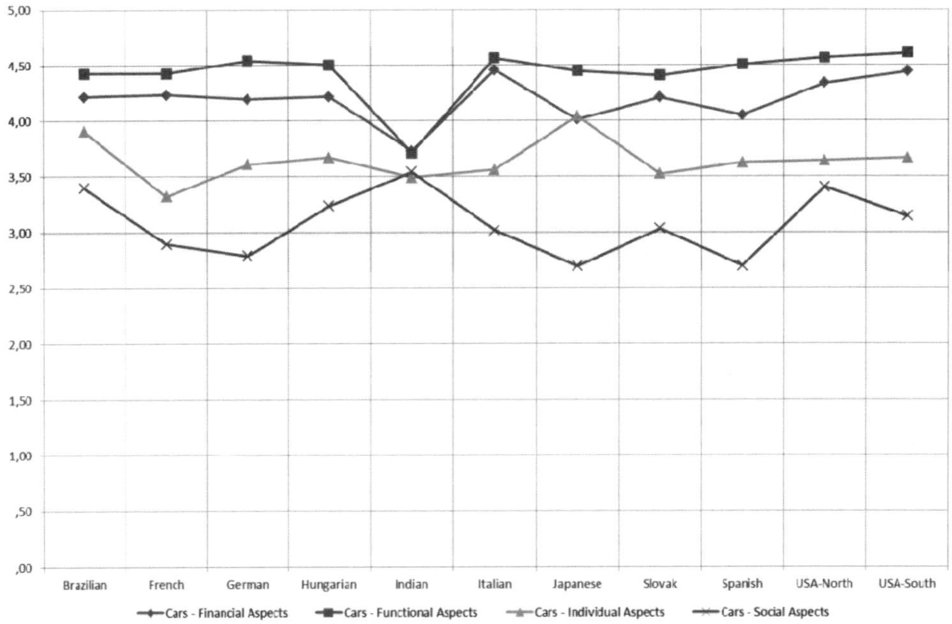

Figure 5.4 illustrates that in the domain of fashion and accessories mean scores differed significantly between countries (see **Table 5.4**). Apart from the perceived less importance of social aspects with values that ranged between 2.66 (Spain) and 3.47 (USA), the other three dimensions of luxury value were rated as being important to a comparable degree by each country. Again, India had the lowest mean scores for all dimensions except the social aspects of luxury value.

Table 5.4 ANOVA Results: Fashion and Accessories

	Financial Aspects	Functional Aspects	Individual Aspects	Social Aspects
Brazil	4.03	4.28	4.16	3.35
France	3.91	4.28	3.83	3.26
Germany	3.84	4.30	4.01	3.07
Hungary	4.05	4.24	4.07	3.27
India	**3.57**	**3.64**	**3.69**	3.39

	Financial Aspects	Functional Aspects	Individual Aspects	Social Aspects
Italy	4.19	**4.31**	4.02	3.14
Japan	**4.20**	4.23	**4.25**	2.71
Slovakia	4.08	4.15	3.63	3.04
Spain	3.90	4.14	3.85	**2.66**
USA-North	4.10	4.19	4.08	**3.47**
USA-South	3.86	4.20	3.67	3.14
F	4.923	6.789	5.462	6.461
Sig	.000	.000	.000	.000

Figure 5.4 Country Comparison: Fashion and Accessories

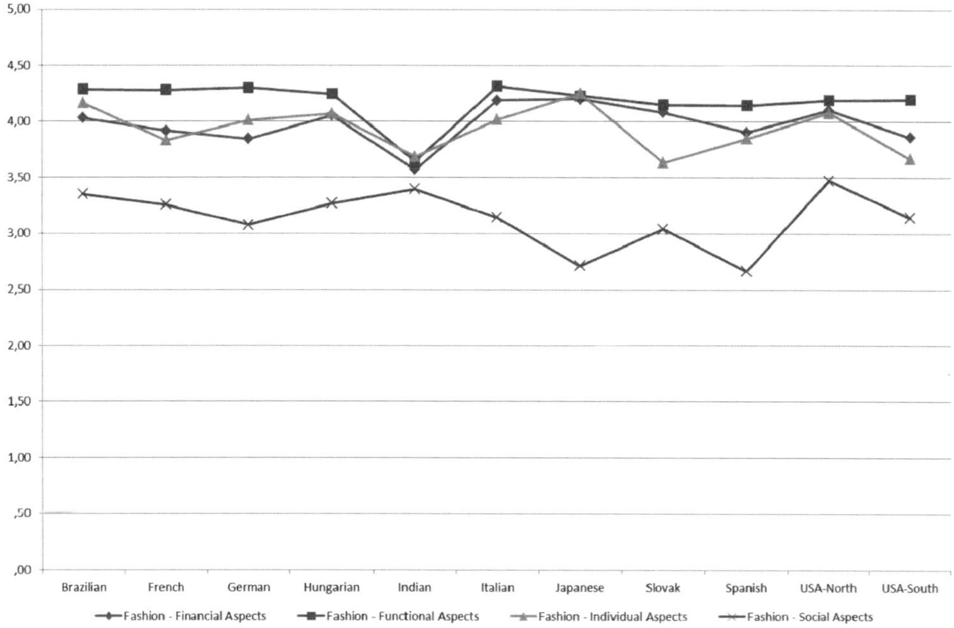

Referring to jewelry and watches, the mean scores for the perceived importance of all luxury value aspects differed significantly between countries (see **Table 5.5**). As shown in **Figure 5.5**, functional and financial aspects were rated as being comparatively important by India, Italy, Japan, Spain, and the USA. In contrast to this, Brazilian respondents associated jewelry and watches mainly with functional and individual aspects (4.37; 4.28), French, German and Hungarian consumers perceived financial and individual aspects to be equally important. India showed similar values for all aspects of luxury value, whereas the social value dimension was perceived to be less important by all other countries.

Table 5.5 ANOVA Results: Jewelry and Watches

	Financial Aspects	Functional Aspects	Individual Aspects	Social Aspects
Brazil	3.99	**4.37**	**4.28**	3.68
France	3.94	4.28	3.87	3.32
Germany	3.70	4.31	3.81	2.79
Hungary	3.97	4.22	3.86	3.22
India	**3.58**	**3.55**	3.81	**3.69**
Italy	4.13	4.19	3.80	3.09
Japan	3.96	4.04	3.87	2.67
Slovakia	4.03	4.04	3.57	3.05
Spain	3.82	3.96	**3.46**	**2.62**
USA-North	4.11	4.13	3.69	3.36
USA-South	**4.19**	4.38	3.84	3.24
F	**4.594**	**8.564**	**4.352**	**11.221**
sig	**.000**	**.000**	**.000**	**.000**

Figure 5.5 Country Comparison: Jewelry and Watches

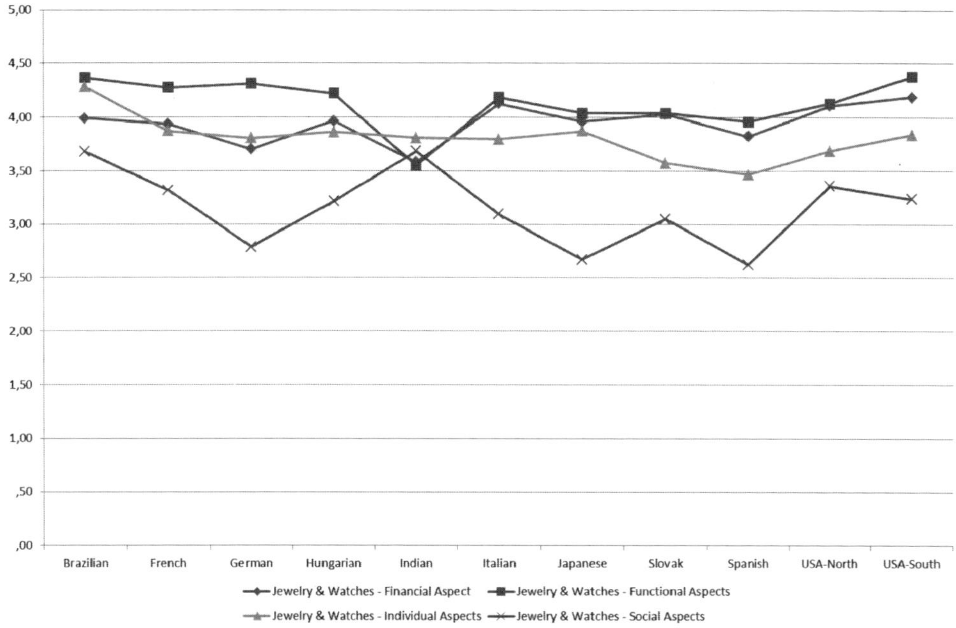

Referring to experiential luxury and the dinner in an exclusive restaurant, mean scores for functional and social aspects differed significantly between countries (see **Table 5.6**). While India showed similar values for all dimensions of luxury value at a moderate level, all other countries differed in their evaluation of the perceived importance of the four value aspects. As illustrated in **Figure 5.6**, all countries perceived the functional aspect to be most important (Italy: 4.61), followed by financial (USA: 3.99), individual (Brazil: 3.75) and, with less importance, the social dimension (India: 3.36).

Table 5.6 ANOVA Results: Dining in an Exclusive Restaurant

	Financial Aspects	Functional Aspects	Individual Aspects	Social Aspects
Brazil	3.82	4.41	**3.75**	3.20
France	3.74	4.40	3.32	2.96
Germany	3.64	4.45	3.59	2.48
Hungary	3.93	4.47	3.35	2.80
India	**3.62**	**3.74**	3.57	**3.36**

	Financial Aspects	Functional Aspects	Individual Aspects	Social Aspects
Italy	3.89	4.61	3.38	2.61
Japan	3.97	4.36	3.45	**2.35**
Slovakia	3.89	4.15	3.19	2.76
Spain	3.92	4.34	3.54	2.53
USA-North	3.92	4.21	**3.14**	2.92
USA-South	**3.99**	**4.56**	3.27	2.97
F	**2.043**	**10.707**	**2.898**	**8.302**
sig	**.026**	**.000**	**.001**	**.000**

Figure 5.6 Country Comparison: Dining in an Exclusive Restaurant

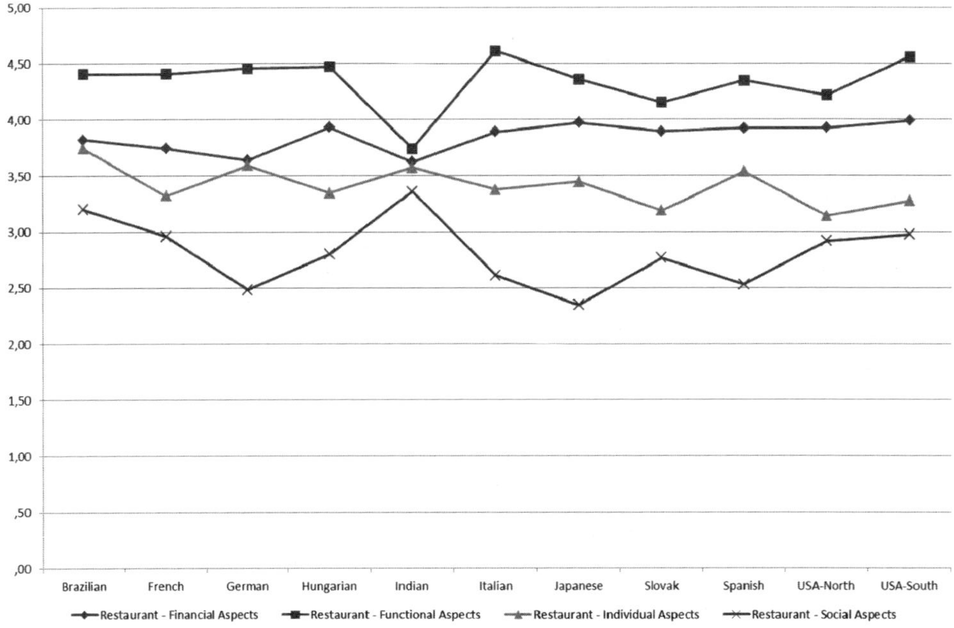

In the context of consumer electronics, country mean scores differed significantly in the perception of all value dimensions (see **Table 5.7**). **Figure 5.7** shows that mean scores for functional and financial aspects were rated in a comparable manner by all countries with slightly higher importance for the functional evaluation. Again, Indian consumers rated these aspects less important than all other countries. While the social dimension was perceived to be less important by all countries except India (3.09), particularly Japanese and Brazilian consumers put emphasis on the individual aspects of consumer electronics (3.95; 3.61).

Table 5.7 ANOVA Results: Consumer Electronics

	Financial Aspects	Functional Aspects	Individual Aspects	Social Aspects
Brazil	4.03	4.59	3.61	2.98
France	4.13	4.40	**2.87**	2.51
Germany	4.24	**4.70**	3.17	2.28
Hungary	4.17	4.66	3.00	2.40
India	**3.47**	**3.62**	3.32	**3.09**
Italy	4.31	4.66	3.09	2.45
Japan	4.22	4.63	**3.95**	**2.18**
Slovakia	4.05	4.46	3.04	2.50
Spain	4.06	4.46	3.30	2.37
USA-North	3.99	4.23	2.88	2.76
USA-South	**4.27**	4.46	3.04	2.58
F	**9.536**	**22.190**	**8.442**	**7.653**
sig	**.000**	**.000**	**.000**	**.000**

Figure 5.7 Country Comparison: Consumer Electronics

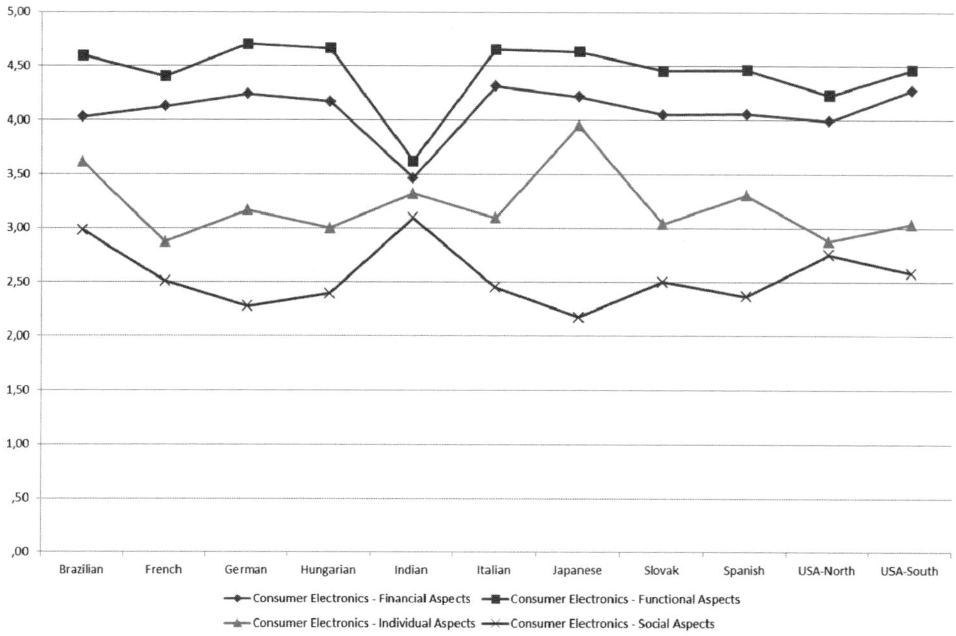

Referring to home luxury and exclusive furniture, as shown in **Table 5.8**, significant differences were found across countries in all luxury value aspects. As illustrated in **Figure 5.8**, apart from India where results revealed similar scores for all dimensions of luxury value, the other countries rated the functional (Italy: 4.63), financial (USA: 4.27), and individual aspects (Japan: 3.94) in a similar pattern with falling tendency. While Indian respondents rated comparatively high importance to social aspects (3.48), all other countries perceived the social dimension of luxury value to be less important.

Table 5.8 ANOVA Results: Home Furnishing

	Financial Aspects	Functional Aspects	Individual Aspects	Social Aspects
Brazil	3.99	4.41	3.88	3.16
France	3.89	4.49	3.57	3.04
Germany	4.04	4.44	3.71	2.53
Hungary	4.06	4.54	3.93	2.64
India	**3.65**	**3.60**	3.56	**3.48**

	Financial Aspects	Functional Aspects	Individual Aspects	Social Aspects
Italy	4.11	**4.63**	3.77	2.95
Japan	4.25	4.59	**3.94**	**2.23**
Slovakia	4.04	4.26	3.52	2.89
Spain	4.02	4.35	3.42	2.36
USA-North	3.92	4.23	**3.39**	2.69
USA-South	**4.27**	4.52	3.57	2.68
F	**4.266**	**16.542**	**3.417**	**12.971**
sig	**.000**	**.000**	**.000**	**.000**

Figure 5.8 Country Comparison: Home Furnishing

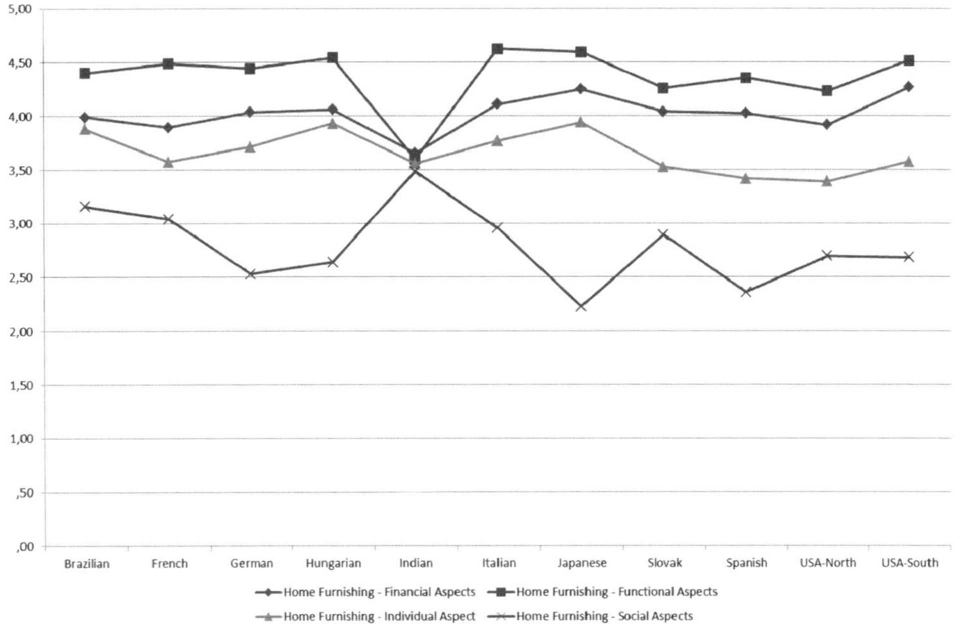

Comparing all four value dimensions in terms of their importance in the context of the different product categories, **Figure 5.9** shows that even if the ANOVA results revealed significant differences across country mean scores, there are similar patterns recognizable.

Figure 5.9 Importance of Value Dimensions across Countries and Industries

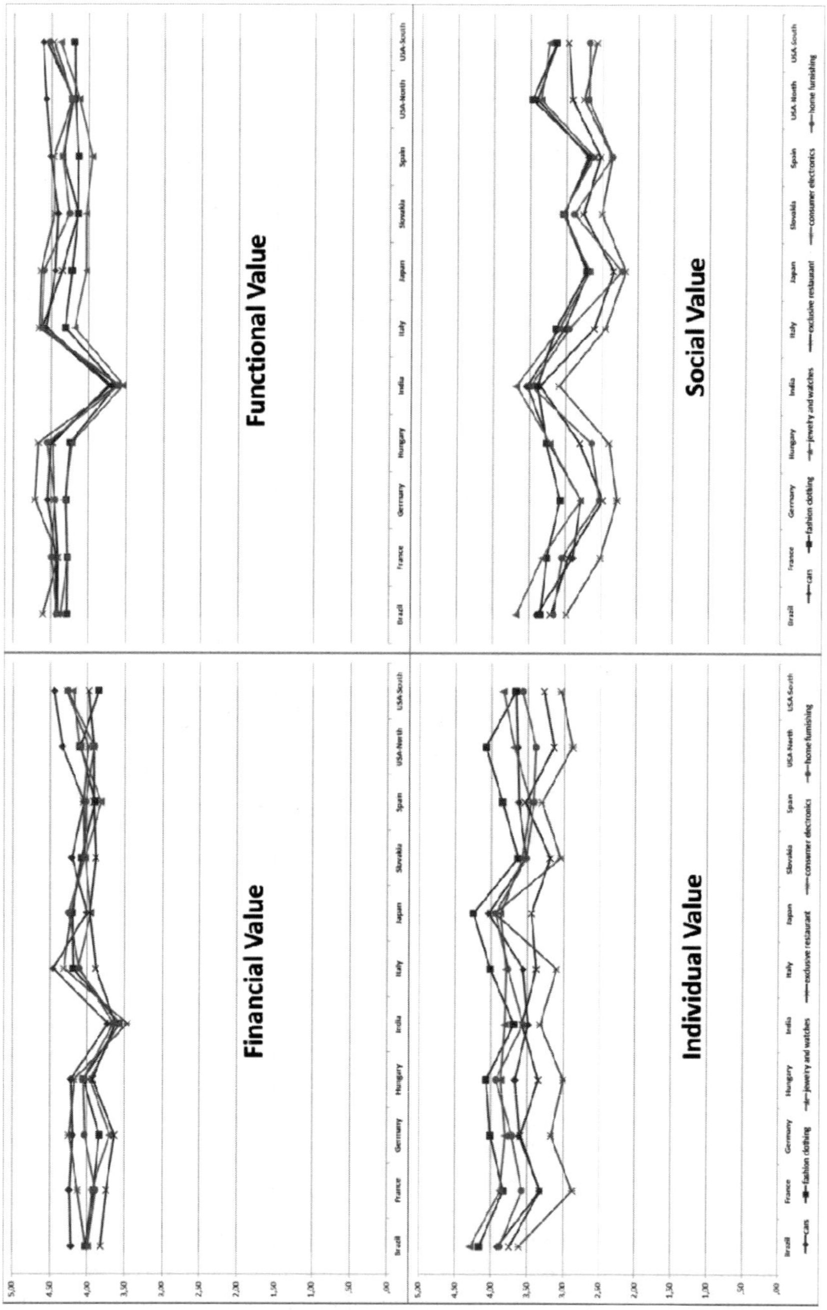

In sum, the ANOVA results provide evidence to confirm significant cross-cultural differences but also similarities between the luxury value dimensions as perceived by consumers from different countries in the context of diverse product categories. Even if these insights are only first steps toward a conclusive picture of key dimensions of the cross-cultural and cross-industry luxury value perception, we believe that the results of this study are worthy of further consideration in future research and business practices.

5.6 Conclusions and Implications

In the interplay between increasing mobility, global communication and media distribution on the one hand, but distinct economic, political, legal, and educational environments on the other hand, the question arises of whether the market for luxury goods can be treated as a single market and whether between-country differences are relevant in luxury marketing. The main objective of the present study was to empirically explore possible similarities and differences related to the luxury value as perceived by consumers from different countries and with reference to diverse luxury product categories. The study results give evidence that the luxury value perception encompassing financial, functional, individual, and social aspects of luxury products varies significantly across countries and industries. Nevertheless, the basic structure of the perceived importance of these four value dimensions was shown to be cross-cultural and cross-industry stable.

Given that this study relies on explicit answers and self-reported survey data, further research should concentrate on implicit measures as well. To overcome existing limitations of explicit measures such as social desirability and the respondent's inability to report correctly, the combination of the measurement of both conscious and unconscious responses to a luxury brand or product category is worth focusing on. Such a holistic approach might lead to an enhanced understanding and interpretation of the value aspects of luxury brand perceptions as a basis to adequately address questions of sampling and country selection in consideration of the variety of cultures across the world.

From a managerial perspective, given that the concept of luxury is subjective and multidimensional in nature, the orientation on an empirically verified system of consumer perceived values is of special importance. As the results revealed, there are cross-cultural differences and similarities in the perceived importance of financial, functional, individual, and social aspects of diverse luxury products. The translation of these individual evaluations to the management luxury brands is a key challenge in a global economy in which low-cost counterfeits are easily available. In fact, to appeal to the cognitive needs and affective desires of luxury consumers it is not so much a question of whether culture is driving luxury consumption, but rather how the concept of luxury is understood and managed:

"Luxury is a culture,
which means that you have to understand it to be able to practice it with flair and spontaneity."

Kapferer and Bastien ([24], p. 313)

References

[1] Agrawal, J./Grimm, P./Kamath, S./Foscht, T. (2011): A cross-country study of signals of brand quality, in: Journal of Product & Brand Management, Vol. 20, No. 5, pp. 333-342.

[2] Atwal, G./Williams, A. (2009): Luxury brand marketing – the experience is everything!, in: Journal of Brand Management, Vol. 16, No. 5, pp. 338-346.

[3] Bastien, V./Kapferer, J.-N. (2009): Luxury marketing plays by a different set of rules, in: Financial Times, June 15th.

[4] Bernstein, L. (1999): Luxury and the Hotel Brand: Art, Science or Fiction?, in: Cornell Hotel and Restaurant Administration Quarterly, Vol. 40, February.

[5] Bernstein, L. (2007): Luxury and the Hotel Brand: it's the experience, only the experience, nothing but the experience, in: 4Hoteliers, Hospitality, Hotel and Travel News, 20th February 2007.

[6] Berthon, P./Pitt, L./Parent, M./Bertho, J.-P. (2009): Aesthetics and Ephemerality: Observing and Preserving the Luxury Brand, in: California Management Review, Vol. 52, No. 1, pp. 45-66.

[7] Chevalier, M./Mazzalovo, G. (2008): Luxury brand management. A world of privilege, Singapore, Wiley.

[8] Calder, B. J./Phillips, L. W./Tybout, A. M. (1981): Designing research for applications, in: Journal of Consumer Research, Vol. 18, September, pp. 197-207.

[9] Cornell, A. (2002): Cult of luxury: The new opiate of the masses, in: Australian Financial Review, p. 47.

[10] Danziger, P. (2005): Let Them Eat Cake: Marketing Luxury to the Masses – As well as the Classes, Chicago, Dearborn Trade Publishing.

[11] Davidson, J. (1898): Luxury and Extravagance, in: International Journal of Ethics, Vol. 9, No. 1, pp. 54-73.

[12] Dawar, N./Parker, P. (1994): Marketing universals: consumers' use of brand name, price, physical appearance, and retailer reputation as signals of product quality, in: Journal of Marketing, Vol. 58, No. 4, pp. 81-95.

[13] Douglas, S. P./Craig, S. S. (1983): International Marketing Research, NJ: Prentice-Hall Inc.

[14] Dubois, B./Czellar, S./Laurent, G. (2005): Consumer Segments Based on Attitudes Toward Luxury: Empirical Evidence from Twenty Countries, in: Marketing Letters, Vol. 16, pp. 115-128.

[15] Dubois, B./Duquesne, P. (1993): The market for luxury goods: Income versus culture, in: European Journal of Marketing, Vol. 27, No. 1, pp. 35-44.

[16] Dubois, B./Paternault, C. (1995): Observations: Understanding the World of International Luxury Brands: The "Dream" Formula, in: Journal of Advertising Research, Vol. 35, No. 4, pp. 69-76.

[17] Durvasula, S. J./Andrews, C./Netemeyer. R. G. (1997): A cross-cultural comparison of consumer ethnocentrism in the United States and Russia, in: Journal of International Consumer Marketing, Vol. 9, No. 4, pp. 73-93.

[18] Furrer, O./Liu, B. S./Sudharshan, D. (2000): The relationships between culture and service quality perceptions: basis for cross-cultural market segmentation and resource allocation, in: Journal of Service Research, Vol. 2, May, pp. 355-371.

[19] Grossman, G. M./Shapiro, C. (1998): Counterfeit-product trade, in: The American Economy Review, vol. 78, pp. 59-75.

[20] Hofstede, F. T./Steenkamp, J.-B. E. M./Wedel, M. (1999): International Market Segmentation Based on Consumer-Product Relations, in: Journal of Marketing Research, Vol. 36, pp. 1-17.

[21] Hume, D. ([1752] 1965): "Of refinement in the arts". In: Lenz, J.W. (Ed.): Of the Standard of Taste and OtherEssays, Bobbs-Merrill, Indianapolis.

[22] Kapferer, J-N. (1997): Managing luxury brands, in: Journal of Brand Management, Vol. 4, No. 4, pp. 251-260.

[23] Kapferer, J.-N./Bastien, V. (2009a): Luxury Strategy: Break the Rules of Marketing to Build Luxury Brand, Milford, CT, Kogan Page.

[24] Kapferer, J.-N./Bastien, V. (2009b): The specificity of luxury management: turning marketing upside down, in: Journal of Brand Management, Vol. 16, pp. 311-322.

[25] Kemp, S. (1998): Perceiving luxury and necessity, in: Journal of Economic Psychology, Vol. 19, pp. 591-606.
[26] Madden, T./Hewett, J. K./Roth, M. S. (2000): Managing images in different cultures: a cross-national study of color meanings and preferences, in: Journal of International Marketing, Vol. 8, No. 4, pp. 90-107.
[27] Merriam-Webster's (2009): Merriam-Webster's Collegiate Dictionary, Eleventh Edition.
[28] Mitchell, V. W./Vassos, V. (1997): Perceived risk and risk reduction in holiday purchases: a cross-cultural and gender analysis, in: Journal of Euromarketing, Vol. 6, No. 3, pp. 47-77.
[29] Nueno J. L./Quelch J. (1998): The mass marketing of luxury, in: Business Horizon, Vol. 41, No. 6, pp. 61-68.
[30] Oxford (1992): Oxford Latin Dictionary, Oxford University Press.
[31] Phau, I./Prendergast, G. (2000): Consuming luxury brands: The relevance of the "rarity principle", in: Journal of Brand Management, Vol. 8, No. 2, pp. 122-138.
[32] Raaij, V./Fred, W. (1978): Cross-cultural research methodology as a case of construct validity, in: Advances in Consumer Research, Vol. 5, pp. 693-701.
[33] Rae, J. ([1834] 1965): Statement of Some New Principles on the Subject of Political Economy, in: James, R. W. (Ed.): John Rae: Political Economist, Vol. II., University of Toronto Press, Toronto.
[34] Reich, C. (2005): Luxusmarken- Eine empirische Untersuchung der Determinanten der Begehrlichkeit im Hinblick auf Luxusmarken und der resultierenden Wirkung auf die Kaufabsicht, Rainer Hampp Verlag.
[35] Sekora J. (1977): Luxury: the concept in Western thought, The Johns Hopkins University Press Baltimore, USA.
[36] Silverstein, M. J./Fiske, N. (2003): Luxury for the Masses, in: Harvard Business Review, April, pp. 48-57.
[37] Shukla, P. (2011): Impact of interpersonal influences, brand origin and brand image on luxury purchase intentions: Measuring interfunctional interactions and a cross-national comparison, in: Journal of world business, Vol. 46, No. 2, pp. 242-253.
[38] Truong, Y./Simmons, G./McColl, R./Kitchen, P. J. (2008): Status and Conspicuousness- Are They Related? Strategic Marketing Implications for Luxury Brands, in: Journal of Strategic Marketing Vol. 16, No. 3, pp. 189-203.
[39] Valtin, A. (2008): Der neue Luxus und die Konsequenzen für die Markenführung, in: Bauer, H. H./Huber, F./Albrecht, C.-M. (Eds.): Erfolgsfaktoren der Markenführung, München, pp. 247-258.
[40] Vickers, J. S./Renand, F. (2003): The marketing of luxury goods: An exploratory study – three conceptual dimensions, in: The Marketing Review, Vol. 3, pp. 459-478.
[41] Vigneron, F./Johnson, L. W. (2004): Measuring perceptions of brand luxury, in: Journal of Brand Management, Vol. 11, No. 6, pp. 484-506.
[42] Webster's (2002): Webster's Third New International Dictionary, Unabridged, Merriam-Webster.
[43] Wiedmann, K.-P./Hennigs, N./Siebels, A. (2007): Measuring consumers' luxury value perception: a cross-cultural framework, in: Academy of Marketing Science Review, Vol. 7, pp. 1-21.
[44] Wiedmann, K.-P./Hennigs, N./Siebels, A. (2009): Value-Based Segmentation of Luxury Consumption Behavior, in: Psychology & Marketing, Vol. 26, No. 7, pp. 625-651.
[45] Zeithaml, V. A. (1988): Consumer Perceptions of Price, Quality and Value: A Means-End Model and Synthesis of Evidence, in: Journal of Marketing, Vol. 52, July, pp. 2-22.

Part 3
Luxury and Luxury Consumption from a Theoretical Perspective

6 Luxury Brands from a Psychoanalytic Perspective

Taro Koyama

6.1 Luxury Brands and Sublimation

What is the difference between luxury brands and commodity goods? Luxury brands improve quality of life because they are the fruits of elaborate design projects that require time and money. In order to improve the quality of life, one must distinguish what the goods do for the person from what that person does with them. Goods do various things for people and various states of being can be realized by goods. In other words, not from the side of human beings, but from the side of goods, we have to consider the idea that goods could have plural potential states simultaneously before one realizes their utility: whether a particular state is chosen and realized is in the hands of the individual [1]. For example, a sofa can realize a state in which people can talk cheerfully with friends or snooze. Goods offer us something to enable us to have a comfortable life. In a highly elaborate design project, the project leader is sure to think about the plural potential states of goods that correspond to various life scenes. The more states of being the goods have the potential to cause, the more the goods approach the status of luxury brands. For example, the famous project designer Bruno Munari [2] sought out the comfortable state that a sofa could realize for him (**Figure 6.1**). He did not try to invent a new function for a sofa but a new way of sitting.

How can we think about the plural potential states of goods?

(A) First, if the goods are luxurious ones, their **acoustic properties** should be good. Successive inorganic sounds made by machines or motor engines should be avoided because people will become exhausted by being constantly forced to hear such sounds. For example, by using an Alessi kettle that makes the sound of a bird, people are able to enjoy the twitter of small birds in order to have a pleasant coffee or tea break. Leading product designers need to consider what kind of emotion their goods will evoke because the goods offer some benefit and must make us happy.

(B) Second, taking into consideration the positive acoustic environment that goods could generate, if the goods are luxurious, (a) **long-wearing** and (b) **natural** materials must be adopted as much as possible because (a) luxury goods are not nondurable and must be repairable (they must endure so that one can remember one's childhood/family memories through them) and (b) cheap-looking plastic or bare concrete should be avoided. When people suffered from a shortage of goods after World War II, plastic and concrete were expected to resolve the problem of poverty; however, recent research shows that if people in jail or hospital come in contact with a picture of a fine view that includes animals and vegetation, they can lower their stress levels more easily and get well more rapidly. In addition, natural materials improve quality of life because they reduce one's level of tiredness. When people sit down on a luxurious sofa, they can relax, surrounded perhaps by wallpaper with a calming plant pattern. (From an evolutionary perspective, people in ancient times are thought to have preferred landscapes like savannas rather than jungles, because in the savanna it is easy to detect hazardous carnivorous animals from a distance [3]).

Figure 6.1 Munari's examination of comfortable states

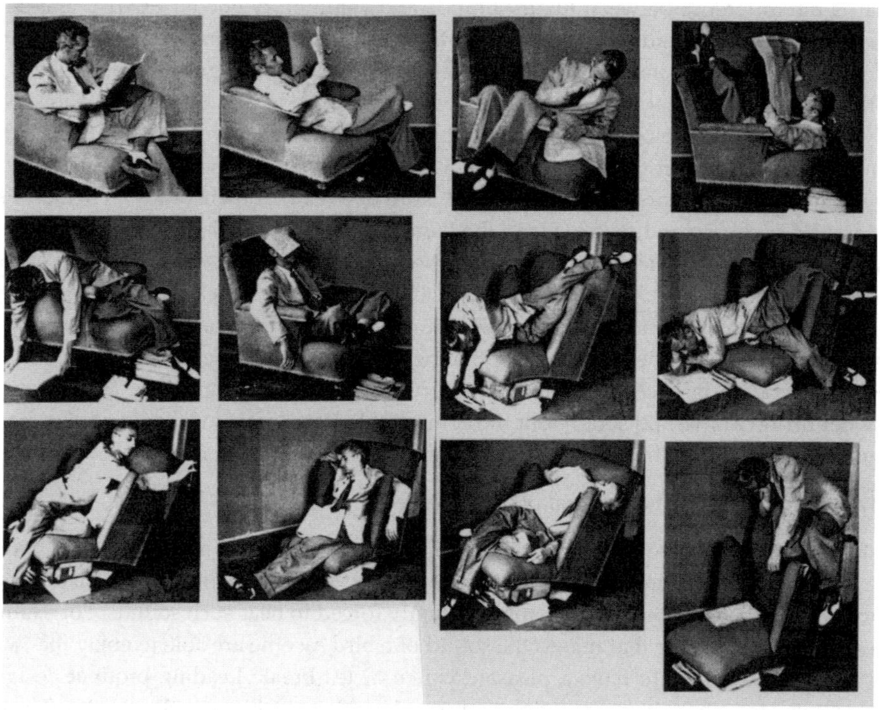

Source: Munari,1977 [2]

(C) Thirdly, one must pay attention to **ergonomics**. Ideally speaking, project designers measure each person's body size and motion pattern and adjust the goods to suit the person's movement, but not vice versa. Keeping track of a ballerina's movement is useful for studying the transition from stillness to movement, as well as balance. There must also be a suitable distance between goods and people. For example, in order for people to recognize each other's facial expressions, urban designers must plan the size of public squares to be within 25 meters; if not, familiarity ("amicizia" in Italian) would be lost in that square. Luxury goods must therefore be adjusted to suit each person's situation from an ergonomic perspective.

(D) Fourth, the most **elegant and streamlined form** of the product must be found through trial and error. Kitsch (meaning "vulgar" in German) decoration is furthest removed from luxury brands. Project designers are apt to pursue three-dimensional aesthetics and flowing form rather than to attach many functions to goods that do not always assure quality of life. By stripping goods of nonessentials and eliminating redundancy, the essential geometric framework, or form, remains. Seen from a distance, well-balanced, non-awkward geometric forms can be examined and determined. In a highly elaborate design project, elegant forms are found by examining the shape of organisms (e.g., the whirlpool pattern of a spiral shell) or by overlapping and combining various kinds of geometric figures, deforming both the angles and the sides of those figures at the same time. Only through trial and error can beautifully shaped goods be found. For example, designer Salvatore Ferragamo attempts to find elegant form, composition, and color pattern (**Figure 6.2**)[4]. He invented a new way of putting on shoes, which proposed that people take into account the beautiful combination of a shoe with a kind of Japanese sock (tabi), according to the situation. For example, he suggested wearing a black suede sandal with tobacco-colored leather socks in the morning, wearing black or white satin socks in the afternoon, and wearing gold leather or satin socks with sequins in the evening. Whether this kind of combination is beautiful or not becomes apparent only through experimentation. Ferragamo's trial and error process is not for the sake of sales, but for the sake of his passion for bringing aesthetics into everyday life, which according to the Greek philosopher Epicurus can generate delight or pleasure. It is well known that design for the sake of sales is called "styling," and styling is bad design because its aim is to create a variety of style patterns and a fleeting trend and it does not examine thoroughly what the goods do for the person [5].

Figure 6.2 Ferragamo's design project

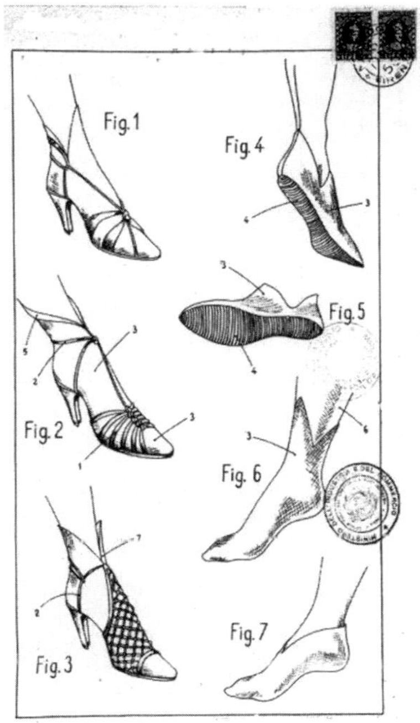

Source: Ricci, 2004 [4]

In pursuing high quality, the project designer must pay attention to unconscious perceptions. For example, Enzo Mari, a pupil and a colleague of Bruno Munari, emphasized that project designers can examine geometric figures, of which it is difficult for people not to be directly conscious (**Figure 6.3**)[6]. In **Figure 6.3**, people might not be immediately aware of the entirety of form (12) because that form emerges as the result of subtraction. If people subtract forms (2), (3), and (9) from the total composition, they get form (12). Enzo Mari thereby attempted to examine the aesthetic combination of geometric figures by considering unconscious perception.

Figure 6.3 Composition of a geometric figure and its components

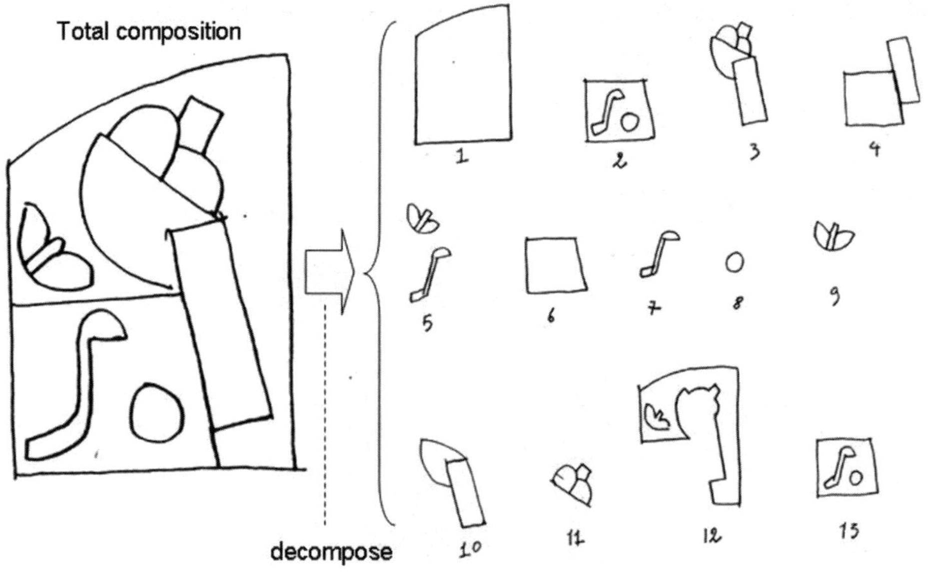

Source: Mari, 2001 [6]

(E)Fifth, Project designers must pay attention to the idea that people's environments are meant to bring them joy; therefore, designers must consider not human beings but their environments as the main character. A luxurious sofa, for example, could be a part of the layout and furnishing of a livable interior. In other words, it could be the part of the interior space (place) that is unique and peculiar to a family. If one family's interior space is replaced with that of another, then the unique and harmonious relationship between the family and the interior space will be destroyed. The utility of goods, which emphasize comfort and convenience, can be easily recognized. However, if only the consumer's desire for efficiency is maintained as the focal point, then changeable environments such as the scenery of a monotonous suburb or a shopping mall will be created without the influence of ethnic individuality. In other words, project designers must respect a **genius loci** which means the special atmosphere of a place and is not able to be substituted for an atmosphere of the other place. If a place is replaceable, the unique relationship between people and the place will be destroyed and people will be also replaceable so that human **dignity** will be lost forever.

(A) through (E) are the laborious processes of **sublimation**. French psychoanalyst Jacques Lacan [7] defines sublimation as a process that "elevates an object to the dignity of the Thing," which means that an artist elevates an object/material to a work of art that can be worshiped or admired because an aura of divinity dwells in it.

In psychoanalysis, it is hypothesized that an impossible enjoyment exists outside of the world in which one is living [8]. In other words, there is enjoyment ("jouissance" in French) between a baby and its mother, and the baby feels a very strong attachment to this enjoyment. Sexual energy (libido) originates in this attachment from the psychoanalytic perspective (Sexuality is attachment), and the happiest situation for the baby will be lost forever because the baby has to grow up and separate from the mother. Therefore, one cannot restore the original happy situation in which enjoyment is supposed to exist: enjoyment is lost forever and is therefore **impossible**. Based on the assumption that enjoyment is an absolutely happy situation, it is usually assumed to be outside the world in which one is living; in other words, it exists in another world. Similarly, an artist attempts to create a work of art in which an aura of divinity dwells—a glimpse or presentiment of enjoyment that exists in another world. The image of enjoyment is the energy of repeating expansions and contractions. If an artist attempts to elevate his/her work of art to the height of sophistication, a glimpse of enjoyment will dwell in that work of art and people will be deeply impressed by it. People will feel that vitality has been put into this lifeless, inanimate world through such a fine work of art.

Project designers are not the same as artists. On the basis of their personal inclination, artists create works of art without taking practical use into account. Ancient Etruscan vases were not objects of art appreciation, but were made for the practical purpose of preserving cooking oil. Daily life and the arts were not separated in ancient times; no work of art existed only for appreciation and no handy tool only for utility [9]. In subsequent years, everyday life and the arts gradually became independent of each other; however, Renaissance artists made a show of their talents in clothes, ceramics, dishes, and daily necessaries in addition to private and public architecture [10]. Renaissance artists combined beauty with everyday goods, as did William Morris, a founder of industrial design, during the Industrial Revolution.

Like Renaissance artists, leading project designers try to passionately invent a new way of life through the goods that they make in cooperation with skilled artisans. To create a luxury brand, both project designers and skilled artisans are required. In India, for example, although there are many artisans who use traditional techniques, there is a lack of leading project designers for the creation of luxury brands [11]. They propose a new way of life so that the quality of life can be dramatically elevated. Elaborately designed products (luxury brands) are a kind of work of art with a practical use, and the design process, as discussed in points A through D above, is laborious in its pursuit of sublimation. To make perfect works of art is to pave the way for sublimation.

Where does aesthetic value fit in psychoanalysis? We find an example in the Greek tragedy *Antigone,* in which the aesthetic area or region crosses paths with the moral area (**Figure 6.4**). In the moral area (α) people had to distinguish right from wrong and obey the laws that King Creon decreed. As long as people did not defy the law and remained within its limits, they were allowed to satisfy their desires. Antigone, however, violated the law that prohibited her from holding a funeral for her elder brother Polyneikes, who had invaded his mother country. Antigone did not obey the law and was therefore sentenced to death. On the basis of fraternal love, Antigone dared to hold a funeral for Polyneikes because she knew that if her husband or child died she could find another husband or bear another child, but if her parents died it would be impossible to have another brother. She was so affectionate toward her brother as to go beyond the law (moral area (α)) and she shone beautifully in the aesthetic area (β). Her blind, intense fraternal love caused her to advance toward her ruin, the land of the dead (γ).

Figure 6.4 Aesthetic area in psychoanalysis

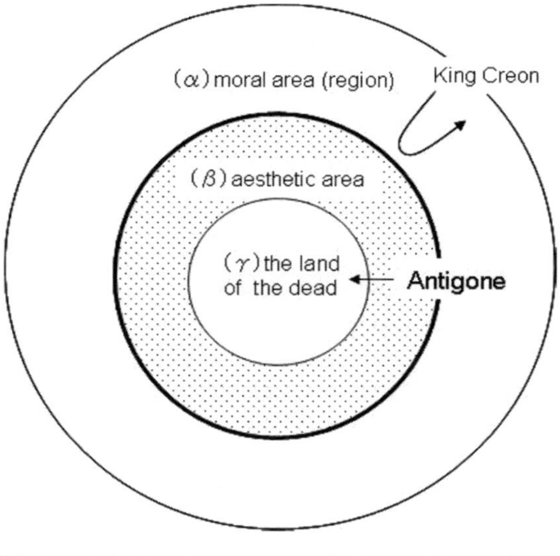

Source: Author

In a marketing context, figuratively speaking, leading project designers have to go beyond economic principles, which involve cost-benefit analysis and at the same time satisfy consumer's desires, so that an aesthetic aura of divinity can dwell in luxury goods. In order to improve quality of life and make people happy, designers may try to elevate an object or material to a kind of work of art in a passionately fanatic and blind manner, like Antigone. In short, to create a luxury brand, it is necessary for project designers to leave behind economic principles. Project designers must obey their internal impulse to create a new way of aesthetic and dramatic life through the sophisticated goods(luxury brands) as a part of stage setting/decoration and must not worry about consumer's evaluation, reputation, or desires because consumers are thought not to be able to create a new way of aesthetic life[12]. Luxury brands must be the cause of consumer's desires which do not exist until s/he sees them. A work of art (or luxury goods) that is seemingly difficult to perfect reflects a glimpse of enjoyment, which may be lost forever.

6.2 Luxury Brands as a Fetish

Consumers are thought to enjoy and gather luxury brands as a fetish. In a psychoanalytic context, a fetish is a substitute for **an ideal ego**, which is inherent in small children. During a child's happy infancy, his/her affectionate mother loves him/her in an unconditional way, which brings about omnipotent feelings in the child. In other words, in his/her happy infancy, the world is made up of only child and mother and it is assumed that there is not pleasure but absolute enjoyment between mother and child. As the child grows up, this enjoyment is lost forever and there is no prospect for its revival (i.e., enjoyment is impossible in principle), because the child as s/he is would not be loved by the mother unless s/he achieved certain accomplishments. A mother who was once satisfied with the child as s/he was and did not desire anything other than the child itself is thereby lost forever. In psychoanalysis, if a person desires something, s/he is regarded as incomplete or lacking because if s/he is complete, s/he should be self-sufficient. In the same way, if a mother wants something other than the child as s/he is, she is lacking and incomplete or she does not seem to be omnipotent in the child's eyes. Confronted with a lacking/incomplete mother, the child tries to revive the omnipotent mother by giving the lacking/incomplete mother something as a gift. This something is known as a fetish and it is tinged with the enjoyment that the once-omnipotent mother was supposed to give to her child. If, in a child's fantasy, the mother became omnipotent, she would love her child unconditionally with an ideal ego, just like it used to be. In other words, as a child grows up, the actual mother does not approve or accept the child's ideal ego in itself and so the child attempts to provide a substitute (fetish) for his/her ideal ego to his/her mother in a fanciful way.

Luxury brands have many favorable attributes as fetishes because their quality is so high that they are admired or worshiped. Luxury brands as fetishes are a favorable signs of lost and impossible enjoyment and deserving gifts for lacking/incomplete mothers. The more luxury brands become brilliant and radiant, the more they act as substitutes for the ideal ego and become desirable because they resemble the omnipotent ideal ego that was once loved by the complete mother but is now not appreciated. (The child's omnipotent ideal ego generally declines with age and the ego ideal, with a moral sense—namely the superego—starts to develop.) In other words the favorable nature of luxury brands is omnipotent and therefore it is desirable that their appearance be very beautiful or elegant, they outperform other goods and have a long lifespan, so that they are able to bring back memories, and they are able to fit a great variety of situations and enlarge the scope of action in an autonomous way. In short, they are able to improve quality of life.

6.3 Identification and Luxury Brands

In a luxury marketing context, the level of congruence between brand personality and self-image are often measured and are supposed to affect the level of consumer satisfaction. From the perspective of psychoanalysis, consumers are thought to identify the image or personality of luxury brands with themselves in either an imaginary or symbolic way. In the case of **imaginary identification**, consumers identify favorable images as seen only by them. This means that the images that other people expect a person to choose are not taken into consideration. For example, the female model in the Prada ad shown in **Figure 6.5** shows simplicity, as opposed to an image of a voluptuous or sensual woman, and emphasizes that women who have less eventful days do not put on clothes that are vibrant with life, but put on modest clothes that do not evoke romantic relationships [13]. (This contrasts with Gucci's fashion, which evokes romance.) Women living in ordinary life can remove their anxiety by identifying the image of the Prada model as lacking in vividness in an imaginary way, because a graceful or feminine image imposes a burden on women in the modern world. Anyone who can identify a really elegant or sensual image of a woman is strong. Forcing women in the modern world to imitate the image and incorporate the characteristics of elegant, sensual models can bring about anxiety.

Figure 6.5 Non-sensual image of a female model in Prada ad

Source: Codeluppi and Galoforo, 2000, p.120 [13]

On the contrary, if one identifies images that other people expect of you, even if one does not like the images that other people perceive highly, **symbolic identification** can occur. For example, let us assume that there is a heavy person who likes to wear clothes with vertical stripes because they seem to make him/her look more slender. If s/he then changes his/her mind and decides to put on clothes with horizontal stripes, people will think that s/he cannot possibly be as heavy as s/he appears. In that situation, symbolic identification takes place because the person has adopted a third party's opinions that contradict his/her intention.

In a given brand community (e.g., Harley Davidson), consumers have the same ego and ego ideal and it is observed that brand images double and redouble. Identification could occur not only on the individual level but also on the group level. Before group identification can occur, the individual's ego ideal and ego are different (**Figure 6.6**). As a child grows up, his/her ego does not always meet his/her parents' expectations and the child regenerates his/her self-respect not in the ego itself but in the ego ideal, which controls his/her behavior based on social norms. In other words, the child feels that his/her ego is too incomplete to meet the parents' expectations and s/he tries to stake his/her self-respect on the developing ego ideal. At that moment, if a remarkably attractive person appears, the ego ideal is substituted for the omnipotent or very attractive person's image (**Figure 6.7**). As this happens, the minds of all of the group members have the same ego ideal, which functions as a major thought. This substituted ego ideal, namely the highly attractive person's image, could have a favorable nature (omnipotence) and could compensate for an incomplete or weak ego. Similarly, the omnipotent image of luxury brands could be substituted for users' ego ideal in a brand community. It is thereby observed that mutual imitation among the members could accelerate and original brand images could intensify.

Figure 6.6 Before group identification

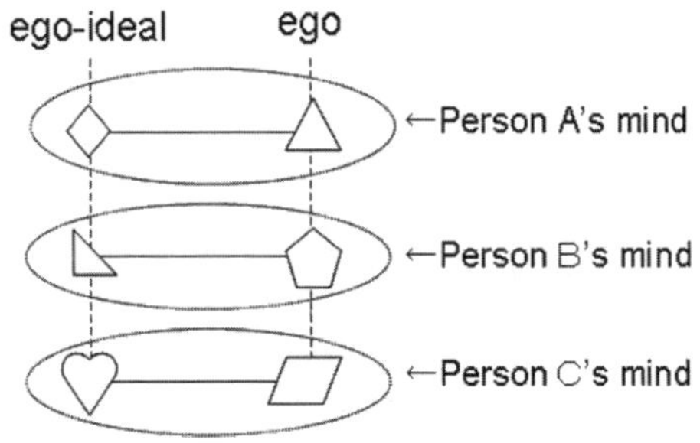

Source: Author

Figure 6.7 After group identification

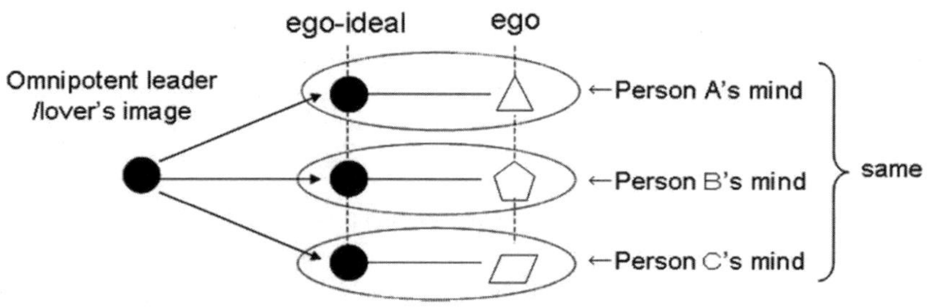

Source: Author

6.4 Borromean Rings and Luxury Brands

Figure 8 shows an image of Borromean rings, which consist of three topological circles. These rings are linked in such a manner that removing any ring results in two unlinked rings. This diagram is useful to distinguish luxury brands from other brands.

Figure 6.8 Borromean rings

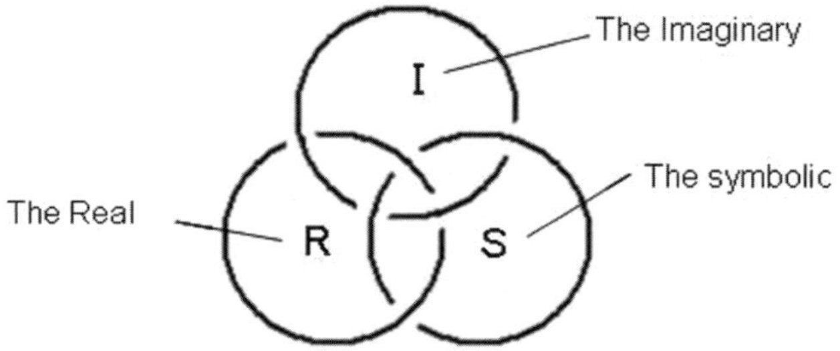

Source: Lacan, 2005 [14]

In the psychoanalytic theory of Jacques Lacan, the mind of human beings is consisted of these three circles [14]. The imaginary means the emotional, namely affection, hatred etc., which originates in emotional exchanges between mother and child. (This imaginary world consists only of mother and child; there are no other people present, even the father.) Each time a child looks at him/herself in a mirror, s/he gradually obtains her total self-image, from which self-consciousness and self-awareness (ego) originates.

⌜...one learns to see oneself as the (mother) sees one, to know oneself as the (mother) knows one. As the child in front of the mirror turns around and looks to the adult standing behind her for a nod, recognition, a word of approval or ratification,... she comes to see herself as if she were the parent, comes to be aware of herself as if from the outside, as if she were another person. [15]⌟

In short, with the help of the mother's approval, the child identifies his/her image in the mirror, which at first looks unfamiliar to him/her, because before self-consciousness is established the child does not have a complete body image, but only fragmented body images. In contrast, the symbolic world is the world in which law and order (social norms) are established by language. The law, in the person of the father, prohibits people from doing improper or unlawful things and the distinction between man and woman or goodness and evilness is provided for examples of social norms. In the symbolic world, there is objective reason/rules and if one refuses to accept the symbolic absolutely, s/he suffers from psychosis.

The real means the inaccessible and unseizable region/area and there is enjoyment in the real, which was lost forever. However, one must distinguish the real from the realistic: the realistic world is the **actual** world, while the real is **outside** the actual world. For example, women cannot perceive the moment of fertilization because without the help of a sensory organ we cannot perceive the events that occur within our bodies. Thus, the body itself is the real: impossible and beyond perception. Why Lacan named the inaccessible/impossible region in mind **the real**? It is because people do not like to know or see one's **true psychic reality** which is suppressed during waking.

In a marketing context, Borromean rings have been used to analyze famous lasting/permanent brands such as Malboro, McDonald etc. [16]. In general luxury brands are related to the tale of the nobility or royalty, who moved in high society. Ingenious project designers have thereby created luxury brands for the upper classes. Luxury brands tend to be named after myths or people in high society in whose lives the **image** of luxury brands originates. **Figure 6.9** shows the Borromean rings of the Lexus brand, wherein the noble story of a kind of royalty is lacking (the imaginary element is weak) and there is no ingenious project designer such as Armani. In other words, the Lexus is technically a highly elaborate car. Lexus development team members lodged at several high-class hotels in Europe to try to understand an aristocratic life-style in high society but they failed although there is an aristocratic life-style also in Asia. Figuratively speaking, the imaginary shown in Figure 8 means brand myth or story whose contents involve brand association and the real is intangible quality or unperceivable technology which supports the overall quality of luxury brands **from behind**. The symbolic consists of both objectively perceivable quality and brand elements (name, logo, characters, etc.) that are capable of being accounted for by using language. **Figure 6.10**, on the other hand, shows the Borromean rings of Coca-Cola, which we do not think of as a high-quality beverage (the symbolic element is weak) but we know the story of its founder, John S. Pemberton, who first created Coca-Cola as a kind of medicated liquor but this story is not related to a kind of the noble tale or a genius loci that expresses American ethnic individuality if it were. Therefore the imaginary element is also not enough. It is impossible to know Coca-Cola's recipes, or industrial secrets, which correspond to the real.

Figure 6.9 Toyota's Lexus

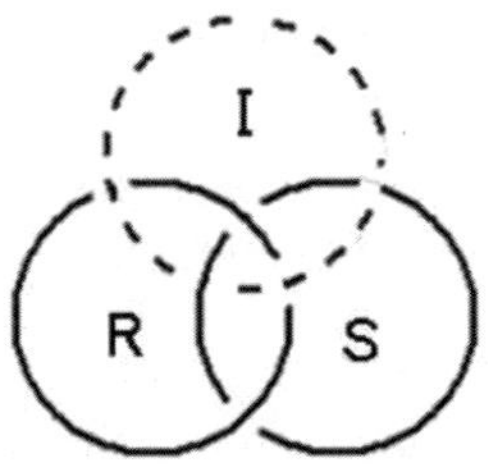

Source: Author

Figure 6.10 Coca-Cola

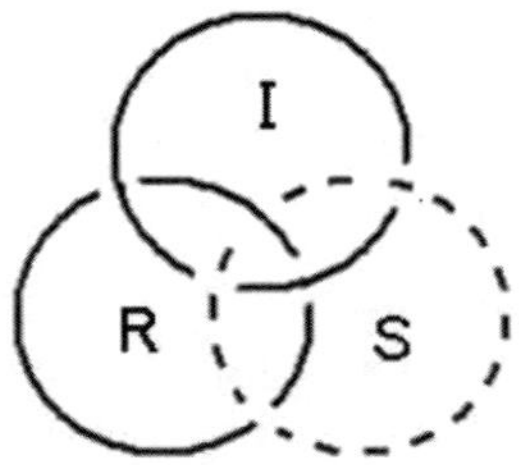

Source: Author

Figure 6.11 shows long-established but low-quality and negligent brands that are content with their present position and no longer introduce technological innovations. In the realm of psychoanalysis, **Figure 6.12** shows "psychosis," or falling into ruin. In psychosis, there is no continuity or consistency. People suffering from psychosis lose coherence and cannot comprehend the thought that tomorrow will be the same as today or that what was good in yesterday will be the same today (Everything is discontinuous.). In a marketing context, for example, it is impossible to combine the noble story of a certain royal family (the imaginary) with incorrect naming (the symbolic) and sewing by cutting corners for putting finishing touches under the hood (the real). To create luxury brands, we must give rich contents to these rings in a coherent manner.

Figure 6.11 Long-established brand lacking high quality

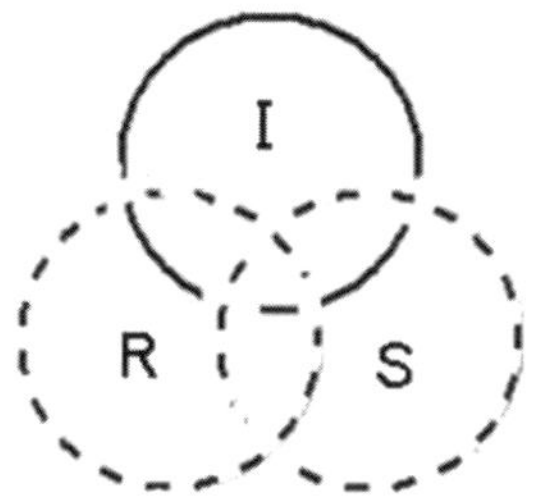

Source: Author

Figure 6.12 Inconsistent / Illogical brands

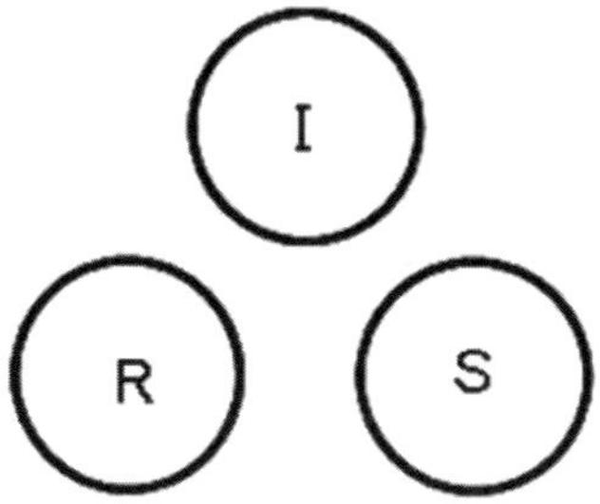

Source: Author

References

[1] Cohen, G.A. (1993): Equality of What? On Welfare, Goods and Capabilities. In Nussbaum, M./ Sen, A. (Eds.): The Quality of Life. USA, Oxford University Press.
[2] Munari, B. (1977): Fantasia [Fantasy]. Roma-Bari, Gius. Laterza & Figli Spa.
[3] Kellert, S.R./Wilson, E.O. (Eds.). (1993): The biophilia hypothesis. Washington, Island Press.
[4] Ricci, S. (Ed.). (2004): Museo salvatore ferragamo Idee, modelli, invenzioni [Museum salvatore ferragamo Ideas, models, inventions], Sillabe.
[5] Bellini, M. (2005): Tra syling e design [Between styling and design]. In Martino, C./Farina, C. (Eds.): Made in Italy. Il design degli Italiani [Made in Italy. The design of Italian]. Roma, R design press.
[6] Mari, E. (2001): Progetto e Passione [Project and Passion]. Torino, Bollati Boringhieri editore s.r.l .

[7] Lacan, J. (1986): Le Séminaire de Jacques Lacan Livre VII — L'Éthique de la Psychanalyse [The Seminar of Jacques Lacan: The Ethics of Psychoanalysis], Seuil.
[8] Juranville, A. (1984): Lacan et la philosophie [Lacan and the philosophy]. PUF.
[9] Munari, B. (1966): Arte come mestiere [Art as handicraft]. Bari, Editori Laterze.
[10] Corbellini, E./Saviolo, S . (2004): La scommessa del Made in Italy e il futuro della moda italiana[The betting of Made in Italy and the future of italian fashon].ETAS
[11] Kapferer, J.N./Bastien, V. (2009): The Luxury Strategy. Kogan Page Ltd.
[12] Verganti, R. (2009): Design-driven innovation: Changing the rules of competiton by radically innovating what thing mean. Boston: Harvard Business School Press.
[13] Codeluppi, V./Galoforo, G. (2000): Mondi di moda: la pubblicità dell' abbigliamento [Fashion worlds: the advertisement of clothes], in: Curicio, A. M. (Ed.): La dea delle apparenze [The goddess of signs], Franco Angeli.
[14] Lacan, J. (2005): Le Séminaire de Jacques Lacan Livre XXIII — Le sinthome [The Seminar of Jacques Lacan: The sin, holy man/Saint Thom], Seuil.
[15] Fink, B. (2004): Lacan To The Letter: Reading Ecrits Closely. University Of Minnesota Press, p.108.
[16] Michon, C. (2002): Un modello lacaniano per lo sviluppo duraturo di una marca forte [A Lacanian model for the lasting development of a strong brand]. Micro & MacroMarketing, n. 3, dicembre, pp. 577-589.

7 Indulging the Self Positive Consequences of Luxury Consumption

Liselot Hudders, Mario Pandelaere

7.1 Introduction

People not only purchase brands for the product behind the brand but also for what they symbolize [69]. As such, luxury brands are widely desired because – compared to their cheaper counterparts – they offer a functional, hedonic and symbolic value to their owners in addition to their utilitarian product value [20], [34], [56], [117], [122], [123]. While the functional value refers to the premium quality and the hedonic value refers to the extraordinary aesthetic components, the symbolic value of luxury brands lies in their ability to signal a consumer's success, wealth and social achievement to others [36], [86], [101], [116]. As luxury brands are associated with a premium quality, an aesthetically appealing design, exclusivity, expensiveness and rarity [20], many consumers incur debt to purchase such a premium priced luxury brands [40].

Spending resources on luxury consumption hinders individuals to spend resources on more rewarding activities, such as spending time with friends and family, work less, or spend more money on inconspicuous consumption that may enhance the quality of our lives to a greater extent [40]. In addition, both their high cost and the existence of several mechanisms that may limit their long-term impact on subjective well-being (e.g. hedonic treadmill, [42]) make it puzzling why consumers continue to spend large amounts of their income on luxury brands. We propose that that luxury consumption may have specific short-term rewards that serve as positive reinforcements for luxury consumption [26]. In particular, we argue that obtaining a luxury brand may (temporarily) enhance one's self-esteem. As the temporary satisfaction of material wants further increases the probability that such wants re-emerge after a short while [82], [118], the enhanced self-esteem may serve to reward the pursuit of luxurious possessions and reinforce this behavior.

Self-esteem is the subjective evaluation of one's self-identity and it consists of all positive and negative thoughts and feelings individuals have about themselves [19]. The need for a high self-esteem is considered to be a universally held need and a fundamental human motive [24], [38], [63]. It is already present in children and also active when lower-order needs are not fulfilled [24]. Some theorists argue that this need has a genetic basis [54], [74], while others see it as the result of learning and experience [1]. In general, self-esteem is considered to be both a stable trait and an unstable state [55], [59]. More specifically, individuals appear to have a stable baseline of self-esteem and a fluctuation of current self-appraisals around this baseline [6], [18], [60]. The current chapter focuses on these fluctuations of self-appraisals and hence, investigates how luxury consumption affects an individual's state self-esteem.

Having a positive sense of self appears to be beneficial [87], [93]. In general, high self-esteem individuals are more able to cope with adversity as they are more persistent in the face of failure, negative feedback and stress. Moreover, self-esteem is positively correlated with various measures of well-being, such as positive affect and satisfaction with life. In contrast, low self-esteem is positively linked with a number of mental health problems, such as depression or social anxiety and with various dark side consumption behaviors, such as binge eating and compulsive spending. Not surprisingly, people are more motivated to achieve increases and avoid decreases in their self-esteem state [22].

To defend a high self-esteem against threats, various studies investigated individuals' (un)conscious cognitive, affective and behavioral strategies such as (in)direct self-affirmations, flattering downward social comparisons and mnemic neglect (see e.g., [4], [51], [61], [111], [120]). In general, high self-esteem individuals tend to overestimate their positive qualities and the successes they achieved in their lives, while they tend to neglect their negative qualities and failures [47]. Accordingly, individuals who provide positive feedback to a high self-esteem individual are highly valued, while individuals who provide negative feedback are ignored and even considered to be incompetent [77].

However, the pursuit of self-esteem not only affects an individual's self and his/her interpersonal relationships directly, it is also a highly motivational force for many consumer decisions as consumers compensate for a bruised self-esteem by inflating the value of certain possessions [5]. More specifically, previous studies on compensatory consumption found that specific aversive states, such as powerlessness, threatened self-confidence, mortality salience and low income generate a higher preference for luxury and high status brands as a means to restore global self-worth [41], [44], [100], [107]. These luxury brands are preferred because they offer an additional symbolic value to their owners, above their utilitarian value [20], [56]. This symbolic value of a luxury brand lies in its ability to signal a consumer's success to others. In this respect, luxury brands may positively increase the sense of self by emphasizing the social value of their owners to relevant others. However, although consumers may prefer luxury brands to restore their global self-worth, it remains unclear if obtaining luxury brands actually enhances an individual's self-esteem.

The current chapter investigates the impact of obtaining luxury brands on an individual's self-esteem. In particular, a survey study shows that luxury consumption enhances an individual's self-esteem. In addition, two experimental studies provide further evidence for the directionality of the relationship between luxury consumption and self-esteem. The results of these studies suggest that acquiring a luxury branded fast-moving consumer product (i.e., chocolate or fruit juice) positively affects an individual's self-esteem level.

7.2 Conceptual Framework and Hypotheses Development

According to *sociometer* theory, a positive sense of self – which is considered to be the evaluative component of an individual's self-identity – functions as a sociometer and serves as an indicator of an individual's social inclusion [66]. Correspondingly, self-esteem fluctuates as a function of the degree to which an individual feels valued by others [110]. Consequently, an individual's self-esteem may decrease when experiencing social rejection and increase when experiencing social appreciation and praises [25], [50], [64], [81], [110]. Individuals who experience acceptance and social inclusion appear to have a higher self-esteem compared to individuals who experience rejection by others (e.g., Leary et al. [66]). In addition, a diary study shows that on the intra-individual level, individuals appear to have a higher self-esteem on days on which they report higher quality interactions and spend more time on such interactions [25].

As people are highly motivated by the need to maintain and enhance their self-esteem [22], [99], they aim to create a favorable self-identity and project this to both oneself and others [17], [37]. Accordingly, projecting these positive self-perceptions to others may lead to a higher social approval and – as suggested by sociometer theory – these positive social appraisals may in turn enhance an individual's self-esteem [66], [110]. This positive effect on an individual's self-esteem might occur because individuals actually experience positive appraisals from others (i.e., perceived social appraisal) or because individuals expect others to evaluate them positively (i.e., imagined social appraisal). In addition, the positive effect on self-esteem might also occur because the brand meaning may rub off on the owner of these brands (e.g., Belk [11]; Park and John [90]). As such, individuals may feel better about themselves because they obtained a luxury brand that reflects success. This brand meaning may rub off on the individual who – in turn – feels successful.

An individual's self-identity consists of all the thoughts and feelings they have about themselves and individuals may use various elements to build a favorable self-identity [39], [99], [125]). In addition to these thoughts about for example their personal history, perceived traits, appearance, and accomplishments [96], individuals use their possessions to construct their identity [31]. By consuming products, an individual integrates the meaning of these products into the self-concept (i.e., Meaning Transfer Model, [76]) and accordingly, the individual shapes his/her identity [21]. Consequently, these products become part of that individual's extended self [11]. However, not all products can be used to construct one's identity. Individuals prefer those products that fit with their (desired) identity as they can fulfill their need for self-consistency [16], [69], [88], [106]. Moreover, especially products with a widely shared meaning that can be easily perceived by others are used for identity-construction [11], [15], [30], [33], [75], [96], [103].

As a consequence, especially luxury brands are used to construct one's identity because these brands are often consumed conspicuously and they have a high symbolic value that is widely known [7], [8], [10], [31], [52], [95], [117]. As such, Puntoni [92] shows that self-identity is a significant predictor of purchase intention for luxury brands. Individuals incorporate the symbolic meanings of luxury brands into their self-concept, thereby approaching an ideal self-identity.

In addition to identity construction, individuals use their luxury possessions to communicate this favorable self-identity to their reference and aspiration groups in order to gain social approval [10], [11], [12], [13], [14], [15], [30], [31], [33], [49], [67], [104], [108], [120], [121]. In this respect, people often purchase luxury brands because their peers also consume them and they *expect* that conforming to the group norms will lead to acceptation and appreciation [29], [68], which, in turn, can positively affect self-esteem. Secondly, owners of luxury brands may demonstrate that they are better off than their peers, which might provide them with a higher status, not only within their own reference groups, but also within their aspiration groups [70]. As a higher perceived status position is positively correlated with self-esteem [62], [114], owners of luxury brands may gain prestige and consequently, achieve a more positive sense of self. As such, luxury brands function as indicators of social affiliation and distinction and – as sociometer theory suggests – this may enhance an individual's self-esteem when he/she holds an imaginary external perspective [5], [23], [44], [105].

Moreover, a study of Sivanathan and Pettit [107] shows that the possession of high status goods might provide psychological benefits and protect individuals against self-threats suggesting that consuming luxuries might enhance an individual's self-esteem. More specifically, their study shows that individuals who ruminated about owning a high-status car were threatened less by negative feedback compared to individuals who ruminated about owning a low-status car. Following this and considering the fact that luxury brands are often used to create and communicate a favorable self-identity to others and that these positive self-perceptions may lead, in turn, to an increased self-esteem, the current chapter proposes that luxury consumption may lead to an enhanced self-esteem. This leads to the first hypothesis that luxury consumption might be used as a self-enhancement strategy:

> *H1: Obtaining luxury brands has a positive impact on an individual's self-esteem*

In addition, the current chapter wants to clarify whether the positive effect of obtaining a luxury brand on self-esteem is affected by the direct appreciation experienced from the experimenter, as sociometer theory suggests that self-esteem is driven by perceived or imagined social appraisal [66]. The fact that respondents in the experimental studies did not have to pay to acquire the (non-) luxuries and the experimenter handed out the luxuries/non-luxuries to the respondents in the experiments may imply that the positive effects of acquiring a luxury brand on an individual's self-esteem is to be moderated by the perceived appreciation of the experimenter who handed out the (non-)luxury brand to the respondent.

Therefore, the main aim of the second experimental study is to test if the positive effect on self-esteem remains significant if the luxury brand is obtained by luck and is thus not related to immediate social appraisal. To disentangle luck versus appreciation, an additional manipulation (i.e., lottery condition versus selection by researcher) is introduced. If self-esteem is boosted because individuals experience a higher appreciation from others, the effect on self-esteem should be moderated by manipulating the appreciation of the choosing agent (i.e., lottery condition, cf. no appreciation but luck versus selection by experimenter, cf. appreciation, instead of luck). However, if immediate social appraisal is not crucial, the self-esteem boost must also occur when there is no direct link to immediate appreciation from others (i.e., in the lottery condition):

> *H2: Both individuals for whom the luxury brand reflects direct appreciation by others and individuals for whom the luxury brand reflects no direct appreciation by others will have a higher self-esteem compared to individuals who obtained the non-luxury brand*

7.3 Study 1: Relationship between Luxury Consumption and Self-Esteem

The main purpose of this exploratory study is to explore the relationship between luxury consumption and self-esteem. Given the fact that self-esteem may depend on the degree to which an individual feels valued by others and that luxury brands have a high symbolic value which reflects success and social achievement, consuming luxuries may lead to a higher self-esteem.

7.3.1 Participants and Procedure

We conducted a large scale survey in the Flemish part of Belgium to investigate the relationship between luxury consumption and self-esteem. Our survey was collected by 327 students in exchange for course credit. Each student filled in the questionnaire and distributed the questionnaire to three other (non-student) people. We strived for a proportional representation in terms of gender and age. 1308 respondents participated in this survey. To check for outliers, we first conducted a multivariate outlier analysis. We calculated the Mahalanobis Distance for the responses on the two central variables in our dataset, namely luxury consumption and self-esteem. On the basis of this analysis we identified 4 outliers for which the observed Mahalanobis distance exceeded the 99.99 % quantile. The remaining data set consists of 1304 valid cases (654 males, $M\_age = 41.08$, $SD = 17.57$).

7.3.2 Measures

Luxury consumption. Previous scales which are related to luxury consumption and ownership do not measure consumers' tendency to purchase luxury brands, regardless of the underlying motives for this behavior [35], [117]. More objective measures of luxury ownership [34] do not account for the fact that products or brands are not inherently luxurious but are perceived as more or less luxurious by an individual. Accordingly, the subjective character may be more important in shaping self-esteem than objective features.

To measure self-perceived luxury consumption, we used the scale of Hudders & Pandelaere [53] that measures consumers' tendency to choose luxury brands (i.e., brands that are perceived as luxurious by an individual) in various product categories.

Self-esteem. To measure self-esteem, we used the self-esteem scale of Rosenberg [98]. This scale consists of ten items that measure people's feelings of self-worth and has a high reliability score ($\alpha = .877$).

Materialism. As materialism is positively related to luxury consumption [53] and negatively related to self-esteem [91], this might produce a spurious negative relationship between luxury consumption and self-esteem. Therefore, we added materialism to the model in order to exclude materialism as a factor. We used the eighteen item material values scale of Richins and Dawson ($\alpha = .83$)[97].

7.3.3 Results and Discussion

The results of a linear regression analysis (R^2 = 1.0 %; $F(1, 1275)$ = 14.06; p < .001) show that luxury consumption predicts self-esteem positively (β = .104, p < .001). These results confirm the first hypothesis. Adding materialism to the model improves the model (R^2-change = 7.0 %, $F(1, 1274)$ = 82.77, p < .001) and strengthens the positive relationship between luxury consumption and self-esteem (β = .185, p <.001), while materialism is negatively related to self-esteem (β = -.258, p <.001). This provides further evidence for the directionality of the relationship.

These results imply that high luxury consumers appear to have a higher self-esteem compared to low luxury consumers. However, the results of this survey study are only correlational in nature. To investigate the directionality of this positive relationship, we conducted an experimental study.

7.4 Study 2: Impact of Luxury Consumption on Self-Esteem

The main aim of this experimental study is to investigate the impact of obtaining a luxury brand on an individual's self-esteem. As consumers derive their self-esteem from positive evaluations of others [66] and they expect luxury consumption to provide them with praise and higher status ([56], we suggest that respondents who obtain a luxury brand will report a higher self-esteem compared to respondents who obtain a non-luxury brand. This study uses a between-subjects single factor experimental design with two conditions (i.e., luxury and non-luxury condition) to test if consumption of luxury brands enhances an individual's self-esteem (cf. hypothesis 1).

7.4.1 Method

7.4.1.1 Materials and Procedure

All respondents received an identical chocolate bar from the Belgian (luxury) chocolate brand *Belcolade. Belcolade* sells its chocolate to the wholesale business, where it is further processed in luxury preparations, such as enrobed pralines or pastries. Consequently, most end consumers do not know this brand. Respondents who recognized the brand are excluded from the analyses in order to avoid confounding effects of brand knowledge and preferences. Respondents were randomly assigned to one of the conditions. Although they all received an identical chocolate, respondents in the two conditions received a different, written description of the chocolate bar.

In the luxury condition, the chocolate bar was presented as a very luxurious, refined, high quality, expensive (i.e., 25 Euros per kilogram) and an exclusive piece of chocolate. In the non-luxury condition, the chocolate bar was presented as an ordinary piece of chocolate which is mass-produced and not expensive (i.e., 3.95 Euros per kilogram). Subsequently, respondents filled in another questionnaire (which was presented as being part of another study), in which respondents' self-esteem level was measured along with some filler questions. To measure self-esteem, the self-esteem scale of Rosenberg [98] was used. This scale consists of ten items that measure people's feelings of self-worth and has a high reliability score (α = .88). At the end of the experimental session, respondents had to indicate how satisfied they were with the chocolate (measured with one item) and had to rate its luxuriousness (measured with one item; manipulation check). To control for materialism, we used the eighteen item material values scale of Richins and Dawson (α = .81)[97]. We measured participants materialism level before they participated in this experiment.

7.4.1.2 Participants

127 respondents participated in this experiment in exchange for 7 Euros. However, eight respondents are excluded because they knew the brand *Belcolade.* 119 valid cases remained (60 males, M_{age} = 21.75, *SD* = 1.09).

7.4.2 Results and Discussion

Respondents report no significant differences in satisfaction level between the two conditions (M_{lux} = 5.64, M_{nonlux} = 5.37, $t(116)$ = -1.48, p = .142). However, respondents in the luxury condition (M = 4.98) rate the chocolate as more luxurious than respondents in the non-luxury condition (M = 4.32; $t(113)$ = 2.66, p = .009). In addition, respondents in the luxury condition (M = 4.28) report higher self-esteem scores compared to respondents in the non-luxury condition (M = 3.95, $t(116)$ = 2.00, p = .048). These results confirm the first hypothesis. Moreover, when controlling for materialism, the positive impact of luxury consumption on self-esteem slightly increased: F(1, 114) = 4.74, p = .032.

However, while these results may be attributed to ownership of a luxury (versus non-luxury) brand, it is also possible that participants in the luxury condition felt better about themselves because they had received something relatively expensive from the experimenter. As such, participants' increase in self-esteem might be due to the perceived social appraisal rather than to ownership of the luxury brand. After all, to the extent that self-esteem is based on perceived value to others [66], it may be directly affected by the direct appreciation experienced from the experimenter (i.e. the chocolate they received). Therefore, the main aim of study 3 is to test if the positive effect on self-esteem remains significant if the luxury brand is obtained by luck and is thus not related to immediate social appraisal. To disentangle luck versus appreciation, an additional manipulation (i.e., lottery condition versus selection by researcher) is introduced.

7.5 Study 3: Alternative Explanation of Direct Appreciation

The aim of this third experimental study is to extend the results of the previous studies by using another product category (i.e., fruit juice instead of chocolate). Secondly, this study wants to test whether direct social appraisal moderates the positive effect of obtaining a luxury brand on an individual's self-esteem. Therefore, an additional manipulation is introduced in which either the experimenter or a lottery determined whether respondents received the luxury or non-luxury brand. In this respect, this study investigates if the effect of luxury consumption on self-esteem depends on the mode in which people receive the luxury brand (i.e., choosing agent) and thus the immediate social appraisal an individual experiences.

7.5.1 Method

7.5.1.1 Materials and Procedure

This study uses a two (i.e., luxury condition: luxury versus non-luxury condition) by two (i.e., choosing agent: selection by lottery versus selection by experimenter condition) between-subjects experimental design to investigate if direct social appraisal moderates the impact of obtaining a luxury brand on self-esteem. All respondents received a glass of fruit juice and they were first instructed to evaluate the juice without tasting it. Next, they had to complete the self-esteem and positive and negative affect scales that are presented as being part of another study. Finally, respondents were instructed to evaluate the taste of the juice (this is only to avoid suspiciousness about the real goal of the experiment).

At the start of the experimental session, two different brands of fruit juice were presented to all respondents, that is, *Pure Fruit* and *Ethno Bar*. Respondents first received some brief information about both brands. *Pure Fruit* was presented as a cheap fruit mix sold in a well-known hard discounter, while *Ethno Bar* was presented as a more exclusive, luxurious and exotic fruit mix available in a well-known upscale store. Next, participants were instructed to evaluate one of both brands, either *Pure Fruit* or *Ethno Bar*. These brands are new in the market and consequently, to emphasize the importance of their evaluation, we told participants that their evaluations would determine whether these brands would be maintained in the store assortments. Next, respondents were randomly assigned to one of the conditions.

As both juices are displayed to all respondents (in contrast to the previous study, where respondents were exposed to either the luxury or non-luxury chocolate and they did not knew that others received a (non-)luxury chocolate), the juices were actually different (in contrast to the second study, where respondents received the same chocolate in both conditions) and had different prices.

In the lottery condition, respondents had to pick a lottery ticket (out of a bag with several tickets) that determined if they had to taste either the *Ethno Bar* or the *Pure Fruit* juice. In the experimenter condition, the experimenter decided conspicuously (*'I want you to evaluate the luxury fruit juice/ non-luxury fruit juice'*) whether respondents had to taste either the *Ethno Bar* or the *Pure Fruit* juice.

In the luxury condition, respondents received the *Ethno Bar* fruit mix and a written description of this juice (i.e., expensive and unique mix of exotic fruits (e.g., litchi), spices and flowers (e.g., lotus) with a limited availability (only available at the end of the year) and excellent quality). In the non-luxury condition, respondents received the *Pure Fruit* fruit mix and a written description of this juice (i.e., cheap mix of ordinary fruits (e.g., apple, orange, pear) which is diluted with sugar water and produced on a large scale).

7.5.1.2 Measures

The Rosenberg's self-esteem scale measures respondents' self-esteem level (α = .91). As manipulation check, respondents had to rate the fruit juice on luxuriousness, expensiveness, and mass production before tasting the fruit juice and on liking and willingness to pay after tasting the juice. These items are measured with a seven-point likert scale ranging from 1 strongly disagree to 7 strongly agree. To control for materialism, we used the eighteen item material values scale of Richins and Dawson (α = .90) [97]. As affective well-being and self-esteem are positively related [27], [78], it is possible that respondents in the luxury condition experience a higher self-esteem, because they feel better after receiving a luxurious chocolate. To control for this alternative account related to happiness and self-esteem, respondents completed the PANAS-scale (α_{PA} = .89; α_{NA} = .83; Watson et al.) [119] and the life satisfaction scale ([28]; α = .79).

7.5.1.3 Participants

151 respondents participated in this experiment in exchange for 7 Euros (52 males, M_{age} = 22.83, *SD* = 5.67).

7.5.2 Results and Discussion

Respondents in the luxury condition rate the juice as more luxurious (*M* = 4.91) compared to respondents in the non-luxury condition (*M* = 2.53; $F(1, 135)$ = 129.17, $p < .001$). Furthermore, respondents in the non-luxury condition (M_{cheap} = 5.53; M_{mass} = 6.07) are more convinced that the juice is cheap and mass-produced than respondents in the luxury condition (M_{cheap} = 2.24, $F(1, 135)$ = 329.59, $p < .001$; M_{mass} = 4.38, $F(1, 135)$ = 55.06, $p < .001$). After tasting the fruit juice, respondents in the luxury condition liked the juice more (*M* = 4.71) and are willing to pay more for the juice (*M* = 1.29) compared to respondents in the non-luxury condition (M_{liking} = 2.24, $F(1, 146)$ = 119.94, $p < .001$; M_{pay} = .80, $F(1, 144)$ = 14.92, $p < .001$). There are no main effects of choosing agent (luck vs. experimenter) and there are no interaction effects.

A two-way ANOVA analysis revealed that there is a main effect of luxury condition on self-esteem ($F(1, 147)$ = 9.35, p = .003). Respondents who acquired the luxurious *Ethno Bar* juice

(M = 5.29) reported higher self-esteem scores than respondents who acquired the non-luxurious *Pure Fruit* juice (M = 4.93). These results again confirm the first hypothesis. Furthermore, there is a main effect of choosing agent on self-esteem ($F(1, 147) = 5.12$, $p = .025$). Respondents who had to pick a lottery ticket (M = 4.93) have a lower self-esteem compared to respondents for whom the researcher decided which juice they had to evaluate (M = 5.24). The main effect of luxury condition ($F(1, 136) = 4.06$, $p = .046$) on self-esteem remains significant when controlling for the alternative accounts of materialism and subjective well-being by including materialism, positive and negative affect and life satisfaction as covariates in the model. Hence, variations in materialism and well-being cannot fully account for this positive effect of acquiring luxuries on self-esteem. On the contrary, the main effect of choosing agent ($F(1, 136) = 1.83$, $p = .179$) condition becomes insignificant after controlling for materialism, life satisfaction and mood.

Finally, the interaction effect between luxury condition and choosing agent on self-esteem is not significant ($F(1, 147) = .039$, $p = .844$). This implies that respondents in the luxury condition experience higher self-esteem compared to respondents in the non-luxury condition, regardless whether the luxury brand reflects direct appreciation from others or not. As such, respondents who obtained a luxury brand that reflects social appraisal (M = 5.46) appear to have a higher self-esteem compared to respondents who obtained a non-luxury brand that reflects social appraisal (M = 5.02, $t(63) = -2.31$, $p = .024$). Similarly, respondents who obtained the luxury fruit juice by luck (M = 5.12) appear to have a higher self-esteem compared to respondents who obtained the non-luxury fruit juice by luck (M = 4.74, $t(57) = -1.95$, $p = .056$), however, this effect is only marginally significant. These results confirm the second hypothesis.

7.6 General Discussion and Conclusions

Scholars from various disciplines conducted research on self-esteem because it has a strong impact on several aspects of human behavior [65],[87]. While low self-esteem appears to be linked to negative behaviors, such as delinquency and antisocial behavior [32], high self-esteem appears to be linked to positive behavioral aspects. High self-esteem individuals appear to be more resistant against self-image threats [109], perceive negative feedback as a challenge rather than a threat [102] and are more capable to savor positive affect [124]. Humans have an innate drive to strive for more positive self feelings and they developed several strategies to restore a negative sense of self. In this respect, several studies found that people with a bruised self-esteem showed to have a higher preference for luxury and high status goods (e.g., Sivanathan and Pettit [107]). The current chapter investigated whether acquiring luxury brands would effectively raise self-esteem.

The results of three studies indicate that this may be the case. In a first study, luxury consumption appears to be positively related to self-esteem. In a second study, participants who acquired a luxury brand reported higher self-esteem compared to respondents who acquired a non-luxury brand. In study 3, this chapter showed that linking a brand to direct appreciation does not moderate the effect on self-esteem. Individuals who obtained a luxury brand reported higher self-esteem ratings compared to respondents who obtained a non-luxury brand, regardless whether this luxury brand reflected immediate social appraisal or not. This may imply that imagined social appraisals may be more important. As such, the positive effect of luxury consumption on self-esteem may be the result of an expected reward due to social appraisal.

In addition, this chapter builds on previous research that shows that brand meanings may rub off on the owners of these brands (e.g., [11], [46], [90]) and this might cause the positive effect on self-esteem. These studies show that consumers not only prefer brands that fit with their sense of self and with their personality, they also enhance their self-identity in line with the brand's meaning. This process can be explained by the Meaning Transfer Model [76] which states that while meaning arises in culture, this meaning can be transferred to brands through advertising and the use of endorsements. Hence, by obtaining these brands, meaning may be transferred to the individual consumer.

This chapter provides further evidence for this meaning transfer of brands to the owners of these brands. However, the occurrence of this meaning transfer might depend on the personality of the owner. In this respect, Park and John [90] show that, while individuals who believe that their personal qualities are invariable (i.e., entity theorists) are influenced by using a brand with an appealing personality, individuals who believe that they can improve their personal qualities (i.e., incremental theorists) are not. In future research, it would be interesting to investigate if the positive effect of owning luxury brands on self-esteem may be moderated by these implicit self-beliefs.

Moreover, future research should investigate if these positive effects of acquiring luxury brands on self-esteem are stronger for materialistic consumers. As such, the negative relationship between materialism and self-esteem may be diminished by materialists' preference for luxurious possessions [97], [115]. As materialists consider luxury possessions as a path to personal happiness, they might experience a stronger effect on their self-esteem compared to low materialistic consumers. However, these positive effects on their self-esteem may reinforce materialistic consumers in their pursuit for luxurious possessions, which might, in turn, hinder them to invest resources in activities that are more rewarding for their self-esteem (e.g., spending time on close friends and family).

The current chapter also has several limitations. First, in the experimental studies, respondents did not have to pay for the fruit juice or the chocolate. In future research, it would be interesting to let respondents actually pay for the luxury good. Furthermore, the hedonic value of the luxuries used in these studies prevails over the status value. The effects on self-esteem might be stronger when choosing a luxury where the status value prevails over the hedonic value. Moreover, the experimental studies only investigated the effect of obtaining a luxurious fast-moving consumer good on self-esteem. In future research, it would be interesting to focus on durable luxury goods, because such durables may enhance self-esteem to a greater extent. Finally, future research should benefit from incorporating the factor public/private luxury consumption. As consumers' self-esteem depends on the evaluation of others, consuming luxury brands in a public setting might enhance self-esteem more than consuming the same brands in a private setting.

The positive effect of owning luxury brands on an individual's self-worth might reinforce one's pursuit of luxurious possessions. However, this does not imply that luxury consumption is beneficial for an individual's self-esteem in the long run. Furthermore, the enhanced self-esteem by status consumption may be illusory. In this respect, many individuals are envious of individuals consuming luxury brands (e.g., many individuals are unfavorably disposed towards men who are showing off their Porsche). Hence, this chapter merely claims that some aspect of owning luxury brands may reinforce buying luxury products but *not* that it is a fruitful way to strive for enduring high self-esteem.

7.7 References

[1] Adler, A. (1956): The Individual Psychology of Alfred Adler. H.L. Ansbacher/R.R. Ansbacher (Eds.). New York: Harper Torchbooks.

[2] Allred, A., Valentin, E. K./Chakraborty, G. (2010) : Pricing risky services: Preference and quality considerations, in: Journal of Product and Brand Management, Vol. 19, No. 1, pp. 54-60.

[3] Ariely, D./Norton, M.I. (2009): Conceptual consumption, in: Annual Review of Psychology, Vol: 60, pp. 475-499.

[4] Baccus, J.R./Baldwin, M. W./Packer, D.J. (2004): Increasing implicit self-esteem through classical conditioning, in: Psychological Science, Vol. 15, No. 7, pp. 498-502.

[5] Banister, E. N./Hogg, M. K. (2003): Possible selves? Identifying dimensions for exploring the dialectic between positive and negative selves in consumer behavior, in: Advances in Consumer Research, Vol. 30, No. 1, pp. 149-150.

[6] Baumeister, R.F. (1998): The self, in: Gilbert, D.T./Fiske, S.T./Lindzey, G. (Eds.): Handbook of social psychology (4th ed., pp. 680-740). New York: McGraw-Hill.

[7] Bearden, W. O./Etzel, M. J. (1982): Reference group influence on product and brand purchase decisions, in: Journal of Consumer Research, Vol. 9, No. 2, pp. 183-481.

[8] Bearden, W. O./Netemeyer, R. G./Teel, R. J. (1989): Measurement of consumer susceptibility to interpersonal influence, in: Journal of Consumer Research, Vol. 15, No. 4, pp. 473-481.

[9] Beggan, J. K. (1992): On the social nature of nonsocial perception: The mere ownership effect, in: Journal of Personality and Social Psychology, Vol. 62, No. 2, pp. 229–237.

[10] Belk, R. W. (1985): Materialism: Trait aspects of living in the material world, in: Journal of Consumer Research, Vol. 12, No. 4, pp. 265-280.

[11] Belk, R. W. (1988): Possessions and the extended self, in: Journal of Consumer Research, Vol. 15 No. 3, pp. 139-168.

[12] Belk, R. W./Bahn, K. D./ Mayer, R.N. (1982): Developmental recognition of consumption symbolism, in: Journal of Consumer Research, Vol. 9, No. 1, pp. 4-16.

[13] Berger, J./Heath, C. (2007): Where consumers diverge from others: Identity signaling and product domains, in: Journal of Consumer Research, Vol. 34, No. 2, pp. 121-134.

[14] Berger, J./Heath, C. (2008): Who drives divergence? Identity signaling, out-group similarity, and the abandonment of cultural tastes, in: Journal of Personality and Social Psychology, Vol. 95, No. 3, pp. 593-607.

[15] Berger, J./Ward, M. (2010): Subtle signals of inconspicuous consumption, in: Journal of Consumer Research, Vol. 37, No. 4, pp. 555-569.

[16] Birdwell, A. E. (1968): A study on the influence of image congruence on consumer choice, in: Journal of Business, Vol. 41, No. 1, pp. 76-88.

[17] Bosnjak, M./Rudolph, N. (2008): Undesired self-image congruence in a low-involvement product context, in: European Journal of Marketing, Vol. 42, No. 5-6, pp. 702-712.

[18] Campbell, J. P. (1990): Modeling the performance prediction problem in industrial and organizational psychology, in: Dunnette, M.D./Hough, L.M. (Eds.): Handbook of Industrial and Organizational Psychology (pp. 687-732): Palo Alto, CA: Consulting Psychologists Press, Inc.
[19] Campbell, R. L./Eisner, S./Riggs, N. (2010): Sources of self-esteem: From theory to measurement and back again, in: New Ideas in Psychology, Vol. 28, No.3, pp. 338-349.
[20] Chevalier, M./Mazzalovo, G. (2008): Luxury brand management: A world of privilege. Singapore: John Wiley & Sons.
[21] Choi, S. M./Rifon, N. J. (2007): Who is the celebrity in advertising? Understanding dimensions of celebrity images, in: The Journal of Popular Culture, Vol. 40, No. 2, pp. 304-324.
[22] Crocker, J/Park, L. E. (2004): The costly pursuit of self-esteem, in: Psychological Bulletin, Vol. 130, No. 3, pp. 392-414.
[23] Csikszentmihaliyi, M. (1999): If we are so rich, why aren't we happy?, in: The American Psychologist, Vol: 54, No. 10, pp. 821-827.
[24] Csikszentmihalyi, M. (2000): Beyond Boredom and Anxiety. San Francisco: Jossey-Bass.
[25] Denissen, J. J. A./Penke, L./Schmitt, D. P./Van Aken, M. A. G. (2008): Self-esteem and reactions to social interactions: Evidence for sociometer mechanisms across days, people, and nations, in: Journal of Personality and Social Psychology, Vol. 95, No. 1, pp. 181-196.
[26] DiClemente, D.F./Hantula, D.A. (2003): Applied behavioral economics and consumer choice, in: Journal of Economic Psychology, Vol. 24, No. 5, pp. 589–602.
[27] Diener, E./Diener, M. (1995): Cross cultural correlates of life satisfaction and self-esteem, in: Journal of Personality and Social Psychology, Vol. 68, No. 4 , pp. 653–663.
[28] Diener, E./Emmons, R./Larsen, J./Griffin, S. (1985): The Satisfaction With Life Scale, in: Journal Personality Assessment, Vol. 49, No. 1, pp. 71-75
[29] Dittmar, H. (1994): Material possessions as stereotypes: Material images of different socioeconomic groups, in: Journal of Economic Psychology, Vol. 15, No. 4, 561-585.
[30] Dittmar, H. (2004): Understanding and diagnosing compulsive buying, in: R. Coombs (Ed.), Addictive disorders: A practical handbook. New York: John Wiley & Sons.
[31] Dittmar, H. (2008): Consumer culture, identity and well-being. The search for the 'good life' and the 'body perfect'. New York: Psychology Press.
[32] Donnellan, M. B./Trezesniewski, K. H./ Robins, R. W./Moffitt, T. E./Caspi, A. (2005): Low self-esteem is related to aggression, antisocial behavior, and delinquency, in: Psychological Science, Vol. 16, No. 4, pp. 328-335.
[33] Douglas, M./Isherwood, B. (1978): The world of goods: Towards an anthropology of consumption. London: Allen Lane.
[34] Dubois, B./Laurent, G. (1996): The functions of luxury: A situational approach to excursionism, in: Advances in Consumer Research, Vol. 23, No. 1, pp. 470-477.
[35] Dubois, B./Czellar, S./Laurent, G. (2005): Consumer Segments Based on Attitudes Toward Luxury: Empirical Evidence from Twenty Countries, in Marketing Letters, Vol. 16, p. 115-128.
[36] Eng, T-Y./Bogaert, J. (2010): Psychological and cultural insights into consumption of luxury western brands in India, in: Journal of Customer Behavior, Vol. 9, No. 1, pp. 55-75.
[37] Escalas, J. E./Bettman, J. R. (2005): Self-construal, reference groups, and brand meaning, in: Journal of Consumer Research, Vol. 32, No. 3, pp. 378-389.
[38] Faunce, W.A. (2003): Work, status, and self-esteem: A theory of selective self investment. Lanham: University Press of America.
[39] Fiske, A./Kitayama, S./Markus, H. R./Nisbett, R. E. (1998): The cultural matrix of social psychology, in:Gilbert, D/Fiske, S./Lindzey, G. (Eds.): The Handbook of Social Psychology (4th ed., pp. 915–981): San Francisco: McGraw-Hill.
[40] Frank, R. H. (1999): Luxury fever. Money and happiness in an era of excess. Princeton: Princeton University Press.
[41] Fransen, M. L./Smeesters, D./Fennis, B.M. (2011): The role of social presence in mortality salience effects. Journal of Business Research, 64, No. 1, pp. 29-33.

[42] Fredrick, S./Loewenstein, G. (1999): Hedonic adaptation. In D. Kahneman, E. Diener/N. Schwarz (Eds.), Well-being: The foundations of a hedonic psychology (pp. 302–329): New York: Russell Sage Foundation.
[43] Frederick, S./Loewenstein, G./O'Donoghue, T. (2002): Time discounting and time preference: A critical review. Journal of Economic Literature, Vol. 40, No. 2, pp. 351-401.
[44] Gao, L./Wheeler, S.C./Shiv, B. (2009): The "shaken self": Product choices as a means of restoring self-view confidence. Journal of Consumer Research, Vol. 36, No. 1, pp. 29-38.
[45] Gawronski, B./Bodenhausen, G.V. /Becker, A.P. (2007): I like it, because I like myself: Associative self-anchoring and post-decisional change of implicit evaluations. Journal of Experimental Social Psychology, Vol. 43 , No. 2, pp. 221-232.
[46] Govers, P./Schoormans, J.P.L. (2005): Product personality and its influence on consumer preference. Journal of Consumer Marketing, Vol. 22 , No. 4, pp. 189-197.
[47] Green, J. D./Sedikides, C., Pinter, B./Van Tongeren, D.R. (2009): Two sides of self-protection: Self-improvement strivings and feedback from close relationships eliminate mnemic neglect. Self and Identity, Vol. 8, No. 2, pp. 233-250.
[48] Greenwald, A.G. /Banaji, M.R. (1995): Implicit social cognition: Attitudes, self-esteem, and stereotypes. Psychological Review, Vol. 102, No. 1, pp. 4-27.
[49] Grewal, R./Mehta, R./Kardes, F.R. (2004): The timing of repeat purchases of consumer durable goods: The role of functional bases of consumer attitudes. Journal of Marketing Research, Vol. 41, No. 1, pp. 101-115.
[50] Gyurak, A./Ayduk, Ö. (2007): Defensive physiological reactions to rejection: The effect of self-esteem and attention control on startle responses. Psychological Science, Vol. 18, No. 10, pp. 886-892.
[51] Hepper, E.G./Gramzow, R. H./Sedikides, C. (2010): Individual differences in self-enhancement strategies: An integrative analysis. Journal of Personality, Vol. 78, No. 2, pp. 781-814.
[52] Holt, D. B. (1995): How consumers consume: A typology of consumer practices. Journal of Consumer Research, Vol. 22, No. 1, pp. 1-16.
[53] Hudders, L./Pandelaere, M. (2011): The silver lining of materialism: The impact of luxury consumption on subjective well-being. Journal of Happiness Studies, doi: 10.1007/s10902-011-9271-9.
[54] James, W. (1890): The principles of psychology (Vol. 1): New York: Henry Holt.
[55] Janis, I.L./Field, P.B. (1959): Sex differences and factors related to persuasibility, in: C.I. Hovland/I.L. Janis (Eds.), Personality and persuasibility (pp. 55-68): New Haven, CT: Yale University Press.
[56] Kapferer, J./Bastien, V. (2009): The luxury strategy. London: Kogan Page Ltd.
[57] Kahneman, D./Tversky, A. (1979): Prospect theory: An analysis of decision under risk. Econometrica, Vol. 47, No. 2, pp. 263-291.
[58] Kashdan, T. B. (2004): The assessment of subjective well-being (issues raised by the Oxford Happiness Questionnaire): Personality and Individual Differences, Vol. 36, pp. 1225-1232.
[59] Kernis, M. H. (1993): The roles of stability and level of self-esteem in psychological functioning. In R.F. Baumeister (Ed.), Self-esteem: The puzzle of low self-regard (pp. 167-182): New York: Plenum Press.
[60] Kernis, M. H. (2005): Measuring self-esteem in context: The importance of stability of self-esteem in psychological functioning. Journal of Personality, Vol. 73, No. 6, pp. 1569-1605.
[61] Knowles, M. L./Lucas, G. M./Molden, D.C./Gardner, W.L./Dean, K.K. (2010): There's no substitute for belonging: Self-affirmation following social and non-social threats. Personality and Social Psychology Bulletin, Vol. 36 , No. 2, pp. 173-186.
[62] Leary, M. R./Cottrell, C. A./Phillips, M. (2001): Deconfounding the effects of dominance and social acceptance on self-esteem. Journal of Personality and Social Psychology, Vol. 81, No. 5, pp. 898-909.
[63] Leary, M. R./Downs, D. L. (1995): Interpersonal functions of the self-esteem motive: The self-esteem system as a sociometer. In M.H. Kernis (Ed.), Efficacy, agency, and self-esteem (pp. 123–144): New York: Plenum Press.

[64] Leary, M. R./Haupt, A. L./Strausser, K. S. /Chokel, J. T. (1998): Calibrating the sociometer: The relationship between interpersonal appraisals and the state self-esteem. Journal of Personality and Social Psychology, Vol. 74, No. 5, pp. 1290-1299.
[65] Leary, M.R./MacDonald, G. (2005): Individual differences in self-esteem: A review and theoretical integration. In M.R. Leary/J.P. Tangney (Eds.), Handbook of self and identity (pp. 401-418): New York: Guilford Press.
[66] Leary, M.R./Tambor, E.S. /Terdal, S.K./Downs, D.L. (1995): Self-esteem as an interpersonal monitor: The sociometer hypothesis. Journal of Personality and Social Psychology, Vol. 68, No. 3, pp. 518-530.
[67] Ledgerwood, A./Liviatan, I./Carnevale, P. J. (2007): Group-identity completion and the symbolic value of property. Psychological Science, Vol. 18, No. 10, pp. 873–878.
[68] Leibenstein, H. (1950): Bandwagon, snob and veblen effects in the theory of consumers' demand. The Quarterly Journal of Economics, 64 (2), 183-207.
[69] Levy, S. J. (1959): Symbols for sale. Harvard Business Review, Vol. 37, No. 4, pp. 117–124.
[70] Linssen, R./Van Kempen, L./Kraaykamp, G. (2011): Subjective well-being in rural India: The curse of conspicuous consumption. Social Indicators Research, Vol. 101, No. 1, pp. 57-72.
[71] Lucas, R.E./Diener, E./Suh, E. (1996): Discriminant validity of well-being measures. Journal of Personality and Social Psychology, 71, No. 3, pp. 616-628.
[72] Lyubomirsky, S. (2001): Why are some people happier than others: The role of cognitive and motivational processes in well-being. American Psychologist, Vol. 56, No. 3, pp. 239-249.
[73] Lyubomirsky, S./Tkach, C./DiMatteo, M.R. (2006): What are the differences between happiness and self-esteem? Social Indicators Research, Vol. 78, No. 3, pp. 363-404.
[74] Maslow, A.H. (1970): Motivation and personality. New York: Harper & Row.
[75] McCracken, G. (1986): Culture and consumption: A theoretical account of the structure and movement of the cultural meaning of consumer goods. Journal of Consumer Research, Vol. 13, No. 1, pp. 71–84.
[76] McCracken, G. (2002): Culture and consumption: New approaches to the symbolic character of consumer goods and activities. Bloomington (Ind.):Indiana university press.
[77] Mezulis, A.H./Abramson, L. Y./Hyde, J.S./Hankin, B.L. (2004): Is there a universal positivity bias in attributions? A meta-analytic review of individual, developmental, and cultural differences in self-serving attributional bias. Psychological Bulletin, Vol. 130, No. 5, pp. 711–747.
[78] Mond, J./Van den Berg, P./Boutelle, K./Hannan, P. (2011): Obesity, body dissatisfaction, and emotional well-being in early and late adolescence: Findings from the project EAT study. Journal of Adolescent Health, Vol. 48, No. 4, pp. 373-378.
[79] Morewedge, C.K./Shu, L.L./Gilbert, D.T./Wilson, T.D. (2009): Bad riddance or good rubbish? Ownership and not loss aversion causes the endowment effect, in: Journal of Experimental Social Psychology, Vol. 45, No. 4, pp. 947-951.
[80] Nayakankuppam, D./Mishra, H. (2005): The endowment effect: Rose-tinted and dark-tinted glasses. Journal of Consumer Research, Vol. 32, No. 3, pp. 390-395.
[81] Nezlek, J.B. (2002): Causal relationships between perceived social skills and day-to-day social interaction: Extending the sociometer hypothesis, in: Journal of Social and Personal Relationships, Vol. 18, No. 3, pp. 386-403.
[82] Norris, J.I./Larsen, J.T. (2010): Wanting more than you have and its consequences for well-being, in: Journal of Happiness Studies, doi: 10.1007/s10902-010-9232-8.
[83] Nuttin, J.M. (1984): Motivation, planning, and action: A relational theory of behavior dynamics. Belgium: Leuven University Press.
[84] Nuttin, J.M. (1987): Affective consequences of mere ownership: The name letter effect in twelve European languages, in: European Journal of Social Psychology, Vol. 17, No. 4, pp. 381–402.
[85] Nuttin Jr., J.M. (1985): Narcissism beyond Gestalt and awareness: The name letter effect, in: European Journal of Social Psychology, Vol. 15, No. 3, pp. 353–361.
[86] O'Cass, A./McEwen, H. (2004): Exploring consumer status and conspicuous consumption, in: Journal of Consumer Behavior, Vol. 4, No. 1, pp. 25-39.

[87] Owens, T.J./Stryker, S./Goodman, N. (Eds.): (2001): Extending self-esteem theory and research: Sociological and psychological currents. New York: Cambridge University Press.
[88] Oyserman, D. (2009): Identity-based motivation: Implications for action-readiness, procedural-readiness, and consumer behavior, in: Journal of Consumer Psychology, Vol. 19, No. 3, 250-260.
[89] Parducci, A. (1995): Happiness, pleasure, and judgment: The contextual theory and its applications. Hillsdale, N.J.: Erlbaum.
[90] Park, J. K./John, D. R. (2010): Got to get you into my life: Do brand personalities rub off on consumers?, in: Journal of Consumer Research, Vol. 37, No. 4, pp. 655-669.
[91] Park, J. K./John, D. R. (2011): More than meets the eye: The influence of implicit and explicit self-esteem on materialism, in: Journal of Consumer Psychology, Vol. 21, No. 1, pp. 73-87.
[92] Puntoni, S. (2001): Self-identity and purchase intention: An extension of the theory of planned behavior, in: European Advances in Consumer Research, Vol. 5, pp. 130-134.
[93] Pyszczynski, T./Greenberg, J./Solomon, S./Arndt, J./Schimel, J. (2004): Why do people need self-esteem? A theoretical and empirical review, in: Psychological Bulletin, 130, No. 3, pp. 435-468.
[94] Reb, J./Connolly, T. (2007): Possession, feelings of ownership and the endowment effect, in: Judgment and Decision Making, Vol. 2, No. 2, pp. 107-114.
[95] Richins, M. L. (1994): Special possessions and the expression of material values, in: Journal of Consumer Research, 21, No. 3, pp. 522-533.
[96] Richins, M. L. (1999): Possessions, materialism, and other-directedness in the expression of self. In M. Holbrook (Ed.), Consumer value: A framework for analysis and research (pp. 85-104): New York: Routledge.
[97] Richins, M. L./Dawson, S. (1992): A consumer values orientation for materialism and its measurement: Scale development and validation, in: Journal of Consumer Research, Vol. 19, No. 3, pp. 303-316.
[98] Rosenberg, M. (1965): Society and the adolescent self-image. Princeton: Princeton University Press.
[99] Rosenberg, M. (1979): Conceiving the self. New York: Basic Books.
[100] Rucker, D.D./Galinsky, A.D. (2008): Desire to acquire: Powerlessness and compensatory consumption, in: Journal of Consumer Research, Vol. 35, No. 2, pp. 257–267.
[101] Rucker, D.D./Galinsky, A.D. (2009): Conspicuous consumption versus utilitarian ideals: How different levels of power shape consumer behavior, in: Journal of Experimental Social Psychology, Vol. 45, No. 3, pp. 49-555.
[102] Seery, M.D./Blascovich, J./Weisbuch, M./Vick, S.B. (2004): The relationship between self-esteem level, self-esteem stability, and cardiovascular reactions to performance feedback, in: Journal of Personality and Social Psychology, Vol. 87, No. 1, pp. 133–145.
[103] Shavitt, S. (1990): The role of attitude objects in attitude functions, in: Journal of Experimental Social Psychology, Vol. 26 , No. 2, pp. 124–148.
[104] Shavitt, S./Nelson, M.R. (2000): The social identity function in person perception: Communicated meanings of product preferences, in: G.R. Maio/J.M. Olson (Eds.), Why we evaluate: Function of attitudes (pp. 37-58): Mahwah, NJ: Lawrence Erlbaum.
[105] Sirgy, J. (1982): Self-concept in consumer behaviour: A critical review, in: Journal of Consumer Research, Vol. 9, No. 3, pp. 287-300.
[106] Sirgy, J./Johar, J.S. (1999): Toward and integrated model of self-congruity and functional congruity, in: European Advances in Consumer Research, 4, 252-256.
[107] Sivanathan, N./Pettit, N.C. (2010): Protecting the self through consumption: Status goods as affirmational commodities, in: Journal of Experimental Social Psychology, Vol. 46, No. 3, 564-570.
[108] Solomon, M. R. (1983): The role of products as social stimuli: A symbolic interaction perspective, in: Journal of Consumer Research, Vol. 10, No. 3, pp. 319-329.
[109] Spencer, S.J./Josephs, R.A./Steele, C.M. (1993): Low self-esteem: The uphill struggle for self-integrity, in: R.F. Baumeister (Ed.), Self-esteem: The puzzle of low self-regard (pp. 21-36): New York: Plenum Press.

[110] Srivastava, S./Beer, J. S. (2005): How self-evaluations relate to being liked by others: Integrating sociometer and attachment perspectives, in: Journal of Personality and Social Psychology, Vol. 89, No. 6, pp. 966-977.

[111] Tesser, A./Crepaz, N./Collins, J. C./Cornell, D./Beach, S. H. R. (2000): Confluence of self-esteem regulation mechanisms: On integrating the self-zoo, in: Personality and Social Psychology Bulletin, Vol. 26, No. 12, pp. 1476–1489.

[112] Thaler, R. (1980): Toward a positive theory of consumer choice, in: Journal of Economic Behavior & Organization, Vol. 1, No. 1, No. 39-60.

[113] Tuominen-Soini, H./Salmela-Aro, K./Niemivirta, M. (2008): Achievement goal orientations and subjective well-being: A person-centred analysis, in: Learning and Instruction, Vol. 18, No. 3, pp. 251-266.

[114] Twenge, J. M./Campbell, W. K. (2002): Self-esteem and socioeconomic status: A meta-analytic review, in: Personality and Social Psychology Review, Vol. 6, No. 1, pp. 59-71.

[115] Van Boven, L. (2005): Experientialism, materialism, and the pursuit of happiness, in: Review of General Psychology, Vol. 9, No. 2, pp. 132-142.

[116] Van Kempen, L. (2007): Status consumption and ethnicity in Bolivia: Evidence from durables ownership, in: International Journal of Consumer Studies, Vol. 31, No. 1, 76-89.

[117] Vigneron, F./Johnson, L.W. (2004): Measuring perceptions of brand luxury, in: Brand Management, Vol. 11, No. 6, pp. 484-506.

[118] Vohs, K.D./Baumeister, R. F. (2007): Can satisfaction reinforce wanting? A new theory about long-term changes in strength of motivation, in: J. Shah/ Gardner, W. (Eds.), Handbook of Motivational Science (pp. 373-389): New York: Guilford.

[119] Watson, D./Clark, L. A./Tellegen, A. (1988): Development and validation of brief measures of positive and negative affect: The PANAS scales, in: Journal of Personality and Social Psychology, Vol. 54, No. 6, pp. 1063-1070.

[120] Wattanasuwan, K. (2005): The self and symbolic consumption, in: Journal of American Academy of Business, Vol. 6, No. 1, pp 179-185.

[121] Wicklund, R. A./Gollwitzer, P. M. (1981): Symbolic self-completion, attempted influence, and self-deprecation, in: Basic and Applied Social Psychology, Vol. 2, No. 2, pp. 89-114.

[122] Wiedmann, K./Hennigs, N./Siebels, A. (2009): Value-based segmentation of luxury consumption behavior, in: Psychology and Marketing, Vol. 26, No. 7, pp. 625-651.

[123] Wong, N. Y. C./ Ahuvia, A. C. (1998): Personal taste and family face: Luxury consumption in Confucian and Western societies, in: Psychology and Marketing, Vol. 15, No. 5, pp. 423-441.

[124] Wood, J. V./Heimpel, S. A./Michela, J. L. (2003): Savoring versus dampening: Self-esteem differences in regulating positive affect, in: Journal of Personality and Social Psychology, Vol. 85, No. 3, pp 566-580.

[125] Zhang, Y./Fishbach, A. (2005): The role of anticipated emotions in the endowment effect, in: Journal of Consumer Psychology, Vol. 15, No. 4, pp. 316-324.

[126] Zinkhan, G. M./Hong, J. W. (1991): Self-concept and advertising effectiveness: A conceptual model of congruency, conspicuousness, and response mode, in: Advances in Consumer Research, Vol. 18, pp. 348-354.

8 How Do Exclusivity Perceptions of Independent and Interdependent Consumers Influence Their Desires for Luxury Products

Hyunjoo Oh

8.1 Introduction

The high signaling nature of a luxury brand exerts a strong impact on consumers' desire to own the brand. In fact, marketing and advertising for luxury brands have focused on portraying the ultra-indulgence of lifestyles that only the privileged '"few"' can afford. Exclusivity (i.e., rarity and scarcity) not only signals high social status by limiting common access, but it also enhances signaling quality by increasing its distinctiveness, uniqueness and salience from being small in size of adopters [4], [23], [24], [25]. Luxury brands, therefore, have strived to create and maintain their exclusivity by advertising to specific target segments, using premium pricing strategy to restrict common access, and controlling distribution. At the same time, successful portrayals of aspirations drive consumers to conform and assimilate, resulting in the popularity of a brand [3], [6], [10], [11], [12], [27]. For instance, the popularity of Louis Vuitton handbags is exemplified by the 60% of young Japanese women who own it. However, this popularity, even in a remote location of Japan, caused a decrease in French women's desire for the big-brand handbag [13]. The paradox of balancing between exclusivity and popularity in luxury brand marketing is an important, yet unexplored, area for academics to understand the mechanisms of luxury product consumption [29].

Using self-construal differences reflected in individualist cultures and collectivist cultures [28], [32], [37], the current study examines how varying adoption levels, which signal exclusivity or popularity, interacts with independent vs. interdependent self-construal in driving the desire for owning luxury products. We examined the interaction effect in a small, peer in-group setting (study 1) and in a broad, large in-group setting where self-identity expression and feedback is not strongly facilitated (study 2).

8.2 Theoretical Background

8.2.1 Self-Construal

People with an interdependent self-construal view themselves as being related to others, whereas people with an independent self-construal view themselves as being different from others [28], [40]. These two types of self-construal correlate with the degree to which a culture is collectivist or individualistic [28], [32], [37]. Previous studies have shown that assimilation motives (conformity) are stronger among people in collectivistic cultures, whereas differentiation (uniqueness) motives are stronger among people in individualistic cultures [2], [12], [18], [25], [33], [19], [20]. The culturally-based propensity to value assimilation or differentiation has been used to account for product preferences. In collectivist cultures, product popularity (i.e., assimilation) communicates attractiveness, assures quality, and reduces the risk of adoption [39], [31]. In individualistic cultures, product exclusivity (i.e., differentiation) signals uniqueness, distinctiveness, and quality [4], [19], [23], [24], [25], [34].

Self-construal is also conceptualized as a mindset, and thus, is sensitive to contextual factors that make one of the two self-construals more accessible [1], [15], [38]. For example, contextual priming tasks have been used to induce independent versus interdependent self-construal which, in turn, influenced risk-taking behaviors [26], self-regulatory goals in information processing [1], information processing modes for product evaluations [41], spatial judgments [21], and even retrieval of chronic individualist and collectivist values [38], [16], [15]. In each of these cases, it has been argued that self-construal alters the relative importance of assimilation and differentiation motives.

Wherein some researchers have posited that a self-construal mindset determines the relative importance of self-identity motives, others have put forth a more complex argument. Although the relative importance of assimilation motives is posited to vary by mindset, the importance of differentiation motives is posited to be strong in both mindsets [9], [40]. In this conceptualization, self-construal impacts the approach people take to achieving differentiation motives. According to Vignoles et al. (2000) [40], people with an independent self-construal express a distinctive self-identity by establishing autonomy from others, whereas people with an interdependent self-construal express a distinctive self-identity by establishing their position within an affiliated group. Thus, different strategies for establishing distinctiveness should contribute to the positive self-identity of people with independent and interdependent self-construals. If this is so, we expect that the different reasons (i.e., establishing autonomy versus establishing a position in a social group) for establishing distinctiveness would influence the ways of how people with independent versus interdependent self-construal change their desires to own luxury products in response to perceived product adoption levels.

8.2.2 Optimal Distinctiveness for Interdependent and Independent Self-Construal

Brewer's (1991) [8] optimal distinctiveness theory (ODT) proposes that a person's optimal distinctiveness in a social group is determined at the point where both a differentiated identity (i.e., be different from the group) and an assimilative identity (i.e., be part of the group) can be expressed [8], [30]. Because luxury brands and products are a means to maintain and enhance social self-identity, assimilation and differentiation motives should contribute to a desire for product ownership. Previous studies have associated the driver of luxury good consumption of collectivist cultures and interdependent self with strong assimilation motives [2], [12], [18]. However, prior studies have neglected how the differentiation motives of people with interdependent self-construal operate to distinguish a position of the self in a social network in the consumption of luxury products. Considering that people both assimilate to and differentiate from others who adopt luxury products, we examine how the optimal distinctiveness is constructed as an interactive function of self-construal and adoption level in the consumption of luxury products.

8.3 Hypotheses

The adoption level signals popularity or exclusivity. As the adoption level increases, the assimilation value rises from the perceived popularity whereas the differentiation value falls from the diminished exclusivity. Since the assimilation value and differentiation value works opposite to each other along with the adoption level, their influence on desire for luxury products should be considered separately before and after reaching the moderate adoption level, where the tension to reconcile between two competitive motives begin to arise. Before passing the moderate adoption level, the desire for luxury goods is driven by stronger assimilation values for interdependent individuals and stronger differentiation values for independent individuals as consistent with previous studies [2]. Therefore, the increasing popularity (from low to moderate adoption rate) would increase the desire to own a luxury product for interdependent individuals but decrease the desire for independent individuals. This is because it would fulfill the need to assimilate and thus achieve the desired position for interdependent individuals but it would not fulfill the need to establish difference from others for independent individuals.

After passing the moderate adoption level, interdependent (vs. independent) individuals become more sensitive to a product's differentiation value. This is because individuals with interdependent self-construal construct their optimal distinctiveness by establishing their position in a social group, whereas individuals with an independent self-construal do it by establishing autonomy from others. As the high adoption level loses the function of signaling a desired position in a social group, the ownership desire would diminish for interdependent individuals. However, the same high adoption level information would signal popularity and thus would increase the desire of independent individuals. Across the low, moderate, and high adoption levels, the desire change would show the inverted U pattern for interdependent individuals and the opposite U pattern for independent individuals.

> H1a: People with an interdependent self-construal will most prefer luxury products that have a moderate level of adoption by important social groups.
>
> H1b: People with an independent self-construal will most prefer luxury products that have a low or high level of adoption by important social groups.

8.4 Study 1

Experiment 1 employed a scenario involving a small social group to test the proposed hypotheses. Participants were asked to review a selection of fashionable handbags and indicate the handbag that they would most like to own. Then they wrote down the amount they would be willing to pay for the chosen handbag. Next, the self-construal and adoption level manipulations were administered. Finally, participants once again indicated the amount they were willing-to-pay for their preferred handbag. Thus, the key dependent measure was the change in the willingness to pay (WTP) for the most desired handbag as a consequence of the self-construal and adoption level manipulations.

8.4.1 Method

Participants and Design. The experiment used a 2 (*self-construal priming*: independent, interdependent) x 3 (*adoption level*: low, moderate, high) between-subjects design. A total of 100 female undergraduate students participated in return for extra credit. Participants were randomly assigned to one of six between-subject conditions.

Procedure. The procedure began with a cover story that the Luxury Marketing Council was looking for ideas on a new strategy that offers affordable luxury products to young consumers. Participants were asked to imagine a series of scenarios designed to create the relevance of luxury good in a peer in-group setting, where 10 people from a personally important social group were invited to an event and entered a raffle on luxury handbags. With the raffle, participants chose the most preferred handbag from three choices and wrote down WTP as the pre-manipulation measure of desirability. Next, the self-construal prime was administrated by using the modified priming procedure of Mandel (2003) [26]. Participants in the independent self-construal condition were asked to list five personal qualities that make them different from their friends and allow them to express their individuality. Participants in the interdependent self-construal condition were asked to list five personal qualities that make them similar to their friends and allow them "fit in" to a social group that they value. Participants in both conditions were asked to describe something nice (luxurious) that they recently purchased to express self-identity.

Next, the adoption levels were manipulated using additional scenario in which participants learned about the preferred choices of their social group members. The manipulated adoption level was 0 (**0**, 5, 5), 2 (**2**, 4, 4), or 6 (**6**, 2, 2) for the handbag that participants selected as the most preferred product. After learning about social group members' preferences for their chosen product, participants were asked to indicate how much they were willing to pay for the product. This WTP was the post-manipulation measure of desirability. Next, participants were asked to evaluate the chosen product in response to statements describing assimilation and differentiation values on a 7-point scale ranging from strongly disagree (1) to strongly agree (7). The statements were developed by modifying items by Berger and Ward (2010) [7] and Tian, Bearden, and Hunter (2001) [34].

Finally, manipulation checks were administered. The effectiveness of adoption level manipulation was assessed by asking participants to estimate how many people were likely to purchase the preferred handbag on a scale ranging from 0 (nobody) to 100 (everybody). The effectiveness of the self-construal manipulation was assessed by selecting four items from previous self-construal studies [32], [36]. Participants indicated their level of agreement with two items (α = .77) of an independent self-construal (My personal identity, which is independent from others, is very important to me; My unique personal characteristics define who I am) and two items (α =.58) of an interdependent self-construal (Belonging to social groups is an important part of my self-image; My relationships to others and my roles define who I am) on 7-point scales. Finally, participants were asked to indicate their experience of purchasing a luxury purse as yes or no and family income on a 5-point level (1: below $25,000, 2: $25,000 - $49,999; 3: $50,000- $74,999, 4:$75,000- $99,999, 5: $100,000 and over). The entire procedure was administered using the Qualtrics software program and personal computers.

Pretests. The pretests were used to select a set of luxury goods and to identify low, moderate, and high adoption levels that were relevant to a sample of participants (female undergraduate students in business school). Pretest 1 ($N = 25$) was conducted to choose the product category. Participants read the same introductory scenario as in the main study and participated in the raffle event. For each of product categories (designer clothing, shoes, handbags, and sunglasses), pictures of eight products (brand names for designer clothing) were presented for participants to choose as their final choice for the raffle entry. The majority of participants (73%) chose a handbag as their product choice for the raffle.

Pretest 2 was conducted to identify adoption levels for fashionable handbags. A total of 134 undergraduate students were told that the study would assess their perceptions regarding the number of people who had the same handbag as them. The participants provided estimates in response to scenarios that describe Just Right and Too Many for product adoption. For each scenario, participants were to write down a number between 0 and 10. Means (SDs) of perceived adoption levels were 2.19 (2.42) for Just Right and 4.47 (2.95) for Too Many. We chose 0 as the low level adoption level, 2 as the moderate level adoption level, and 6 as the high level of adoption level.

8.4.2 Results

Manipulation Checks. To check the effectiveness of the self-construal manipulation, the independent self-construal score was submitted to a self-construal x adoption level ANCOVA analysis with the covariates of prior luxury purse purchase (no/yes) and family income. Participants primed with an independent self-construal ($M = 6.26$) reported being more independent than participants primed with an interdependent self-construal ($M = 5.89$; $F(1, 93) = 3.38$, $p = .07$). The same analysis conducted on the interdependent self-construal score did not yield a significant effect from self-construal manipulation ($M_{\text{interdependent}} = 4.48$, $M_{\text{independent}} = 4.23$; $F < 1$) because a social group setting activated an interdependent self-construal.

To check the effectiveness of the adoption level manipulation, the perceived adoption level was submitted to a self-construal x adoption level ANCOVA analysis with the covariates of prior luxury purse purchase and family income. The expected effect of adoption level manipulation was significant, $F(2, 93) = 4.28$, $p = .02$. The perceived adoption level did not significantly change as the manipulated adoption level changed from low ($M = 50.0$) to moderate ($M = 48.5$), $F(2, 93) = 1.15$, $p = .29$, but significantly changed as the manipulated adoption level went from moderate to high ($M = 59.5$; $F(2, 93) = 7.49$, $p = .007$). People with an interdependent self-construal generated higher estimates ($M = 55.4$) than people with an independent self-construal ($M = 45.8$), $F(1, 93) = 5.08$, $p = .03$.

Hypothesis 1 Test: Change in Willingness-to-Pay (WTP). The WTP change score was computed by subtracting the pre-treatment WTP amount from the post-treatment WTP amount. The change in WTP was submitted to a self-construal by adoption level ANCOVA analysis with the covariates of prior luxury purse purchase and family income. The covariate-adjusted means are shown in **Figure 8.1**. The interaction between the self-construal prime and the adoption level was significant (*F* (2, 92) = 4.07, *p* = .02). Participants primed with an interdependent self-construal showed a negative quadratic response to the increase in adoption levels (*F* (1, 92) = 4.57, *p* = .04; M_{low}= -59.44, $M_{moderate}$ = -12.26, M_{high}= - 82.73). Participants primed with an independent self-construal showed a positive quadratic response to the increase in adoption levels (*F* (1, 92) = 1.15, *p* = .29; M_{low}= -86.49, $M_{moderate}$ = -92.14, M_{high}= -21.25).

Figure 8.1 Results on the Desire for Luxury Products at Different Levels of Adoption

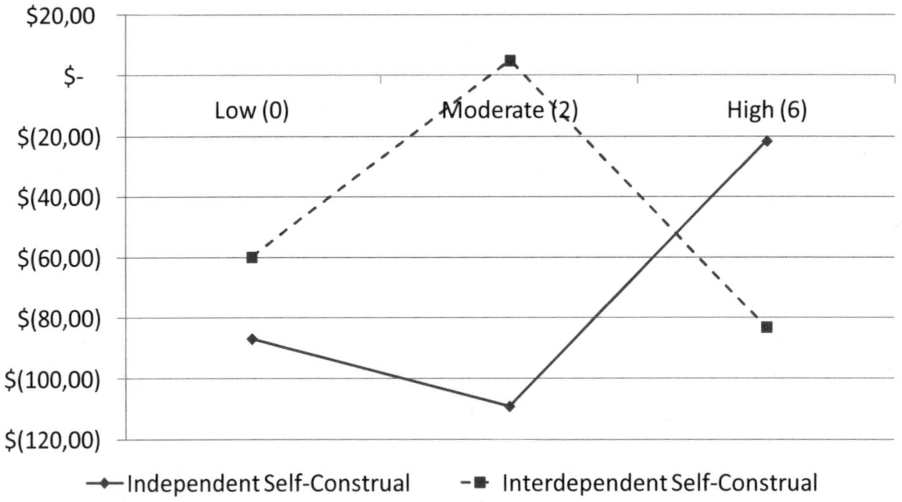

Assimilation and Differentiation Values. We tested our assumption that the perceived assimilation value would increase but the perceived differentiation value would decrease as the adoption level among important social group members increases. The perceived assimilation and differentiation value scores were submitted to a self-construal by adoption level by value type repeated measures MANCOVA with motive type as a within-subject factor and the purchase experience and income covariates. Only the expected value type by adoption level interaction was significant, *F*(2,93) = 8.68, *p* = .000. The assimilation values did not change as the adoption level went from low (*M* = 3.96) to moderate (*M* = 4.01; *F*(1, 96) = .13, *p* = .72) but increased as the adoption level went from moderate (*M* = 4.01) to high (*M* = 5.20; *F*(1,96) = 3.84, *p* = .05). The perceived differentiation value also didn't change as the adoption level went from low (*M* = 4.15) to moderate (*M* = 4.16; *F*(1, 96) = .00, *p* = .99) but decreased as the adoption level went from moderate (*M* = 4.16) to high (*M* = 3.49; *F*(1, 96) = 3.02, *p* = .09). These patterns support our assumption.

8.4.3 Discussion

Results from experiment 1 support our hypothesis that self-construals interact with product adoption levels to determine the desire to own luxury products. We tested our hypothesis in the context of a small peer group with willingness-to-pay for a luxury handbag. As predicted, people primed with interdependent self-construal preferred luxury products at the moderate adoption level reconciling assimilation and differentiation motives. In contrast, for people primed with an independent self-construal, the optimal distinctiveness was constructed at the lowest adoption level that ensures difference from in-group members. Knowing that no one in a peer group selected her preferred handbag produced more positive change in WTP than knowing that two of 10 peer group members selected her preferred bag. But the high adoption level of peer members (6 out of 10) signaled popularity for people primed with an independent self-construal. In response, the change in WTP positively increased. However, the same high adoption level decreased the desire of people primed with an interdependent self-construal, suggesting that more frequent new product introduction cycles would be effective in collectivist cultures.

We propose that the reconciliation between assimilation motives and differentiation motives for optimal distinctiveness is likely to occur in contexts where one's choices and preferences are visible to group members and meaningful to the expression of social identity. This argument implies that a small in-group size in study 1 provides a boundary within which people respond to adoption level information to construct optimal distinctiveness. In study 2, we examine whether a large in-group size (100 people) removes the need for fulfilling differentiation motives in response to increased adoption level information. When in-group is large in size, the motivation for self-expression is meager and thus subsequent social comparison and feedback are not likely to arise. Assimilation motives and differentiation motives do not compete and thus no reconciliation between the two motives is likely to occur. Thus, the increasing adoption level signals only popularity of a product leading to increased ownership desire. This effect will be stronger for individuals with independent self-construal than for individual with interdependent self-construal. This is because the adoption level on a luxury good is not informative to interdependent individuals when its signaling value is diffused in a loose, large social group. Thus, their desire to own a luxury good is not swayed by the adoption levels. We hypothesize the change in WTP for owning luxury products in a broad social context as follows.

> H2a: People with an independent self-construal will increase their desire for luxury goods as the adoption level from a broad social group increases.
>
> H2b: People with an interdependent self-construal will not change their desire for luxury goods as the adoption level from a broad social group increases.

8.5 Study 2

The objective of study 2 was to examine how people with independent and interdependent self-construals respond to adoption level information from a broad in-group, where self-identity expression is not strongly motivated.

8.5.1 Method

Participants and Design. The experiment used a 2 (*self-construal priming*: independent, interdependent) x 3 (*perceived adoption level*: low, moderate, high) between-subjects design. The low, moderate, and high adoption levels provided to participants were 1%, 17%, and 59% of 100 people. A total of 114 female undergraduate students, from a Principles of Marketing class, participated in return for extra credit. Participants were randomly assigned to one of six between-subject conditions.

Procedure. The procedure, stimuli, and measures were the same as in study 1, except the changes made to accommodate the change of social group and, subsequently, the source of adoption level. Instead of attending the even with a group of friends in study 1, participants accepted an individual invitation. The adoption level, which manipulates the popularity of items in a social group, was introduced as the preference survey conducted in the Atlanta Chapter of the National Association of Women's MBAs. The adoption level was presented as 1 (low), 17 (moderate), or 59 of young professionals who voted for the most preferred bag. In addition, self-construal manipulation check measures were expanded. Four items measuring independent self-construal (α = .88) were averaged. Four items measuring interdependent self-construal (α = .80) were averaged.

Pretest. A pretest was conducted to identify adoption levels among 100 aspirational in-group members. Ninety-nine female undergraduate students were told that the study would assess their opinions about fashion goods. Participants were asked to select their most preferred handbag from a set of ten. Next, the participants provided estimates in response to scenarios described as Just Right, Too Few, and Too Common product adoption levels. For each scenario, participants were to write down a number between 0 and 100 to indicate Too Few, Just Right, and Too Common. Means (SDs) of perceived adoption levels were 7.8 (17.4) for Too Few, 13.5 (15.7) for Just Right, and 46.1 (55.6) for Too Common with substantial variance, as indicated in standard deviations. Many participants indicated '0' for both Too Few and Just Right. We chose 1 as the low level adoption level, 17 as the moderate adoption level, and 59 as the high adoption level. These rates roughly correspond to the adoption rates used in study 1.

8.5.2 Results

Manipulation Checks. To check for the effectiveness of the self-construal manipulation, the independent self-construal score was submitted to a self-construal x adoption level ANCOVA analysis with prior luxury purse purchase (no/yes) and family income as covariates.

Because the scenario stressed individuality by asking participants to imagine that they were one of 100 participants attending the same event, the self-construal manipulation did not change their perception of independent self-construal ($F(1, 106) < 1$). However, the manipulation worked on the interdependent self-construal measure. As predicated, participants primed with an interdependent self-construal ($M = 4.98$) reported being more interdependent than participants primed with an independent self-construal ($M = 4.55$; $F(1, 106) = 3.81$, $p = .05$). To check the effectiveness of the adoption level manipulation, the perceived adoption level was submitted to a self-construal x adoption level ANCOVA analysis with the covariates of prior luxury purse purchase and family income. As expected, the estimated numbers of people (out of 100) who would buy the product increased as the manipulated adoption level changed from low ($M = 39.1$) to moderate ($M = 48.5$) to high ($M = 59.5$; $F(2, 106) = 5.50$, $p = .005$).

Change in Willingness-to-Pay. As in Study 1, the change in WTP was submitted to a self-construal by adoption level ANCOVA analysis with the covariates of prior luxury purse purchase and family income. As predicted, we found a self-construal by adoption level interaction, (F (2, 106) = 2.95, $p = .06$). A broadly defined in-group with the size of 100 produced an increasing linear pattern of change in WTP for preferred products for people primed with an independent self-construal, F (2, 107) = 9.85, $p = .000$ (M_{low}= -92.3, $M_{moderate}$ = -7.0, M_{high}= 5.78). For people with an interdependent self-construal, the adoption level effect did not affect the change in WTP, F (2, 106) = .87, $p = .42$ (M_{low}= -31.45, $M_{moderate}$ = -20.95, M_{high}= -2.13).

Assimilation and Differentiation Values. The hypothesis 2 assumes that the perceived assimilation value, but not the differentiation value, will increase as the adoption level increases. Without an opportunity to distinguish oneself in a broad social group, the desire for luxury products would increase based on the assimilation value perceived from others' preferences. The assimilation and differentiation value scores were submitted to a self-construal by adoption level by value type repeated measures MANCOVA with motive type as a within-subject factor and purchase experience and income as covariates. Only the predicted value type by adoption level interaction was significant, $F(2,106) = 5.64$, $p = .005$. The assimilation values continuously increased as the adoption level increased, $F(2,106) = 5.64$, $p = .005$ (M_{low}= 4.48, $M_{moderate}$ = -4.90, M_{high}= 5.20), whereas the differentiation values did not differ across the adoption levels (F (2, 106) < 1, $p = .43$, M_{low} = 4.70, $M_{moderate}$ = 4.05, M_{high} = 4.32),

8.5.3 Discussion

Study 2 examined how the adoption level in a large in-group changed WTP of people with independent and interdependent self-construals. When the adoption level information came from the large group (100) of young professionals as a broadly defined in-group, the change in WTP for their preferred product was influenced by what others liked instead of the opportunity to express optimal distinctiveness for their core self-construal. Supporting our assumption, the increase in the adoption level changed the assimilation values perceived upward but didn't change differentiation value perceived. For people with an independent self-construal, the desire to own luxury products, measured as the WTP change, was guided by the perception of popularity signaled from adoption level. However, for people with an interdependent self-construal, the lack of signaling channel in a broadly defined, large in-group attenuated the effect of popularity on their ownership desire. Thus, study 2 suggests a boundary within which the optimal distinctiveness mechanism operates. It requires certain group features, such as a small in-group within which individuals can effectively use the signal afforded by luxury products to express their self-identity.

8.6 General Discussion and Conclusion

Our research demonstrates that self-construal moderates how product exclusivity/popularity influences the ways that individuals construct the optimal distinctiveness within their in-group through the ownership of luxury products. In study 1, we demonstrated that the optimal distinctiveness was achieved at the moderate adoption level for people with an interdependent self-construal. Their optimal distinctiveness of self-identity was constructed at the adoption level that holds both assimilation and differentiation values. This result suggests that the tension reconciling between the competing assimilation and differentiation motives drives the change in desires for luxury products for consumers with an interdependent self-construal as well as consumers in a collective culture. For people with an independent self-construal, the optimal distinctiveness occurred both at the low level, which conveys the high differentiation value, and at the high adoption level, which signals popularity. The increase in the ownership desire of people with an independent self-construal at the high adoption level resulted from the perceived popularity overriding the activation of differentiation motive. Thus, the optimal distinctiveness of independent participants was determined without reconciling the two competing motives. This result may suggest that consumers with an independent self-construal and in an individualistic culture respond more positively and purchase luxury products both when only a few people have adopted as well as when many people in their in-group also have adopted.

Our study highlights the complexity of how consumers respond to the popularity and exclusivity signaled by varying levels of product adoption within in-group contexts. Previous studies have shown that assimilation values drive desire for products in collectivism cultures, whereas differentiation (uniqueness) values drive desires for products in individualism cultures [2], [25], [33], [19]. These prior findings hold when there are no tensions between the assimilation and differentiation motives. We also observed consistence with these findings when the adoption level increases from the low to moderate level; the desires for luxury products increased for people with an interdependent self-construal but decreased for people with an independent self-construal. However, we found the opposite patterns when the tensions between the assimilation and differentiation motives emerge from the increased adoption levels from moderate to high. The desires for luxury products decreased for people with an interdependent self-construal but increased for people with an independent self-construal. This result suggests that the high adoption level signaled the loss of exclusivity for interdependent people who desire to differentiate their position within in-group, whereas it signaled the popularity for independent people.

Our study proposes that within-group differentiation is a dynamic force that drives continuous desires for luxury products. Considering that within-group differentiation has been rarely studied in previous studies compared to between-group differentiation [7], [17], future work is required to expand our understanding the dynamics of within-group differentiation in the consumption contexts of identity signaling products. The opposite patterns of desire change, the inverted U pattern along for interdependent consumers and the U pattern for independent consumers, were observed in the context of a small peer group (10 people in an important social group) used in study 1, but not in the context of a board, large group (100 young professionals). Therefore, we suggest that the use of luxury products to express optimal distinctiveness requires group contexts where self-identity is effectively managed through acquisition of luxury products.

Findings of our study provide global luxury marketers with the framework to understand the drivers of luxury product consumption in individualistic and collective cultures and psychological mechanisms that how people respond to popularity and exclusivity signaled by product adoption levels. Considering that the growth of the luxury market is coming from the rising segment of affluent global consumers in emerging economies with collective cultures like China, the deeper understanding of psychological mechanisms is valuable to develop effective marketing strategies. In fact, luxury brand marketers have been aggressively expanding into the fastest-growing economies, including BRIC (Brazil, Russia, India and China) in order to capitalize growing market potentials. Since the appeal of luxury brand marketing relies on access to rare, exclusive, and desirable products, brand marketers must understand how consumers in individualistic versus collectivistic cultures perceive popularity and exclusivity signaled by the perceived product adoption levels. Our findings suggest that people with an interdependent self-construal (in collectivist cultures) are sensitive to assimilation values in the early stage of product introduction and the differentiation values in the later stage of product life cycle. Thus, it is recommended to have narrow product lines in collectivist cultures to increase a signaling value for assimilation, followed by more frequent introductions of new products at the later stage of product adoption when the exclusivity signal fades. Considering consumption of counterfeit products flooded in those markets, it would be worthwhile to investigate how authenticity plays a moderating role in consumers' desire for luxury brands in emerging economies.

References

[1] Aaker, J. L./Lee, A. Y. (2001): "I" Seek Pleasures and "We" Avoid Pains: The Role of Self-Regulatory Goals in Information Processing and Persuasion, in: Journal of Consumer Research, Vol. 28, No. 1, pp. 33-49.

[2] Aaker, J. L./Schmitt, B. (2001): Culture-Dependent Assimilation and Differentiation of the Self: Preferences for Consumption Symbols in the United States and China, in: Journal of Cross-Cultural Psychology, Vol. 32, No. 5, pp. 561-76.

[3] Amaldoss, W./Jain, S.. (2005): Pricing of Conspicuous Goods: A Competitive Analysis of Social

Effects, in: *Journal of Marketing Research*, Vol. 42, No. 1, pp. 30-42.
[4] Bearden, W. O./Etzel, M. J. (1982): Reference Group Influence on Product and Brand Purchase Decisions, in: Journal of Consumer Research, Vol. 9, No. 2, pp. 183-94.
[5] Belk, R. W. (1988): Possessions and the Extended Self, in: Journal of Consumer Research, Vol. 15, No. 2, pp. 139-68.
[6] Berger, J./Heath, C. (2007): Where Consumers Diverge from Others: Identity Signaling and Product Domains, in: Journal of Consumer Research, Vol. 34 (August), pp. 121-134.
[7] Berger, J./Ward, M. (2010): Subtle Signals of Inconspicuous Consumption, in: Journal of Consumer Research, Vol. 37, No. 4, pp. 555-569.
[8] Brewer, M. B. (1991): The Social Self: On Being the Same and Different at the Same Time, in: Personality and Social Psychology Bulletin, Vol. 17, No. 5, pp. 475-482.
[9] Brewer, M. B./Pickett, C. L. (1999): Distinctiveness motives as a source of the social self, in: Tyler, T./Kramer, R. /John, O. (Eds.): The psychology of the social self, Hillsdale, NJ: Erlbaum, pp. 71-87..
[10] Corneo, G./Jeanne, O. (1997): Conspicuous Consumption, Snobbism, and Conformism, in: Journal of Public Economics, Vol. 66, No. 1, pp. 55-71.
[11] Escalas, J. E./Bettman, J. R. (2003): You Are What They Eat: The Influence of Reference Groups on Consumers' Connections to Brands, in: Journal of Consumer Psychology, Vol. 13, No. 3, pp. 339-48.
[12] Escalas, J. E./Bettman, J. R. (2005): Self-Construal, Reference Groups, and Brand Meaning, in: Journal of Consumer Research, Vol. 32, No. 3, pp. 378-89.
[13] Friedman, V./Ward, A. (Sunday, June 4): Brands Assess Risks of Extending Global Reach, retrieved on October 2, 2008 from http://us.ft.com/ftgateway/superpage.ft?news_id=fto060420061816551622 .
[14] Gao, Leilei, S. /Wheeler, C./Shiv, B. (2009): The "Shaken Self": Product Choices as a Means of Restoring Self-View Confidence, in: Journal of Consumer Research, Vol. 36, No. 1, pp. 29-38.
[15] Gardner, Wendi L./Gabriel, S./Lee, A. Y (1999): 'I' Value Freedom, but 'We' Value Relationships: Self-Construal Priming Mirrors Cultural Differences in Judgment, in: Psychological Science, Vol. 10 No. 4, pp. 321-26.
[16] Gardner, Wendi L./Gabriel, S./Hochschild, L.(2002): When You and I Are 'We,' You Are Not Threatening: The Role of Self-Expansion in Social Comparison, in: Journal of Personality & Social Psychology, Vol. 82, No. 2, pp. 239-51.
[17] Hornsey, M. J./Jetten, J. (2004): The Individual Within the Group: Balancing the Need to Belong with Need to be Different, in: Personality and Social Psychology Review, Vol. 8, No. 3), pp. 248-264.
[18] Kampmeier, C./Simon, B. (2001): Individuality and Group Formation: The Role of Independence and Differentiation, in: Journal of Personality and Social Psychology, Vol. 81, No. 3, pp. 448-462.
[19] Kim, H. S./Drolet, A. (2003): Choice and Self-Expression: A Cultural Analysis of Variety-Seeking, in: Journal of Personality & Social Psychology, Vol. 85, No. 2, pp. 373-82.
[20] Kim, H. S./ Drolet, A. (2009): Express Your Social Self: Cultural Differences in Choice of Brand-Name Versus Generic Products, in: Personality and Social Psychology Bulletin, Vol. 35, No. 12, pp. 1555-66.
[21] Krishna, A./Rongrong, Z./Shi, Z. (2008): The Effect of Self-Construal on Spatial Judgments, in: Journal of Consumer Research, Vol. 35, No. 2, pp. 337-48.
[22] Liebenstein, H. (1950): Bandwagon, Snob, and Veblen Effects in the Theory of Consumer Demand in: Quarterly Journal of Economics, Vol. 64, pp. 183-207.
[23] Lynn, M. (1989): Scarcity Effects on Value: Mediated by Assumed Expensiveness?, in: Journal of Economic Psychology, Vol. 10, pp. 257-274.
[24] Lynn, M. (1991): Scarcity Effects on Value: A Quantitative Review of the Commodity Theory Literature, in: Psychology and Marketing, Vol. 8, pp. 43-57.
[25] Lynn, M.,/Snyder, C. R./Lopez, S. J. (2002): Uniqueness Seeking, in: Handbook of Positive Psychology. New York, NY US: Oxford University Press, pp. 395-410.
[26] Mandel, N. (2003): Shifting Selves and Decision Making: The Effects of Self-Construal Priming on Consumer Risk-Taking, in: Journal of Consumer Research, Vol. 30, No. 1, pp. 30-40.

[27] Mandel, N./Petrova, P. K./Cialdini, R. B. (2006): Images of Success and the Preference for Luxury Brands, in: Journal of Consumer Psychology, Vol. 16, No. 1, pp. 57-69.
[28] Markus, H. R./Kitayama, S. (1991): Culture and the Self: Implications for Cognition, Emotion, and Motivation, in: Psychological Review, Vol. 98, pp. 224-253.
[29] Phau I./Prendergast, G. (2000): Consuming Luxury Brands: The Relevance of the Rarity Principle, in: Journal of Brand Management, Vol. 8, pp. 122-38.
[30] Pickett, C. L./Silver, M. D./Brewer, M. B. (2002): The Impact of Assimilation and Differentiation Needs on Perceived Group Importance and Judgments of Ingroup Size, in: Personality and Social Psychology Bulletin, Vol. 28, No. 4, pp. 546-58.
[31] Roth, M. S. (1995): The Effects of Culture and Socioeconomics on the Performance of Global Brand Image Strategies, in: Journal of Marketing Research (JMR), Vol. 32, No. 2, pp. 163-75.
[32] Singelis, T. M. (1994): The Measurement of Independent and Interdependent Self-Construals, in: Personality and Social Psychology Bulletin, 20 (October), pp. 580-591.
[33] Snyder, C. R./Fromkin,H. L. (1977): Abnormality as a Positive Characteristic: The Development and Validation of a Scale Measuring Need for Uniqueness, in: Journal of Abnormal Psychology, Vol- 86 (October), pp. 518-527.
[34] Tian, K. T./Bearden, W. O./Hunter, G. L. (2001): Consumers' Need for Uniqueness: Scale Development and Validation, in: Journal of Consumer Research, Vol. 28, pp. 50-66.
[35] Timmor, Y./Katz-Navon, T. (2008): Being the Same and Different: A Model Explaining New Product Adoption, in: Journal of Consumer Behaviour, Vol. 7 (May-June), pp. 249-262.
[36] Triandis, H. C./Gelfand, M. J.(1998): Converging Measurement of Horizontal and Vertical Individualism and Collectivism, in: Journal of Personality & Social Psychology, Vol. 74, No. 1, pp. 118-28.
[37] Triandis, H. C./Bontempo, R./Villareal, M. J./Asai, M./Lucca, N. (1988): Individualism and Collectivism: Cross-Cultural Perspectives on Self-Ingroup Relationships, in: Journal of Personality and Social Psychology, Vol. 54, No. 2, pp. 323-38.
[38] Trafimow, D./Triandis, H. C./Goto, S. G. (1991): Some Tests of the Distinction between the Private Self and the Collective Self, in: Journal of Personality & Social Psychology, Vol. 60, No. 5, pp. 649-55.
[39] Van den Bulte, C./Joshi, Y. V. (2007): New Product Diffusion with Influentials and Imitators, in: Marketing Science, Vol. 26, No. 3, pp. 400-21.
[40] Vignoles, V. L./Chryssochoou X./Breakwell, G. M. (2002): Sources of Distinctiveness: Position, Difference and Separateness in the Identities of Anglican Parish Priests, in: European Journal of Social Psychology, Vol. 32, No. (6), pp. 761-780.
[41] Zhu, R./Meyers-Levy, J. (2009): The Influence of Self-View on Context Effects: How Display Fixtures Can Affect Product Evaluations, in: Journal of Marketing Research (JMR), Vol. 46, No. 1, pp. 37-45.

Part 4

Different Approaches to Understanding Luxury and Luxury Brands from the Customer's Perspective within Specific Cultural Contexts

9 Luxury Buying Behaviour and the Role of Culture: An Indian Context

Kartik Dave, Garima Dhamija

9.1 Introduction

India's tradition of luxury goes back in time when the maharajas were connoisseurs of luxury. They always satiated themselves with beautiful and handcrafted products. After independence in 1947, the country having taken the socialist route, the industrialist families still continued to engage with luxury products. Post the economic reforms in 1991, the Indian consumer has seen a significant increase in income across levels. And over the next decade, India is likely to experience the largest growth in total millionaire wealth. So today, it is no surprise that in the luxury market place, amongst Asian countries, India is being considered as a "promising territory for the future and a long-term player". According to Vadukut (2011) [21] , Buoyed by prosperity, maturing retail experience and well travelled, well informed consumers, the Indian luxury sector is booming.

A study conducted by Deloitte Center for Financial Services titled 'The next decade in global wealth among millionaire households', suggests that emerging markets will see a significantly higher growth rate in millionaire households compared to developed markets. And amongst the emerging markets, India is likely to experience the largest growth in total millionaire wealth (405%) over the next decade. According to the same report, China's wealth among millionaires will grow by 394%, followed by Brazil at 257%, and Russia at 241%.

The growth though, is expected to vary across different wealth segments. So over the next decade,:

- the \$5m-\$30m segment would see the greatest growth, at 161%
- the \$1m-\$5m segment would have a growth rate of 142%
- the \$30m+ segment will follow closely with likely growth rate of 115%
  With this growth, the picture in 2020 is likely to look like.
- 0.51 million millionaire households in \$1m – \$5m wealth segment, (73% of the total)
- 0.13 million millionaire households in \$5m – \$30m wealth segment and
- 0.05 million millionaire households in the \$30m+ segment.

Therefore, India's income demographics are becoming increasingly unrecognisable and would translate into an enormous buying power. The stage is already set.

In the luxury car segment (Average price Rs. 25-30 lakhs) grew at 23% during the recession in the year 2007, though overall passenger car sales declined by 0.5% during this time. This segment accounts for 3-4% of total car sales in India but what is really alluring is that it is growing at a rate of over 20% year on year. (SIAM).

And then there is the Jewellery market in India which is the second largest in the world with a market size of \$16 billion, next only to USA (\$ 40 billion) and larger than that of China (\$13 billion). The gold jewellery market is growing at 15% per annum and diamond jewellery at 27% per annum. (Gems and Jewellery Export Promotion Council) Whether it is cars, jewellery or designer apparel, Indians are embracing the rediscovered luxury.

It is no surprise then that in the luxury market place, amongst the Asian countries, India has been considered as a next big opportunity. For example, a survey conducted by AT Kearney projects India's luxury market at 14.7 billion dollars in the year 2015, from 5.5 billion in 2010 [14].In a new study McKinsey suggests that if India continues on the recent growth path, by 2025 it will become the world's 5th-largest consumer economy .In the process the spending patterns will shift towards discretionary purchases. This is an opportunity no luxury player can afford to ignore.

9.2 Indian Culture

While his volume and growth is being increasingly recognized, the diversity of the vast Indian consumer and the complexities in the culture have intrigued the international luxury houses that want to make an India entry or have already entered. Therefore, it is worth exploring the complexities of culture and their impact on luxury buying behaviour in India.

The oft-quoted book 'A Rough Guide to India' says, "It is impossible not to be astonished by India. Nowhere on Earth does humanity present itself in such a dizzying, creative burst of cultures and religions, races and tongues........ Perhaps the only thing more difficult than to be indifferent to India would be to describe or understand India completely........."

This variety is evident in several aspects of culture in India and it has far reaching effects on the buying behaviour of the people. According to past research, national culture has considerable influence on consumer behaviour (Jaishankar 1998 in Banerjee [3]– Dimensions on Indian Culture...) The relationship between culture and values manifesting themselves through consumption of products is well-documented, with material goods being important to individuals due to their ability to carry and communicate cultural meaning [2].

Take for instance a 100-year-old Phantom Rolls-Royce built for Krishna Raja Wodeyar IV, Maharaja of Mysore (in South India) went under the hammer in August 2011 and fetched £400,000. The car was customized with umbrellas to shield the maharajas, servants from the sun. The maharaja who ordered this model in 1911 called it 'Victoria'. He was one of the world's wealthiest men, being worth the equivalent of £35billion at his death in 1940.

Luxury Rolls-Royces were popular among Indian rulers because their sturdy construction equipped them to negotiate the tricky terrain. He was so keen on Rolls-Royces that he used to order them in batches of seven. Within the company, 'doing a Mysore' became a shorthand for placing bulk orders.

And he wasn't the only one. Legend has it that in 1930, Maharaja Bhupinder Singh felt slighted at the British Rolls Royce company's refusal to accept an order from him for a new Rolls Royce car. Reacting to the refusal, the Maharaja put some of his old Rolls Royce cars (amongst the 24 he had at one point of time) to work hauling garbage, dung and filth in Patiala city to the chagrin of the all-powerful British ruling establishment who quickly prevailed upon the Rolls Royce Company to comply with the Maharaja's wishes.

Figure 9.1 Rolls Royce

The Patiala Necklace was made for the same Maharaja Bhupinder Singh of Patiala in 1928. It had five rows of diamond-encrusted platinum chains and oversized gems. The original Patiala necklace, had 2,930 diamonds and weighed almost a thousand carats. It included as its centerpiece the famous De Beers diamond, nearly as big as a golf ball. The Maharaja was often photographed wearing the necklace. The original, according to Cartier, would be worth about $30 million today.

Figure 9.2 Patiala Necklace

The book 'Made For Maharajas : A Design Diary of Princely India' [15] is full of such instances not limited to any one prince or any area. From Rani Sita Devi who had a collection of designer clothes and was supposed to be one of the best dressed women in the world; to the children of Gaekwar Pratapsinh Rao of Baroda who went from their palace to the nearby royal school in a miniature train made by Royal Locomotives, the tales of the Indian tradition of luxury are unending. Indulging in the ceremonial and the grand has always been a part of the Indian culture. Be it celebrating the innumerable festivals across the year or elaborate wedding ceremonies, some values are very basic to the Indian culture and remain deeply embedded. So, if culture is broadly defined as an integrated pattern of beliefs, values, attitudes and practices of a group then in case of India, this pattern is definitely complex and deep rooted.

9.2.1 The New Luxury

The luxury market is undergoing dramatic changes and the New Luxury that we see in India today is very different from the Old Luxury – that of the Maharajas and royalty. "Post independence period was largely marked by socialist ideology of community living and self- reliance." [6]. The New Luxury emerged after the economic changes initiated in 1991. It was only after 1991 that, with the integration of the Indian economy with the global one, a contemporary form of consumption culture was initiated [16],[18].Past 20 years since then have seen a rapid northward increase in accepting consumption, so much so that the current educated urban generation defines itself through consumption. This 'educated urban India' is part of the Great Indian Middle Class which includes "entrepreneurs, high salaried professionals, engineers, expatriates, and so on" [6]. This middle class has become representative of India's growth and hence the country is coming out of the austerity which was encouraged from 1947 until the 80s. The prosperity is happening mostly through knowledge based professions.

As Chadha and Husband (2007) [4] point out, Forbes India's 40 richest list includes 19 self made people, 11 of the 40 richest made their fortune in technology and 9 in pharmaceuticals. As they further suggest, 'the inherently Indian cultural and social factors will speed up the development of the luxury phenomenon. 'What they allude to as the 'The Indian Weddings, *Izzat* and Hinduism' is actually a combination of multiple factors.

Hofstede's (1980) five cultural dimensions: power distance, individualism and collectivism, masculinity and femininity, uncertainty avoidance, and short and long-term orientation can serve as the basis for understanding Indian culture [8]. Specifically, power distance seems to be playing a role in the Indian society which has had a long tradition of hierarchy (caste system) and hence luxury consumption may distinguish an individual's status in the society [7].

The Hofstede analysis for India shows a large power distance society and all other measures are relatively moderate. This would be indicative of the fact that India is in the midst of change. India has Power Distance (PDI) as the highest Hofstede Dimension for culture, with a ranking of 77 compared to a world average of 56.5. This score indicates a high level of inequality of power and wealth within the Indian society. However, since this represents inequality defined from below, not from above, it suggests that a society's level of inequality is accepted both by the followers and the leaders, which means this phenomenon is endorsed by most of the population as a norm India's Long Term Orientation (LTO) Dimension rank is 61, with the world average at 48. A high score can be indicative of a culture that is perseverant and highly inclined towards savings. India has Masculinity as the third highest ranking Hofstede Dimension at 56, with the world average just slightly lower at 51. A higher score on masculinity would stand for a greater the gap between values of men and women. In the context of luxury, this is evident in the fact that most of the jewellery purchases in India are led by women. India's lowest ranking Dimension is Uncertainty Avoidance (UAI) at 40, compared to the world average of 65. On the lower end of this ranking, the culture may be more open to unstructured ideas and situations. To take this further, this also indicates a mindset open to new ideas. To demystify this we would have to understand some core values of Indian culture because the endurance and longevity of core values has a significant effect on the behaviour of consumers. According to Ramesh Kumar (2009) [19], there are 32 Indian values that are relevant to marketing. Some of these like power distance and uncertainty avoidance have been discussed above. Other than these, there are some values which remain are typical and remain core to the Indian culture. First, there is high family orientation – including extended family like parents, grandparents and siblings.

Luxury brands which communicate family togetherness and sharing tend to go down well with the Indian consumer. According to Chadha and Husband (2009) [4], the 'Gold Collars' who are 35-55 years of age, well educated, have the money but not the mind set to spend it on luxury brands. They take their families on vacations overseas but will not buy a suit for $ 1000. They 'invest' in US/UK college education for their children but wouldn't spend on $ 400 Prada shoes. They go on to add that this group would spend on a Louis Vuitton bag if it was a gift for their wife. Products which communicate feelings and emotions gel with the Indian consumers. A majority of television commercials are based around emotions centering on the family and nurture and care. So much so that even McDonalds is positioned in its commercials as a 'family restaurant'.

And hand in hand with this goes savings orientation. Unlike in developed countries, government doesn't provide medical services or assured pension post retirement. So, Indian families depend on their savings for their expenses post retirement. Besides saving up for old age, Indians also save for weddings in the family and then spend lavishly during these occasions. Consumers across all sections are interested in jewellery and look at it as an investment. They invest a significant amount of their savings in jewellery. According to the World Gold Council, India is the largest market for gold jewellery in the world, representing a staggering 746 tonnes of gold in 2010. Whilst over 50% of gold jewellery is bought for weddings, the wedding anniversary has now become the most aspirational occasion for receiving gold today, extending a couple's relationship with gold beyond the marriage ceremony.

The festival of *Dhanteras,* and *Akshay Tritiya* sees phenomenal and annually increasing success; sales during 2011 grew over 28% over the previous year. Undoubtedly, India's culture embraces gold. Sample this, in 2011, the World Gold Council introduced a campaign to bring alive the significant role that gold plays in our lives. The campaign was drawn from the insight that investing in Gold jewellery gives us an emotive payoff and strengthens relationships. It further drives the realization that gold is actually a smarter investment.

The core creative idea of the campaign – '10 *Diwalis* ago', really stems from the insight that no purchase made 10 *Diwalis* ago would be as valuable as gold. (*Diwali* is the festival of lights and the biggest festival in India) In other words, many festivities are marked by buying or presenting jewellery. And since it is combined with the savings orientation, it is a great fit in the Indian culture.

Other than wedding ceremonies, many other festivities are marked by buying or presenting jewellery. During festivals of *Akshay Tritiya, Dhanteras* and *Diwali,* buying gold coins is considered auspicious. Largely, every celebration is centered around the rituals of prayer, seeking exchanging goodwill, decorating houses, wearing new clothes, music, dance and feasting.

While these traditional values are ingrained in India, some trends are changing, especially in urban areas. Now, the young professional is engaged in knowledge based industries and is strongly influenced by the west. He has high achievement orientation and is conscious of her material success. This segment is diametrically different from the same segment 20 years ago (pre 1991) – both in terms of education and material consumption. The present day young adult has high international exposure owing to technological development, and work experience abroad. He aspires for bigger cars, foreign holidays and quality services like baby-sitters, laundry services, dog walkers and home delivery of food and groceries All in all, besides several kinds of services they would engage in something that would have a symbolic value of having 'arrived'. To this affect, they also possess one or more credit cards , and regularly use this thus displaying a trend towards impulse gratification. So, in terms of luxury, even though India is at the beginning stage' the 'new money' is helping the middle class up its quality of life.This can even be noticed from the ever popular Indian cinema. Indian cinema epitomises thr Indian culture in ways more than one and there has been a sea change in the way the Indian popular cinema portrays wealth. For example, until a few years ago, the rich were regarded suspiciously. However, present movies celebrate materialism and the protagonist tends to a professional or an entrepreneur as compared to a blue collar worker in the 70s. In the 1950s.

Take the example of Devdas a film made in the 50s based on an old Indian novel, the heroine was portrayed as the girl next door, dressed in a simple cotton saree (an Indian wraparound garment for women) in, in a remake the opulence depicted was unbelievable.

Despite the high urbanisation, it is interesting that in terms of the geographical spread of growth in india—the middle class growth is expected to spread beyond the top tier cities. According to a study by McKinsey, the geographic pattern of India's income and consumption growth is seeing a shift too. Today, despite their lower incomes, rural households, due to their majority share of the population, are collectively India's largest consumers—57 percent of current consumption is in rural areas versus 43 percent in cities. Undoubtedly, the income growth would be very fast but average real rural income growth per household is expected to accelerate to 3.6 percent over the next two decades and consumption is expected to reach today's average urban household level by 2018.

Figure 9.3 Devdas

However, by 2025 , urban areas will account for over two-thirds of the future growth in the Indian market despite the fact that they would have only 37 percent of the population. Within urban areas though India's middle tier and smaller cities would come up attractive markets with substantial numbers of middle class customers. The report goes on to add that while India's two largest cities Delhi and Mumbai, will continue to be the country's biggest markets and the top eight cities will remain the important locations for upper-income consumers. However, almost two-thirds of India's middle class opportunity will be in mid-size cities such as Chandigarh, Ludhiana and Amritsar that have average incomes equal to, or higher than, the top-tier cities.

There is also an increasing trend of impulse gratification. So, in terms of luxury, even though India is at the beginning stage' the 'new money' is helping the middle class up its quality of life.

9.3 Discussion

Due to rapid globalization, though we are fast absorbing concept of "no-border" world, yet cultural factors moderate many aspects of consumer's buying behaviour. There is a growing interest in cultural differences of countries and understanding the cultural context of consumer behaviour in a globalized marketplace [17].

According to *Forbes*, India has the fastest-growing population of millionaires in the world. But for Western luxury brands operating in the country, grabbing a piece of the market has proven more difficult than anticipated and many are in the process of re-conceiving their India strategies.

Indian culture is a beautiful example of "unity in diversity", incredibly rich in culture and heritage. Few quotations or statements cannot describe the pedestal that India holds in the world map because of its colorful and unique culture. From the times of Mauryas, Cholas, Mughals till to the period of British Empire, India has always been famous for its traditions and hospitality. The warmth in the relations and euphoria in celebrations make the country stand out distinctively in the clutter. The country's liveliness and generosity attract a number of tourists. The cuisines, festivals, music, literature, and theatre, everything is 'special' in this 'land of gods'. The Indian culture is traditional yet contemporary, it moves on with the time. There are certain things about India that are famous worldwide, like: Indian Marriages, Time has changed but the lavishness has always been an integral and indispensable part of Indian marriages. In India, marriage is still taken as an institution where not two people but two families get united. Wedding is a big occasion in different sub cultures in India, and big opportunity too for luxury brands in categories like apparel and jewellery, In *Hindu* marriages, there is a lot of exchange of gifts across sub cultures, Muslims also have their own special way of celebrating the marriage ceremony, popularly called *Nikaah*. During the auspicious occasion, the groom's family gives *mehar* (gift) to the bride. Every state within country has its own special way of celebrating the marriage ceremony. Luxury Brands in apparels category understand the importance of clothing for beauty of Indian women. Very traditional and ethnic yet contemporary Indian *Saris* are famous worldwide. For men too, there is no dearth of variety. India is a secular state and every citizen enjoys an equal right of choosing and following a particular religion. There are many religions and so many pilgrimages but still people are united. It is the Indian values that unite people. People are not judged by caste, color or creed. Despite so much diversity, people in India are united and they feel proud of their culture and tradition. On the world stage, either through international film festivals or through beauty pageants, India regularly displays its talent and culture. Many rulers came but none could bring harm to its culture. In fact, Indians preserved the good values of their culture. The flexibility and movement with time make Indian Culture fashionable and acceptable too.

It may be argued that deep-rooted cultural diversity in India makes it difficult for a marketer to go for a standardized strategy to do better than competition. It has been found that even after tremendous exposure to globalization, consumers from different cultures have different attitudes, perceptions, tastes, preferences and values, and remain reluctant to purchase foreign products [20]. As a result of consumer acculturation to the country of origin, consumers respond in a manner that is consistent with their culture's norms and values [23]. A correct identification of country-specific cultural and value dimensions and its proper consideration play a critical role in ensuring success in the highly uncertain and competitive market scenario. Indeed, in order to succeed in India, luxury brands need to localize and formulate more culture specific marketing strategies. This goes further than just putting an Indian print on a bag or collaborating with a local celebrity. A multitude of cultures, languages, religions, festivals, colours and tastes make up this land of 1.2 billion people. So, it's about understanding the difference between the flashy nature of a *Punjabi* or *Delhi* customer, and the more reticent nature of a *Gujarati,* and speaking to each of them in the specific cultural register that they respond to. Luxury brand Montblanc is a good example, which successfully operates nineteen retail points across first, second and third tier cities in India, has regionalized all their marketing material. Brands like Louis Vuitton and Rolls Royce have also localized their approaches, identifying important events and celebrations amongst potential clients and arriving with personalized gifts or a surprise car service for the occasion. Getting toehold in the Indian sub-continent has been a challenge for every Western luxury brand that has tried to crack this complex new market.

As discussed in earlier section, Power distance, individualism and collectivism, masculinity and femininity, uncertainty avoidance, and short and long-term orientation are five cultural dimensions, which can serve as the basis for comparing Western and non-Western or Indian cultures. Specifically, power distance is prevalent in a hierarchical society and hence luxury consumption may distinguish an individual's status in the society [7]. A collective culture is representative of Indian values such as behavioural norms of displaying wealth through luxury possessions. Masculine traits of achievement, assertion and performance may be conveyed in terms of success and luxury consumption. Societies with high uncertainty avoidance tend to follow adequate norms and reference groups, which may explain the popularity of some foreign masstige luxury brands. Opposite to the short-term orientation of Western culture, long-term orientation emphasises ordering relationships such as perceived status and class of luxury possession in the society. However, Global demand for luxury goods can be argued as being part of global consumer culture. In this sense, global luxury brands may convey universal recognition and transfer across cultures. Wiedmann et al. (2007, p.3 [22]) note that *"there is no understanding of luxury conceivable which is nationally or regionally bound"* on the world marketplace.

It can be concluded that good understanding and right management of Indian culture would facilitate Luxury Brands to thrive as India is the next big opportunity for marketers across globe. Firms need to workout major issues involved with culture and values.

References

[1] Ablett, J./Baijal, A./Beinhocker, E./Bose,A,/Farrell, D./Gersch, U./Greenberg,E./Gupta, S./Gupta, S. (2007): The "Bird of Gold ": The Rise of India's Consumer Market. Executive summary. Business, May 2007.

[2] Askegaard, S./Kjeldgaard, D. (2002): The water fish swim in? Relations between culture and marketing in the age of globalization", in: Knudsen, T., Askegaard, S. and Jorgensen, A.N. (Eds), Perspectives on Marketing Relationship, Thompson, Copenhagen, pp. 13-35.

[3] Banerjee S. (2008): Dimensions of Indian culture, core cultural values and marketing implications, in: Cross Cultural Management: An International Journal Vol. 15 No. 4, pp. 367-378.

[4] Chadha, R./Husband, P. (2007): The cult of the luxury brand. Inside Asia's love affair with luxury. London: Nicholas Brealey International.

[5] Chaudhuri, H. R. (2006): Of Diamonds and Desires: Understanding Conspicuous Consumption from a Contemporary Marketing Perspective, Academy of Marketing Science Review.

[6] Eng, T.-Y./Bogaert, J., 2010: Psychological and cultural insights into consumption of luxury western brands in India, in: *Journal of Customer Behaviour,* Vol. 9 No. 1, pp. 55-75.

[7] Heisley, D. D./Cours, D. (2007): Connectedness and worthiness for the embedded self: A material culture perspective", Consumption, Markets and Culture, Vol. 10, No. 4, pp. 425-450.

[8] Hofstede, G. (1980): Culture's consequences: International differences in work-related values. Newbury Park, CA: Sage.

[9] http://en.wikipedia.org/wiki/Bhupinder_Singh_of_Patiala
[10] http://www.clearlycultural.com/geert-hofstede-cultural-dimensions/power-distance-index.
[11] http://www.dailymail.co.uk/news/article-2031237/The-Penelope-Pitstop-maharajas-Indian-royals-400-000-cartoon-like-Rolls-Royce-colourful-brolly.html#ixzz1eVKSZ0xK
[12] http://www.deloitte.com/assets/Dcom-United-States/Local%20Assets/Documents/FSI/US_FSI_GlobalWealthExecutiveSummary_050611.pdf)
[13] http://www.patialaweb.page.tl/Made-for-Maharajas.htm
[14] India Luxury Review 2011- A CII-AT Kearney Report
[15] Jaffer, A. (2006): Made for Maharajas: A Design Diary of Princely India
[16] Kurien, C.T. (1995): Aspects of India's Economy. New Delhi: Orient Longman
[17] Maheswaran, D./Sharon, S. (2000): Issues and new directions in global consumer psychology, in: Journal of Consumer Psychology, Vol. 9 No. 2, pp. 59-66.
[18] Mankekar, P. (1999): Screening Culture, Viewing Politics. An ethnography of television, womanhood, and nation in postcolonial India. London. Duke University Press.
[19] Ramesh Kumar , S. (2009): Consumer Behaviour and Branding : Concepts, Readings and Cases.
[20] Suh, T./Kwon, I.-W. (2002): Globalization and reluctant buyers, in: International Marketing Review, Vol. 19 No. 6, pp. 663-80.
[21] Vadukut, S. (2011): National Treasure, Mint Indulge, Mint, p.5
[22] Wiedmann, K. P./Hennigs, N./Siebels, A. (2007): Measuring consumers' luxury value perception: a cross-cultural framework, in: Academy of Marketing Science Review, Vol. 11, No. 7, pp. 1-21.
[23] Zhang, Y./Neelankavil, J. P. (1997): The influence of culture on advertising effectiveness in China and the USA, in: European Journal of Marketing, Vol. 31, No. 2, 131–149.

10 Perceptions, Attitudes and Luxury Brands

An Australian Study of Consumers' Perceptions of and Attitudes to Luxury Brands

Nicole Stegemann, Sara Denize, Kenneth E. Miller

10.1 Luxury Brands in Australia

The Australian market for luxury brands has been largely overlooked, but in the first decade of the new millennium it has quietly developed into a success story for the luxury industry with *"international luxury executives consistently remarking that Australia is a surprisingly sophisticated market"* [23]. Spurred by a growing middle class and increased disposable income, the democratisation of luxury is seen in the Australian market place in much the same way as it is in the rest of the world. This process of democratisation and the underlying strength of the Australian economy have been key drivers of growth in the luxury brand sector in Australia [22].

Access to luxury brands has also improved with newly established inner city shopping destinations for luxury brands in both Sydney and Melbourne [22]. Well-known brands such as Burberry and Louis Vuitton moved into new and bigger flagship stores [4]. Other well-established luxury brands such as Miu Miu and Veneta Botega moved into the new and extravagant Westfield Sydney City shopping mall that opened in July 2010 [4], [38].

Australian consumers are also active online shoppers and are placed third on the luxury online shop Net-A-Porter and spend over AU$1,190 annually [22]. Most of the luxury shoppers are Australian residents making up 60-70% while the balance is attributed to tourists of mainly Asian origin [22].

These developments explain how the Australian luxury fashion and accessories industry reported steady increases in sales (24% and 22% respectively) during the global financial crises in 2008/2009 [22]. The combined market value for fashion, accessories, fine jewellery and Australian fine jewellery luxury brands was about AU$800 million in 2009 and, even with a conservative estimate of an 11% growth rate is expected to move to the $AU1billion mark [22].

The growth in luxury brands in Australia is also underpinned by increasingly positive attitudes towards luxury brands and increased consumer sophistication [22]. This trend to increasingly positive predispositions is consistent with general research findings that consumers' buying processes for luxury brands are influenced by consumers' perceptions and attitudes [7], [10], [16], [36], [37], [39], [40].

The relationship between perceptions and attitudes is well established in the literature [14], [15]. However, consumers exhibit conflicting attitudes and behaviors in respect to luxury and luxury brands. Research has shown that consumers may have negative attitudes to the "excesses" of luxury while simultaneously expressing positive attitudes to particular luxury brands [10]. This potential for ambiguity is something about which little is known. Even less is known about this process in the Australian context.

This chapter therefore examines how consumer perceptions drive attitudes to luxury brands in Australia. We extend the basic model of perceptions and attitudes to include the impact of consumers' general attitudes to luxury and their familiarity with luxury brands on this relationship. We consider the literature on perceptions and attitudes to develop the conceptual framework that forms the basis of the research. This overview includes conceptualisations of the constructs and their relationships with each other, and provides justification for the hypotheses that we test in our empirical study. In the later part of the chapter we describe the results of our research. In particular we consider how the perceived degree of brand luxury, brand familiarity and general attitudes to luxury interact to affect attitudes to luxury brands. We conclude the paper by considering the limitations of the research, by highlighting a number of theoretical and managerial implications, and by outlining future research that may contribute to our understanding of luxury brands.

10.2 Conceptual Framework

Perceptions are a precursor to attitude formation [19]. This is an accepted axiom for all things, but particularly so for brands in general as well as for luxury brands. However, as we have noted, there are also other factors that influence attitude formation. Researchers investigating attitudes to luxury brands have identified both consumer familiarity with brands and their general attitude to the concept of luxury as possible drivers of brand attitude formation [10], [29]. We explore the literature pertaining to attitudes and perceptions and brand familiarity as the basis for the development of a conceptual model of luxury brand attitudes.

10.2.1 Luxury Brand Attitudes

Brand attitudes are the general feeling held by a consumer for a particular brand. For luxury brands these attitudes are influenced by the perceived degree of luxury of a brand [37]. However, as we have noted, Dubois & Laurent [9] and Dubois et al. [10] reported that an individuals' attitudes towards the concept of luxury in general and particular luxury brands may vary greatly. The dual nature of the concept of luxury is highlighted by individuals' avoidance and attraction reactions to luxury. They may simultaneously display strong positive and negative feelings towards luxury although, as these researchers note, relatively few consumers expressed indifference to the concept. Furthermore, this apparent contradiction may lead to inconsistencies when consumers report their attitudes to luxury brands [24].

We seek to distinguish between more general attitudes to luxury (discussed below) and attitudes to specific luxury brand. We note however that much of the extant work exploring brand luxury confounds these two distinct constructs [30].

Attitudes to luxury brands can be described as the cognitive and affective components of the tri-component attitude construct commonly used in attitude research [28]. The emotional and subjective qualities, as highlighted by descriptors such as "beauty" and "refinement" as well as the more negative descriptor "flashy" [10], characterize the affective dimension of luxury brand attitudes (liking and disliking). In contrast, aspects such as quality and experience relate to the cognitive dimension of luxury brand attitudes [29]. The behavioral dimension is expressed in consumers' intention to buy a luxury brand, and is influenced by consumers' affective and cognitive views of the luxury brand. Thus, luxury brand attitudes are central to the decision-making process for consumers and are of the upmost importance for marketers.

10.2.2 Perceptions of Brand Luxury

General consumer decision making literature emphasizes the effect of attitudes on perceptions and vice versa [19]. Keller [14] showed that the perceptions that are unique, favorable and strong result in positive brand evaluations and attitudes. Thus, brand associations relate to brand attributes and characteristics and influence the formation of consumers' brand attitude. Attitudes towards a brand may be expressed as the subjects' liking or disliking of that brand [26].

More specifically, early work on brand luxury established that the luxuriousness of a brand could be measured by various brand attributes including brand conspicuousness, uniqueness, quality, hedonism and self-extension [37]. These dimensions are grouped into non-personal-oriented perceptions (conspicuousness, uniqueness, and quality), and personal-oriented-perceptions (hedonism and extended self). Most individuals perceive brands as multifaceted combinations of these dimensions. Thus when making evaluations of brand luxury consumers will trade off less salient dimensions of luxury for more salient ones [37].

Non-personal perceptions of brand luxury are described as an individual's concerns about how other people may perceive them when using and consuming luxury brands. Personal-oriented perceptions rather relate to an individual's inner desires and needs rather than how they are perceived by others. These dimensions are now briefly outlined.

Perceived conspicuousness is based on the Veblen effect [35], which suggests that individuals are concerned about how they are perceived by others and therefore engage in conspicuous consumption in order to display their wealth, status and prestige. Perceived uniqueness is usually described through the snob effect [18], which occurs when individuals are concerned about social status. In their view, social status is gauged as exclusive consumption [20]. Therefore, these so-called 'snobs' only adopt newly launched goods and brands owned and consumed by very few consumers [21]. Perceived quality relates to consumers emphasizing brand quality when evaluating the degree of luxury a brand [36].

Evaluations of brand quality may be based on indicators such as exclusive product attributes and premium prices [36], [37]. Perceived hedonism captures individuals motivated by the feelings and affective state they achieve through the purchase of a luxury brand. These individuals are therefore not concerned with how they are perceived by others. Perceived self-extension relates to consumers using a brand's symbolic meaning and its ability to reinforce an individual's self [11].

In their research Vigneron and Johnson [36], [37] emphasize that the level of luxury for any given brand may vary from 'very little' to 'a great deal'. There are notable differences in the degree of luxury for any given brand. For Vigneron and Johnson [37], brand variations drive consumers perceptions of brand luxury. This is confirmed by similar research investigating the brand perceptions of, and attitudes to, luxury brands [10] as well as more general attitudes to luxury.

Perceived luxury is generally regarded as a positive indicator of quality and each of the other brand luxury indicators (conspicuousness, uniqueness, hedonism and self-extension). In combination, higher levels of these indicators of brand luxury will have a positive effect on luxury brand attitudes [35], [10]. Our model of brand luxury therefore incorporates the relationship between these two constructs. This resulted in the development of H1.

Hypothesis 1: The greater the degree of luxury perceived in a brand the more positive the attitudes towards that brand.

The consumer decision making literature establishes that an individual's familiarity with the brand [29] and their general attitude to luxury [35], [9], [10] will influence the underlying relationship between luxury brand perceptions and attitudes. Each of these moderating variables will now be considered in turn.

10.2.3 Brand Familiarity

Brand familiarity is "*the number of product related experiences that have been accumulated by the consumer*" [3], and can be further described as the degree of awareness of a specific brand [17]. Alba and Hutchinson [3] report that with increased product familiarity consumers apply more complete and refined cognitive structures to differentiate products. Sujan's (1985) findings concur with these results, reporting that expertise influences evaluations. More specifically, we note that consumers use product and brand knowledge to form attitudes towards a brand by evaluating its advantages over competitive brands [14].

Literature on luxury also suggests that brand attitudes are impacted by the extent to which a person is familiar with a brand [29]. Roux and Boush [29] examined the impact of brand familiarity on brand evaluations, and found that greater brand knowledge and familiarity resulted in more objective brand evaluations and associations.

We contend that the more familiar a person is with a brand the more they will be able to accurately perceive how luxurious it is and thus they develop brand attitudes that more accurately reflect that state of luxury. This is in line with the findings by Roux and Boush [29] who showed that greater brand knowledge improves perceptions of quality. Thus brand familiarity will have a positive moderating effect on the relationship between percieved luxury and luxury brand attitudes. This led to the development of H2.

Hypothesis 2: Familiarity will positively influence the relationship between perceived degree of brand luxury and brand attitude.

10.2.4 General Attitudes to Luxury

The second influencing variable considers the impact of general attitudes to the concept of luxury on the relationship between luxury brand perceptions and attitudes to luxury brands. It is important to remember the distinction between this more general view of luxury and attitudes towards a brand discussed above. There is very little literature addressing how a consumer's general attitude to luxury influences the formation of luxury brand attitudes. However, as we have noted, consumers' attitudes to the concept of luxury in general and to particular luxury brands may be both strongly negative and positive [9], [10]. It is therefore important to consider individuals' general attitudes towards the concept of luxury.

In research exploring these general attitudes associations may range from the positive (upscale, quality, good taste, and class) to the negative (flashiness and bad taste) [9], [10]. As expected, wealthy consumers usually express positive attitudes towards the concept of luxury. Less affluent consumers appear to be more critical and often display mixed attitudes towards the concept of luxury [10]. They are concerned that people buy luxury goods to copy the wealthy, and to be different [10]. However, individuals generally express positive attitudes towards the concept of luxury although they often lack expertise and familiarity when purchasing luxury brands.

We argue that attitudes held by a consumer about specific luxury brands may be influenced by that consumer's attitude to the concept of luxury itself and will therefore influence the relationship between the perceptions of how luxurious the brand is and the formation of attitudes towards that brand. The nature of this influencing effect is particularly interesting. It is expected that individuals with a negative disposition to luxury will produce more negative attitudes to particular luxury brands when they perceive the brand is more luxurious. Conversely, when an individual has positive dispositions to luxury they will have more positive attitudes towards a brand when perceptions of luxury are positive. This led to the development of H3.

> Hypothesis 3: The general attitude to luxury will positively influence the relationship between the perceived degree of brand luxury and luxury brand attitude.

10.2.5 A Model of Attitudes to Luxury Brands

The hypothesized relationships between perceived brand luxury, brand familiarity, and general attitudes to luxury and their impact on attitudes to brand luxury, form the basis of a model of attitudes to luxury brands. With such a model, it is possible to begin to understand how perceptions of luxury influence brand attitude formation of luxury. The following diagram presents a model of luxury brand attitudes, summarizes each of the hypotheses and shows the relationships between them.

As shown in the diagram we add the constructs familiarity, and general attitude to luxury to the conventional model of luxury brand attitudes. We expect that these co-variates will moderate the role that the perceived level of luxury plays in forming brand attitudes for luxury brands. In the following section, we present the results of a study to investigate the proposed model.

Figure 10.1 A model of luxury brand attitudes

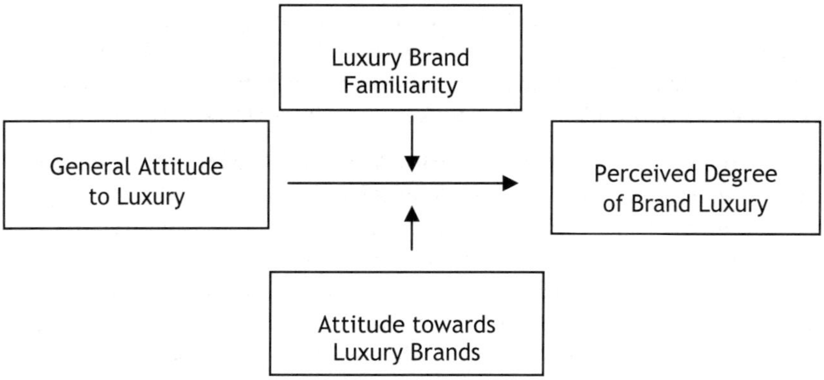

10.3 A Study to investigate Brand Luxury Attitudes

The investigation of luxury brand attitudes requires consideration of specific brands in order to develop a framework for respondent decision-making. In this research, we used an experimental design asking respondents to consider one of six luxury brands (LB) or one of two non-luxury brands (NLB). We chose brands in two product categories: cars (Ferrari LB, Porsche LB, Mercedes LB, and Toyota NLB) and fashion apparel (Prada LB, Ralph Lauren LB, Calvin Klein LB and Levis NLB). Through a pre-test with 541 respondents it was confirmed that the target respondent group had sufficient knowledge of each brand and category.

10.3.1 Approach and Methods

Data was collected using a self-completion questionnaire that included a number of multi- and single item measures. The appendix provides details of the operationalization of each of the main constructs and they are briefly summarized here.

Following Putrevu and Lord [26], we measured luxury brand attitudes using a shortened four-item seven point measure that captures consumers' attitudes towards a brand and relate to their liking or disliking of a particular brand. To measure perceived brand luxury we used a refined seventeen item seven-point scale [31] based on the Brand Luxury Index (BLI) originally developed by Vigneron and Johnson [36], [37]. The scale captures consumer perceptions of the degree of luxury of specific brands and products. Brand familiarity was measured using a three item semantic differential seven-point scale [17]. Lastly, following Homer [12] we used the shortened seven-item semantic differential seven-point measure to reveal consumers' general attitudes to luxury.

We used a convenience sample of 1440 undergraduate and postgraduate business students who were each required to fill in the self-completion questionnaire (with 180 respondents for each brand evaluation). We considered the use of the convenience sample was justified as the purpose of the study was model testing rather than generalization. There are other quantitative research studies that used business student respondents to measure changes in brand related elements. Dubois et al. [10] show that management students are suitable as research subjects when investigating the diversity of consumers' attitudes towards luxury. Kapferer [13] emphasizes that management students are current and potential consumers for luxury products.

10.3.2 Analysis Approach and Cleaning Variables

We used multiple brands to be able to test the model across a wider range of settings and to offer a more robust measure of the constructs. However, the limitation of this approach is that there may be brand-specific relationships between the constructs. To reduce the impact of brand specific distortions we aggregated the results for each brand into a single data file. The instrument's specific measures focused on measuring research constructs for a specific brand. The brand examined varied across execution. Hence, when the data was aggregated into one data file there was an understandable unease that brand specific effects would degrade (confound/bias) the measurement of each psychometric construct in the study. In order solve this potential difficulty we used a classic econometric methodology whereby regression coefficients (cleaning terms) were introduced into each of the main research models to capture these effects. We were then able to partition out these brand specific effects for the main and interaction effects of interest.

The cleaning terms were incorporated in the model as dummy coded variables that reproduce the specific brand in each execution. Ralph Lauren was used as the reference brand (and is therefore not shown in the regression model). The choice of Ralph Lauren was arbitrary. Any significant brand coefficients can be interpreted as the difference in effect of the significant brand from Ralph Lauren. We expect a mixture of significant and non-significant results for all of the cleaning terms. This has no theoretical implication apart from potentially identifying the relationships being examined.

10.4 Results and Findings

The constructs allowed the examination of the impact of perceived level of luxury, familiarity and general attitude to luxury on luxury brand attitudes. We tested the model and each of the hypotheses using regression (see **Table 10.1**). Output is interpreted using a 95 % significance level.

Table 10.1 Regression output perceived degree of brand luxury (PBL) on luxury brand attitude

		β	Std. Error	t	Sig.
	(Constant)	1.964	.308	6.380	.000
B1	Mercedes	.377	.218	1.726	.084
B3	Ferrari	.505	.225	2.246	.025
B2	Porsche	.325	.216	1.506	.132
B4	Toyota	1.062	.207	5.126	.000
B7	Calvin Klein	.123	.212	.580	.562
B5	Prada	-.594	.207	-2.867	.004
B8	Levis	.330	.197	1.670	.095
H1	**Perceived Brand Luxury**	**.196**	**.063**	**3.107**	**.002**
9	BLI*Mercedes	-.039	.042	-.928	.354
10	BLI*Ferrari	-.079	.045	-1.752	.080
11	BLI*Porsche	-.111	.044	-2.491	.013
12	BLI*Toyota	-.226	.037	-6.048	.000
13	BLI*Calvin Klein	-.023	.041	-.561	.575
14	BLI*Prada	.046	.044	1.029	.303
15	BLI*Levis	-.153	.037	-4.138	.000
16	Familiarity	.154	.039	3.951	.000
17	Familiarity*Mercedes	-.015	.019	-.786	.432
18	Familiarity*Ferrari	-.083	.022	-3.779	.000
19	Familiarity*Porsche	.001	.016	.084	.933
20	Familiarity*Toyota	.091	.023	4.038	.000
21	Familiarity*Calvin Klein	-.040	.021	-1.911	.056
22	Familiarity*Prada	-.010	.016	-.628	.530
23	Familiarity*Levis	.079	.021	3.776	.000
24	General Attitude Luxury	.089	.042	2.121	.034
H2	**PBL *Familiarity**	**.019**	**.006**	**3.145**	**.002**
H3	**PBL *General Attitude to Luxury**	**.028**	**.008**	**3.614**	**.000**
	Familiarity*General Attitude Luxury	-.009	.005	-1.789	.074

All terms relevant to testing the research hypotheses are in bold. All other terms are cleaning terms. F = 171.672, Significance p = 0.000, R^2 = .306, Adjusted R^2 = 0.304. BLI = Perceived degree of luxury of a brand. B1 to B8 = Alternative Specific Constants of each brand's effect. 8 to 24 = Cleaning Terms

As can be seen in the regression output, the overall model is significant with F = 171.672 and with the coefficient of determination (R2) equal to 0.306. This output confirms that the model explains sufficient variability in the dependent variable to continue interpretation. Before discussing the results of interest, we briefly explain the various terms in the model. Coefficients B1 to B8 are the alternative specific constants (ASCs) of each brand's effect on brand attitude. These ASCs partition out the effect of each brand from the main effects of interest. As noted above, this was done to remove any brand specific effects for the construct. Likewise, coefficients 9-24 partition out the brand specific effects of perceived level of luxury, brand familiarity and general attitude to luxury on brand attitude. The inclusion of these cleaning terms ensures that we have accounted for any variance in the dependent variable that may be misattributed to a main or interaction effect.

The model shows the main effect of perceived level of brand luxury on luxury brand attitudes, and is labelled H1. We note that the interactions between each of the main terms were included in the regression model for completeness. However, only the interactions between perceived brand luxury and familiarity [H2] and perceived level of brand luxury and general brand attitudes [H3] are relevant to this investigation. The research hypotheses are now considered.

10.4.1 Perceived Brand Luxury

It was hypothesized that the greater the degree of perceived brand luxury the more positive the attitudes towards that brand. As can be seen from the output, the standardized beta for perceived degree of luxury of a brand (H1) is significant (β= 0.196, t = 3.107) indicating a relationship exists. This result is in line with the theoretical discussion whereby perceptions when unique, favorable and strong result in positive brand evaluations and attitudes [14]. Hypothesis 1 is supported.

10.4.2 Familiarity

Theory suggests that familiarity will positively influence the relationship between perceived brand luxury and luxury brand attitude.

The beta for the PBL*familiarity interaction term is significant (β= 0.019, t = 3.145) indicating a relationship exists. The beta coefficient itself is 0.019 indicating that the relationship is positive. Theory suggests that the level of brand familiarity will produce more accurate consumer evaluations [1], [2], [6], [25], [27], [33], [34]. As most respondents were familiar with the eight brands a relatively small correction was expected. This result offers support for Hypothesis 2.

10.4.3 General Attitude to Luxury

The literature suggests that the general attitude to luxury will positively influence the relationship between perceived degree of brand luxury and brand attitude.

The beta for the PBL*general attitude to luxury interaction term is significant (β= 0.28, t = 3.614) indicating a relationship exist. This result is in line with the literature. Most respondents displayed positive attitudes to the general concept of luxury, which explains the magnitude of the impact. This result offers support for Hypothesis 3.

Further main effects were found, which were not hypothesized. The main terms brand familiarity (β= 0.154, t = 3.951) and general attitude to luxury (β= 0.089, t = 2.121) positively influence the relationship between perceived degree of luxury of a brand and brand attitudes. The influencing effect of general attitude to luxury on specific luxury brand attitudes is minimal and does not change the result greatly. However, brand familiarity has a greater impact and should be further researched.

10.5 Conclusion

The general literature suggests that the perceived level of luxury of a brand has a positive effect on a person's attitude to that brand [35], [10]. This study finds that consumers' perceptions of luxury brands directly influence the formation of their attitudes to that brand. It follows that the perceived degree of luxury has a primary impact on the consumer decision making process.

This critical relationship between the perceived degree of brand luxury and its subsequent impact on attitudes is positively influenced by the general attitude to luxury and the familiarity of the consumer with the brand. With regard to the degree of familiarity with the brand, as we have noted the literature suggests that the more familiar a person is with a brand the more accurate the perceptions and brand evaluations [1], [2], [6], [25], [27], [33], [34]. This improved accuracy would work in favor of a luxury brand as luxury brands are generally of higher quality and superior on a number of dimensions compared to other alternatives in the market. Thus, for a luxury brand, familiarity is likely to positively influence the relationship between perceived degree of luxury of a particular brand and brand attitudes. Not only is there a theoretical basis for this effect, this research now provides empirical support for its presence.

The other influencing variable, consumers' general attitude to the concept of luxury, was also thought to have a positive effect on this relationship. Literature had previously shown that consumers' attitudes towards the concept of luxury, and to luxury brands in general, can vary considerably [9], [10]. That is consumers can hold a general liking, disliking or ambivalence towards the ownership of luxury products. Although never previously considered, the general attitude to the concept of luxury is a logical influencing factor between a consumer's perceived level of luxury of a particular brand and their attitude towards that brand. The results of this research give clear evidence for the presence of such an influencing factor. Indeed this research shows that the general attitude to the concept of luxury positively influences this relationship such that if people are generally positively predisposed to the ownership of luxury brands this factor magnifies the impact perceived level of luxury has on the formation of attitudes. The reverse is also true; when they have a negative predisposition to the ownership of luxurious types of brands those perceived to be luxurious attract more negative brand attitudes.

10.5.1 Theoretical Contributions

A model has been developed that integrates the concept of luxury into the theory of consumer decision making processes. This was achieved through the development of a model linking relevant constructs to luxury. This model describes the relationships between the constructs of perceived degree of brand luxury and luxury brand attitudes. The model also identifies two influencing factors as relevant to the concept of luxury. These are consumers' level of brand familiarity and their general attitude to the concept of luxury. This model provides marketers and scholars with further understanding as to how consumers choose luxury brands.

10.5.2 Implications for Practice

The vital question now is if the perceived degree of luxury is a key driver for luxury brand attitudes then luxury brands manufacturers have to build and maintain a desired level of luxuriousness in a brand. This means they have to provide justifications as to how their luxury brands provide a high level of gratification and indulgence in order that consumers develop strong and positive brand attitudes. It is no longer satisfactory to think that consumers will buy luxury brands because they are expensive and of superior quality. Instead, the communication strategy has to incorporate the desired level of luxuriousness of a brand.

10.5.3 Limitations and Future Research

This study has identified a number of limitations and opportunities for future research.

The model developed in this paper has been tested for the first time and should be applied to other brands for generalization. A limitation of this research is the use of a convenience sample of students. The use of a consumer panel would add further insights into the model's applicability by providing longitudinal, rather than cross-sectional, data. This research study included eight brands and two categories (clothes and cars). No category or brand specific analysis was undertaken. Future research will look at brand and category specific differences. An aggregated brand luxury index was used in this study and a disaggregation into its five dimensions will provide further understanding how to design successful communication strategies. Further research will provide new insights that will build on our understanding of the luxury brand decision process, and in particular, on the formation of luxury brand attitudes. This research has established that conventional models do not apply.

References

[1] Aaker, D.A./Joachimsthaler, E. (2000): The brand relationship spectrum: The key to the brand architecture challenge, in: California Management Review, Vol. 42, pp. 8-24.
[2] Aaker, D.A./Keller, K.L. (1990): Consumer Evaluations of Brand Extensions, in: Journal of Marketing, Vol. 54, pp. 27-41.
[3] Alba, J.W./Hutchinson, J.W. (1987): Dimensions of Consumer Expertise, in: Journal of Consumer Research, Vol. 13, pp. 411-454.
[4] Anonymous (2011): A whole new mall game, in: The Sydney Morning Herald. Sydney, Fairfax Media.
[5] Bone, P. F./Ellen, P. S. (1992): The Generation and Consequences of Communication-evoked Imagery, in: Journal of Consumer Research, Vol. 19, pp. 93-104.
[6] Broniarczyk, S. M./Alba, J. W. (1994): The Importance of the Brand in Brand Extensions , in: Journal of Marketing Research, Vol. 31, pp. 214-228.
[7] De Bernier, V./Falcy, S./Valett-Florence, P. (2008): Comment mesurer les perceptions du luxe ? Une comparaison entre les échelles de Kapferer (1998), de Vigneron et Johnson (1999) et de Dubois et al. (2001), International Marketing Trends Congress, Congrès ESCP Paris-Venise.
[8] Dubois, B./Czellar, S./Laurent, G. (2005): Consumer Segments Based on Attitudes Toward Luxury: Empirical Evidence from Twenty Countries, in Marketing Letters, Vol. 16, p. 115-128.
[9] Dubois, B./Laurent, G. (1994): Attitudes toward the concept of luxury: An exploratory analysis, in: Asia-Pacific Advances in Consumer Research, Vol. 1, pp. 273-278.
[10] Dubois, B./Laurent, G./Czellar, S. (2001): Consumer rapport to luxury: Analyzing complex and ambivalent attitudes, in: Working Paper Series: CR 736/2001, Paris, HEC.
[11] Holt, D. B. (1995): How consumers consume: A typology of consumption practices, in: Journal of Consumer Research, 22 June, pp. 1-16.
[12] Homer, P. M. (1995): Ad Size as an Indicator of Perceived Advertising Costs and Effort: The Effects on Memory and Perceptions, in: Journal of Advertising, 24 (Winter), pp. 1-12.
[13] Kapferer, J.-N. (1998): Why are we seduced by luxury brands?, in: Journal of Brand Management, Vol. 6, pp. 44-49.
[14] Keller, K. L. (1993): Conceptualizing, Measuring and Managing Customer Based Brand Equity, in: Journal of Marketing, Vol. 57, pp. 1-22.
[15] Keller, K. L. (2003): Strategic branding management: building, measuring, and managing brand equity, Upper Saddle River, New Jersey, Prentice Hall.
[16] Keller, K. L. (2009): Managing the growth tradeoff: Challenges and opportunities in luxury branding, in: The Journal of Brand Management, Vol. 16, pp. 290-301.
[17] Kent, R. J./Allen, C. T. (1994): Competitive Interference Effects in Consumer Memory for Advertising: The Role of Brand Familiarity, in: Journal of Marketing, Vol. 58, pp. 97-105.
[18] Leibenstein, H. (1950): Bandwagon, snob, and Veblen effects in the theory of consumers' demand, in: Journal of Economics & Business, Vol. 64 (May), pp. 183-207.
[19] Lutz, R. J. (1991): The Role of Attitude Theory in Marketing, in: Robertson, H. H. K. A. T. S. (Ed.) Perspectives in Consumer Behavior, 4th ed., Englewood Cliffs, NJ, Prentice Hall.
[20] Mason, R. (1981) Conspicuous Consumption, New York, N.Y., St Martin's Press.
[21] Mason, R. S. (1992) Modelling the demand for status goods. Working Paper. UK, Department of Business and Management Studies, University of Salford.
[22] O'Rourke, M. (2011a): Australian Luxury Market Nearing 1 Billion Dollars. The MO Down. Sydney, MO Luxury, assessed 17 January 2012, http://www.moluxury.com.au/themodown/?p=2113
[23] O'Rourke, M. (2011b): Sydney, A Surprisingly Sophisticated Market Sydney, Luxury Society, assessed 17 January 2012, http://luxurysociety.com/articles/2011/03/sydney-a-surprisingly-sophisticated-market.
[24] Otnes, C./Lowrey, T./Shrum, L. J. (1997): Toward an Understanding of Consumer Ambivalence, in: Journal of Consumer Research, Vol. 24 (June), pp. 80-94.

[25] Park, W. C./Jaworski, B. J./Macinnis, D. J. (1986): Strategic Brand Concept-Image Management, in: Journal of Marketing, Vol. 50, pp. 135-145.
[26] Putrevu, S./ Lord, K. R. (1994): Comparative and Noncomparative Advertising: Attitudinal Effects Under Cognitive and Affective Involvement Conditions, in: Journal of Advertising, Vol. 23 (June), pp. 77-90.
[27] Reddy, S./Holak, S. L./Bhat, S. (1994): To extend or not to extend: Success determinants of line extensions, in: Journal of Marketing Research, Vol. 31, pp. 243-262.
[28] Rosenberg, M. J./Hovland, C. I. (1960): Attitude Organization and Change: An Analysis of Consistency Among Attitude Components, New Haven, Connecticut, Yale University Press.
[29] Roux, E./Boush, D. M. (1996): The role of familiarity and expertise in luxury brand extension evaluation, in EMAC Conference Proceedings. Budapest, Hungary.
[30] Stegemann, N./Denize, S./Miller, K. E. (2007): Measuring Consumers Attitudes to Luxury, in: Merunka, D. & Warlop, L. (Eds.), The La Londe Seminar, La Londe les Maures, France.
[31] Stegemann, N./Denize, S./Miller, K. E. (2011): Refinement of the Brand Luxury Index, in the proceedings of the 1st International Conference on Luxury and Counterfeiting, Geneva, Wesford Laboratoire de Recherche, 9 – 10 June.
[32] Sujan, M. (1985): Consumer Knowledge: Effects on Evaluation Strategies Mediating Consumer Judgments, in: Journal of Consumer Research, Vol. 12, pp. 31-46.
[33] Tauber, E. M. (1981): Brand Franchise Extension: New Product Benefits from Existing Brand Names. Business Horizons, Vol. 24, pp. 36-41.
[34] Tauber, E. M. (1988): Brand Leverage: Strategy for growth in a cost-control world, in: Journal of Advertising Research, Vol. 28, pp. 26-30.
[35] Veblen, T. B. (1899): The Theory of the Leisure Class, Boston, Houghton Mifflin.
[36] Vigneron, F./Johnson, L. W. (1999): A review and a conceptual framework of prestige-seeking consumer behavior, in: Academy of Marketing Science Review, Marketing Science Institute, http://www.amsreview.org/articles/vigneron01-1999.pdf.
[37] Vigneron, F./Johnson, L. W. (2004): Measuring Perceptions of Brand Luxury, in The Journal of Brand Management, Vol. 11, pp. 484-506.
[38] Westfield (2010): Shopping Directory. Sydney, Westfield, assessed on 17 January 2012, http://www.westfield.com.au/sydney/directory/search/store?category=2180.
[39] Wiedmann, K.-P./Hennigs, N./Siebels, A. (2007): Measuring Consumers' Luxury Value Perception: A Cross-Cultural Framework, in: Academy of Marketing Science Review, http://www.amsreview.org/articles/wiedmann07-2007.pdf.
[40] Wiedmann, K.-P./Hennigs, N./Siebels, A. (2009): Value-Based Segmentation of Luxury Consumption Behavior, in: Psychology and Marketing, Vol. 26, pp. 625-651.

Appendix: Operationalization of constructs

Perceptions of the Degree of Luxury – Brand Luxury Index (BLI) (refined by Stegemann, Denize and Miller, 2011; based on Vigneron and Johnson, 1999 and 2004)

Cronbach alphas between 0.69 and 0.95 (Vigneron and Johnson, 2004), and 0.85 and 0.92 for refined scale (this study). Refined seventeen-item seven-point scale (Strongly Disagree – Strongly Agree) for each brand including six luxury brands (Ferrari, Porsche, Mercedes, Prada, Ralph Lauren, Calvin Klein) and two non-luxury brands (Levis and Toyota) . This brand is:

1. For Elitist
2. Extremely Expensive
3. For Wealthy
4. Very Exclusive
5. Unique
6. Rare
7. Crafted
8. Best Quality
9. Sophisticated
10. Superior
11. Exquisite
12. Attractive
13. Stunning
14. Leading
15. Very Powerful
16. Rewarding
17. Successful

Attitudes toward luxury brands (shortened scale based on Putrevu and Lord 1994)

Cronbach alpha = 0.87 (Putrevu and Lord, 1994), and 0.85 for shortened scale (this study). Refined four item seven-point likert-type scale (Strongly Disagree – Strongly Agree) for each brand including six luxury brands (Ferrari, Porsche, Mercedes, Prada, Ralph Lauren, Calvin Klein) and two non-luxury brands (Levis and Toyota):

- Buying (brand) is a good decision
- I think (brand) is a satisfactory brand
- I think (brand) has a lot of beneficial characteristics
- I have a favorable opinion of (brand)

Attitudes towards the Concept of Luxury (shortened scale based on Homer, 1995)

Cronbach alpha = 0.97 (Homer, 1995), and 0.93 (this study). Shortened seven-item semantic differential seven-point scale. Luxury for me is:

- Negative – Positive
- Unpleasant – Pleasant
- Disagreeable – Agreeable
- Worthless – Valuable
- Bad – Good
- Unfavorable – Favorable
- Dislike a lot – Like a lot

Familiarity/Experience (Kent and Allen, 1994 and Machleit, Allen and Madden, 1993)

Cronbach alpha = 0.85 (Kent and Allen, 1994), and 0.82 (this study). Three-item semantic differential seven-point scale.

- Unfamiliar – Familiar
- Inexperienced – Experienced
- Not knowledgeable – Knowledgeable

Part 5

Luxury and Luxury Consumption in the Context of Societal Change

11 Luxury Marketing in the Age of Cheap An Exploratory Survey on Consumer Behavior in Online Shopping Clubs

Wolfgang Fritz, Wencke Gülow

11.1 Introduction

Price competition in the retail industry, notably in Germany, has reached enormous dimensions. Large retail companies counteract decreasing market shares and aggressive pricing of discounters with spectacular sales promotions [20]. According to Horx (1995) [27], this process is embedded in a „recession culture" which originated in the 1990s. The author regards it as the typical megatrend of the western world. This development implies that an increasing number of people, not only of the lower but also of the middle and upper classes, search for the cheapest products. Therefore, he argues that the retail sector as well as parts of our society suffer a significant loss of luxury orientation [27]. This development may lead to a process of "trading down" in economy and society.

In contrast, it is also noticeable that consumers have become more sophisticated and show a rather polarized consumption behavior [29]. They tend to demand either a high-end luxury brand (e.g. a designer fashion item) or a low-cost good (e.g. generic food at the discounter) to meet their needs best [44]. Ko and Sung (2007) [29] consider this consumer behavior as the "trading up and trading down phenomena" of our society.

This paper will examine the prospects of luxury marketing in light of the unfolding megatrend in today's economy and society known as the "Age of Cheap" [8]. According to Bosshart [8] , there has recently been an increasing polarization between the small premium and the large discount sector. Silverstein and Fiske (2005) [46] argue that consumers trade down in "categories of little importance to them" (p. 6) which enables them to afford premium or luxury products in other categories. Hence, the "Age of Cheap" exists but not in the exceeding dimension for our economy and society as some authors suggest.

In accordance with the research objective, the paper is organized as follows: We give a brief overview of prior research on the "trading down" concept and mainly refer to Bosshart's (2004) [8] study of the "Age of Cheap". Subsequently, we examine the contrary development of "trading up" that leads to the successful establishment of new luxury brands. In this context, we illustrate the transformation of the luxury market that has recently embraced affordable goods for the masses. In this regard, we include results of an empirical study, using a sample of 274 online shoppers. The purpose of this survey was to analyze the consumer behavior regarding new luxury brands sold in online shopping clubs, such as Brands4Friends or BuyVIP. We conclude this paper by summarizing our findings, addressing its limitations, and discussing further research.

11.2 The Phenomenon of "Trading Down"

The intense price competition in the German retail industry serves as evidence for the apparent trading down phenomenon in our society and economy. Aggressive price promotions are expected to counter long-lasting buying resistance, increasing competition, and constantly growing market shares of discounters. Therefore, discount campaigns and special sale promotions have become a noticeable market phenomenon [21].

The German futurologist Matthias Horx already recognized this trading down development throughout the retail industry in the mid-1990s. By using the term "Aldisierung" he stated that in course of the recession culture people of the middle and upper classes search for the cheapest goods, especially in the food sector [27]. Bernd M. Michael concludes that Germany suffers a decline in values and quality as well as a fading fascination for shopping and longing of goods [38]. Referring to Horx, that trend originated in the 1990s and appeared not only in Germany but also across all western nations. In reference to the USA, Kotler (2011, p. 134 [31]) emphasizes that many households have chosen to shift to a "less is more" lifestyle as a result of reduced income in the recent financial meltdown. Besides trading down in the retail industry, Horx also discovered a social dimension characterized by consumer's loss of luxury orientation and a penchant towards frugality and poverty in the western world [27]. Kull (2004) [33] and Fritz et al. (2007) [20] also outline the structural shift in the retail industry towards the discount principle which has an immediate impact on the western society.

Two interdependent factors exist that support the trading down development. First, an increasing intersectoral search for the cheapest offer, even by wealthy consumers, encourages a purchase in discount stores such as Aldi or Lidl. Second, trading down is characterized by aggressive pricing and sales promotion battles among retail companies. Hence, business models with an integrated discount pricing strategy, e.g. discount supermarkets and drugstores, have a competitive advantage in the course of the recession culture [18].

Whereas Horx regards the recession as a typical megatrend of the 1990s, David Bosshart (CEO of the Gottlieb Duttweiler Institute) describes trading down in a more comprehensive and prevailing social context. Referring to Bosshart (2004) [8] this development is embedded in a broader structural change of economy and society, which he refers to as the "Age of Cheap". He states, it is a logical consequence of saturated markets that consumers become increasingly price sensitive and focus on bargain buys. Furthermore, trading down may not only affect shopping habits but also other areas of life, e.g. politics, morals, family, education, and leisure. Trading down, Bosshart claims, evolves into the most important and decisive lifestyle of our time [8]. Some of the trading down areas that had been introduced by Bosshart (2004) [8] and Fritz et al. (2007) [20] are discussed below.

- **Cheap Fashion:** It has become particularly apparent in the fashion industry that there can be reasonably priced offers even for fashion-conscious consumers. The sustained success of vertical fashion retailers, such as H&M, New Yorker, Mango, and Zara, supports this statement [36].
- **Cheap Food:** Discount grocery stores become increasingly accepted to European consumers at all income levels [15]. Similar business models have been established in the United States, such as the so-called Dollar Stores. By offering an assortment of goods at affordable prices even for financially weaker consumers, these stores grow faster than the market leader Wal-Mart [8]. In Europe, the number of discount grocery stores nearly doubled between 1991 and 2005 [15]. Despite the market share of discount brands in-

creasing between 2005 and 2011, the GfK consumer panel (2011) shows a slight trading up trend in favor of premium brands which is due to growing quality awareness, especially in the consumer segment of people aged 50 years and above [22].

- **Cheap Computers:** Due to the saturation of the computer market, it has become necessary to sell hard- and software products at lower prices. Cheaper computers, equipped with freeware operating systems such as Linux, increasingly reduce prices beyond their already subordinate prices [21]. Large discounter chains like ALDI sell modern tablet computers at half of the price of an iPad.
- **Cheap Home Improvement, Cheap Consumer Electronics:** Consumer electronics, home appliances, and furniture gradually become cheaper. Stores like Ikea, Home Depot, Best Buy, Media Markt or Saturn and their price-oriented advertising slogans support this development [21]. Additionally, the online platform "MyHammer", for example, offers all kinds of craftsmen services for prices below average.
- **Cheap Mobility, Cheap Travel:** Budget Airlines such as Southwest Airlines in the USA, Ryan Air and easyJet in Europe, or Tiger Airways in South East Asia have experienced tremendous success in recent years. The price sensitivity in this area has increased more than anywhere else and brings about high competitive pressure [21]. Established airlines such as Lufthansa, have to adjust their offers in order to compete, especially in short-haul flights. The increased price sensitivity in air travelling is an international phenomenon and independent of social classes [8]. The easyGroup appreciates this trend and gradually enlarges its scope of products by launching additional concepts such as easyCruise, easyBus, and easyHotel. These business concepts offer good value for money by waiving costly services and comfort. "Cheap Mobility" is further driven by an enhanced information and communication mobility [21] that enables consumers to compare prices and easily book travel itineraries.
- **Cheap Know-how:** The internet has drastically changed the handling of information. Knowledge has become a commodity that is virtually open to everyone who possesses an internet access [19]. Search engines like Google, the encyclopedia Wikipedia as well as online newspapers or TV channels substitute conventional and nongratuitous sources of information.
- **Cheap Chic:** Cheap is usually associated with negative attributes. However, Bosshart (2004) [8] prophesies "Cheap Chic" a promising future. If everything becomes cheaper, design and taste will be relevant differentiation factors to be competitive [8]. As mentioned earlier, some fashion companies (e.g. Zara, H&M) prove this statement.

In addition to Bosshart's trading down categories, a recent study of the Boston Consulting Group identified mobile telephony, car insurance, internet services, fashion accessories, and food as the main trading down sectors in Europe [43]. In contrast, consumers in emerging markets like China preferably trade down in various food categories, e.g. biscuits, fast foods, desserts, fine dining, chocolate, and drinks. This example shows that there are significant differences between mature and developing markets in which product categories consumers trade down.

Bosshart (2004) [8] identified economic, political, demographic, and technological factors that drive the trading down phenomenon. Furthermore, he defines "consumer democracy" as one important movement that results from a globalized world. Due to innovations in transportation and information technologies, consumers nowadays have the opportunity to purchase globally with little to no additional effort. Internet makes it possible to instantly compare prices and to order goods from a wide range of suppliers located across the globe. Convenient electronic payment options and faster transportation systems contribute to this trend. On the supply side, enhanced production and transportation technologies enable companies to offer higher quality products at lower costs and in a greater variety as well as in a higher frequency [43]. A broader choice and the alignment of quality levels lead to interchangeable products and services. Hence, the price increasingly becomes the decisive factor for a consumer when purchasing a product [8], [54]. Apart from that, higher levels of education, rising real incomes, and the increasing popularity of traveling abroad create more sophisticated consumers with a greater polarization of their spending patterns: saving money in categories that are less important and "splurging" in some others that are meaningful to them [15], [43]. The BCG Consumer Survey (2007, 2008) shows that more than 90% of consumers in developed markets like USA, Europe, and Japan have at least one category in which they try to save by spending less [43].

The accompanying phenomenon of consumers trading up in certain product categories has a substantial impact on the luxury market and is subject of the following chapter.

11.3 Recent Developments in Luxury Marketing

As described above, "trading up" describes the behavior of consumers who shift their spending in favor of premium or luxury goods. Many of these consumers wish to reward themselves with products that make them feel good in their stressful lives or to enjoy an upmarket lifestyle [48]. However, financial pressures force them to focus on their budgets, search for discounts, and find best value for money in categories of less interest [43].

The trading up trend originated from the US middle class and resulted in a global movement [45]. Trading up is usually associated with "new luxury goods", i.e. "goods [that] evoke and engage consumers' emotions while feeding their aspirations for a better life" (Silverstein and Fiske, 2003, p. 48 [44]). New luxury brands attract mostly middle-class consumers because they combine emotional benefits, such as prestige or self-fulfillment with accessible prices [48], [29].

Despite the worldwide economic depression the size of the luxury market rapidly increases. The value of luxury goods was estimated to be about 220 billion dollars in 2009 and is anticipated to augment up to 2 trillion dollars in 2010 due to the consumers in the developing markets [37]. The importance of emerging markets, particularly China, was already evident before the financial crisis, but within the past years the shift towards these markets has increased even more [5]. The demand in emerging markets has been so robust; luxury retail companies such as Prada, Gucci, and Cartier have been able to raise their prices in

China by one-third despite a global crisis [26]. Besides the extension to new markets, fundamental structural changes have occurred in the luxury sector and will be illustrated in the following sections.

11.3.1 Luxury as a Real Experience

As opposed to the traditional luxury market, "new luxury" comprises various sectors and can also have various interpretations. As daily life becomes busier, more and more consumers regard immaterial things as luxury. For example, spa treatments, holidays in health resorts, or leisure time that can be spent with family and friends are perceived as luxurious commodities since they notably enhance the quality of life [35]. According to an IPSOS (2005) study, factors such as "being able to determine one's own schedule", "experiencing something extraordinary", or "having time for myself" are the most important needs of many consumers [30]. Therefore, some people treat themselves with a cleaning service, private chauffeur, personal trainer, or even a cook to have time to spare and add a sparkle to their daily lives. An example of real and exclusive luxury that can be experienced is Gorilla Watching in Ruanda, for which only a limited number of visas is being issued by the government each year [32]. The global turnover in this luxury sector of "real experience" is estimated to increase up to 750 billion Euros in 2010 [4]. As the global economy recovers and the number of potential, affluent tourists of the baby boomer generation and from BRIC countries increases, an ILTM study (2011) [28] anticipates a fast market growth for exclusive travel in the next years.

11.3.2 Transformation into a Mass Market

A significant factor for the growth of the luxury market is the changing purchasing pattern of the middle class. In 2010, the middle class already accounted for 54 percent of the total revenues in the luxury market [5], [4]. Hence, the luxury market gradually transforms into a mass market [48]. The increasing demand for luxury products in the BRIC nations additionally drives the market's growth.

Besides new luxury brands, a lot of the traditional luxury companies extend their brand range by offering accessible product variants that meet the new demands of the middle class. BOSS Orange, Armani Exchange, and Ralph Lauren's Polo are new luxury brands that stand for a profitable brand stretching strategy of their parent company: the new brand benefits from the image of the parent company whereas the brand extension prevents the parent brand from decline in sales [12]. However, along with the "democratization of luxury" (Dubois and Laurent, 1995, p. 69 [16]) luxury brands also lose their exclusiveness.

The consumer's propensity to save on the one hand and to buy luxury goods on the other hand has an immediate effect on products of the middle price range. It becomes more and more difficult for mid-range brands to stay competitive in the market. Therefore, companies can either seek to compete with low-priced products or create an emotional and authentic brand image in order to convince the consumers to buy their products [44]. So-called Masstige brands pursue the latter strategy: they are considered luxury or premium brands but are priced well below traditional luxury goods and above conventional brands in their product category [3]. Porsche Boxter, Lauren by Ralph Lauren, Armani Jeans, and Victoria's Secret are representative examples for Masstige brands that bridge the gap between consumer and super premium brands.

11.3.3 Sustainable Luxury

A growing number of consumers make their purchasing decisions dependent on factors like health, fair trade, or eco-friendliness. This new consumer segment is described by the term LOHAS, an acronym for Lifestyles of Health and Sustainability, that was coined by the US-American sociologist Paul Ray. Consumers who belong to this lifestyle have a significantly higher income at their disposal and are willing to spend more money for products that fulfill their requirements and beliefs [32]. Therefore, LOHAS offer additional opportunities for the luxury industry. They are able to combine the virtual conflict of delightful consumption and responsibility in their lifestyles [6]. Thereby, their "green" lifestyle encompasses almost all consumption areas, e.g. apparel made of natural material, fair trade food, eco-friendly cars, natural cosmetics, or sustainable travelling.

Around 20 to 30 percent of the U.S. population already account for the LOHAS segment (Kotler, 2011, p. 134 [31], [11]). In the meantime, this lifestyle has extended to nearly all European countries, Australia as well as Asia [52]. Norbert Reithofer, CEO of BMW AG, also claims that "in the future, premium will increasingly be defined by sustainability" [41]. Hence, marketers have to revise their practices by adjusting product development, pricing, distribution, and branding according to the consumers rising demand for sustainable products and services [31].

11.3.4 Luxury Brands in the Internet

Another trend is based on the fact that new luxury consumers – compared to consumers of traditional luxury – are relatively young, cosmopolitan, and possess a better brand knowledge than the former generation of luxury shoppers [32]. Furthermore, an increasing internet affinity of the global population leads to an intensified integration of e-commerce in the luxury industry. According to a KPMG study (2009), luxury "newcomers" are particularly open for online shopping. Luxury brands, therefore, have the possibility to extend their customer range because the psychological threshold to purchase luxury in the internet is also much lower [17]. However, traditional luxury brands like Chanel or Hermès have resisted selling their products via internet [39]. Losing customer contact and the exclusive shopping experience – both important image drivers – might be one main reason for their resistance. Another fear of the companies is product piracy and the high number of imitations that are still very common in the luxury sector [37], [39].

According to Bain & Company (2011) [2] there is a particularly fast growth of the e-commerce channel in China because luxury consumers are younger than in other markets.

The consumers in the United States have also widely accepted the internet as a shopping channel for luxury goods [17]. Hence, company owned e-commerce channels gain in importance if the makers want to control the distribution to the end customer ([37]). Brands like Armani, Louis Vuitton, and Gucci have already entered the online shopping market.

In the course of the "trading up" and "trading down" development, online shopping clubs like Vente-Privée or Brands4Friends become more and more popular. As the acceptance of e-commerce in this sector increases, luxury companies even include the shopping club sales in their production planning [34]. The following chapter deals with consumer behavior in these shopping clubs by presenting results of an exploratory online survey.

11.4 Luxury Orientation of Consumers in Online Shopping Clubs

Online Shopping Clubs (or: Private Sales Clubs) are commercial online communities which offer "new luxury" brands (e.g. Calvin Klein, Guess, or GAP) at discounted prices. These communities sell current collections from over-stock productions, off-season articles as well as special collections [7], [25]. Thereby, brand manufacturers benefit from a nearly hassle-free clearance while the turnover increases. These exclusive sales clubs also make it possible for luxury companies to help potential customers getting acquainted with luxury goods and to increase the brand's recognition [53]. Vente-Privée, the world's first online shopping club, and most of its competitors have reported double-digit growth rates in the last years [34]. The club has ca. 800.000 members solely in Germany who generated a turnover of nearly 71 million Euros in 2010 [50].

Customers are required to be registered as members to gain access to sales promotions. In this way, a certain exclusiveness arises which gives promoted brands an additional appeal and also counters a dumping of the brand [34]. In every instance, the promotions are limited in time and stock. Therefore, members constantly receive information on current sales promotions via email.

The three largest private sales clubs Vente-Privée, BuyVIP and Brands4Friends [7] are subject of our questionnaire survey. The following table summarizes relevant facts of these communities.

Table 11.1 Company Facts of Vente-Privée, Brands4Friends, and BuyVIP

	Vente-Privée	Brands4Friends	BuyVIP
Foundation	2001 in Paris (France)	2007 in Berlin (Germany)	2006 in Madrid (Spain)
Employees	> 1.300	> 200	> 130
Markets	Europe (A, D, E, F, GB)	Europe (A, D) Japan	Europe (A, D, E, I, NL, P, PL)

	Vente-Privée	Brands4Friends	BuyVIP
Membership	> 13 million	ca. 4 million	ca. 6 million
Sales (2010)	ca. 969 million Euros	ca. 250 million Euros	ca. 100 million Euros
Promotions (p.a.)	> 3.770	> 1.800	> 520
Brands	> 1.200	> 1.000	> 500
Discount	50-70 %	up to 70 %	30-70 %
Registration	Website Registration; Invitation	Waiting List; Invitation	Website Registration; Invitation
Duration of Promotion	2-4 days	2-3 days	1-5 days
Special Features	VP Lounge VP Gift Mobile App for Smart Phones Forum	Blog Fashion Magazine	VIP status Super Sale BuyVIP Brand Shop Blog Fashion Magazine

Source: Vente-Privée, 2011b; Brands4Friends, 2011a; BuyVIP, 2010, 2011

11.4.1 Study Design

Online Shopping Communities enable consumers to afford new luxury and premium goods for reasonable prices and, hence, give them access to the world of luxury. Prior empirical research on the subject has been limited, for example, Heitmeyer and Naveenthirarajah (2010) [25] published an excerpt on a Brands4Friends customer segmentation, but the essay does not provide any relevant information regarding motives, shopping patterns, or consumer attitudes. Hence, the objective of the questionnaire study was to outline the consumption behavior of shopping community members and to understand their motivation for luxury shopping at discounted prices.

In this study an e-mail supported web-survey was implemented. 1700 members of an online access panel were invited to the survey by e-mail. The link to the questionnaire was additionally posted on social network websites (e.g. Vente-Privée's fan page on Facebook). The survey was carried out over a period of four weeks in August and September 2011. Vouchers for an online shop served as an incentive. Altogether 328 probands took part in the study, whereas 274 questionnaires had been completely filled out and were used for the analysis.

Figure 11.1 Simplified Structure of Standardized Questionnaire

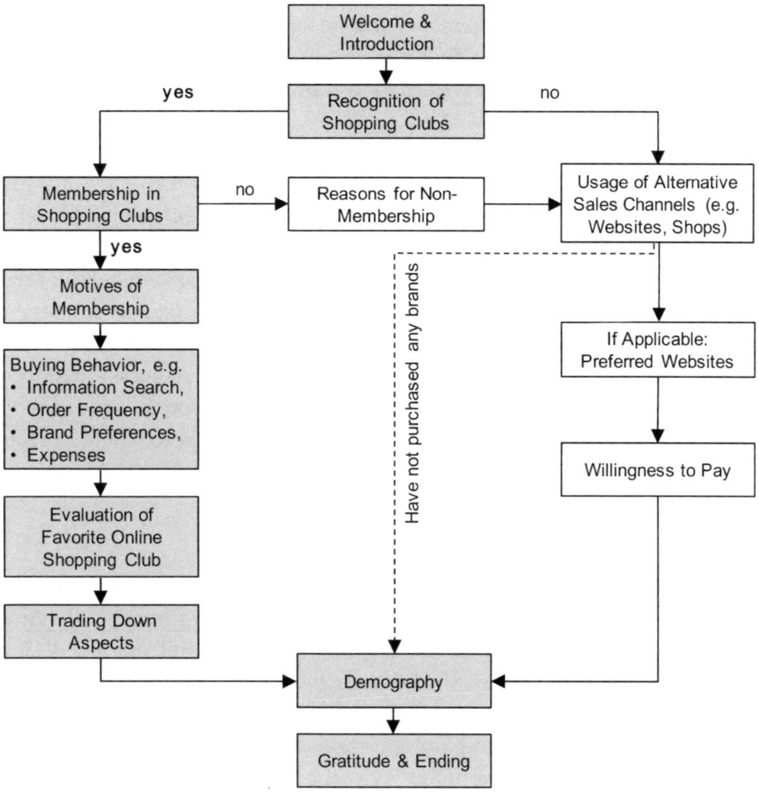

Source: Adapted from Tietjen (2011), p. 53

The questionnaire included several filter questions, for example, regarding recognition of or membership in shopping clubs. Participants who indicated to be shopping club members had to answer far more complex questions and, therefore, needed relatively more time to fill out the questionnaire (9 min. on average). If probands did not know any shopping clubs, they were asked to name the sales channels where they usually purchase premium or luxury brands and to quantify their willingness to pay. If shopping clubs were known but not used as a sales channel, respondents had been additionally asked to reveal their reasons not to join a shopping community. In the case that premium or luxury brands had never been purchased, probands were directly guided to the demographic part in the end of the questionnaire. Figure 1.1 illustrates the structure of the questionnaire. In order to understand behavioral patterns and motives of shopping community members, they were asked to fill out multiple item batteries with five-point rating scales. The items had been collected by literature research and several explorative interviews with shopping club members. The most relevant results will be presented and discussed in the following chapter.

11.4.2 Results and Discussion

Out of 274 analyzable questionnaires, 53 respondents indicated to be a member of at least one online shopping club. Their statements will be subject of the analysis. Due to the small sample, these results can only give a first exploratory impression of the buying behavior in online shopping clubs but cannot be regarded as representative.

Profile of shopping club members

Almost two-third (64%) of the shopping club members are women, 36% are men. The respondents in our sample are between 20 and 62 years old (average age: 31 years). Similar distributions are revealed in the Brands4Friends customer profile. Their customers are described to be 35 years on average; 69% of them are women, 31% men [10]. Regarding the marital status, the shopping club members show a high percentage of singles / unmarried persons (77%). The monthly net income is between 1001-1500 Euros on average which equals that of the whole sample (n = 274). The sample of club members mainly consists of salaried employees (43%) and students (30%). This distribution is similar to that of the whole sample (47% employees, 31% students). However, the percentage of club members who earn less than 500 Euros is peculiarly low (6% compared to 21% of non-members).

Consumer Behavior in Online Shopping Clubs

74% of the members regularly read the club's newsletters in order to be informed about current sales promotions. Nearly one half of the respondents directly receive their information from the club's website. Almost everyone (94%) shops for oneself, 40% of the purchases are done for family members and 34% for the partner. Thereby, apparel is the category that is purchased the most (83%), followed by shoes (49%), and accessories (45%). 85% indicated that they buy premium and luxury brands also in classical online shops or in local stores. Even though, 42% of the respondents do not only buy brands for discounted prices, customers of online shopping communities expect a discount of 40% on average. Only 6% would also accept non-discounted goods in shopping clubs. First and foremost, the respondents seem to be shopping club members because of the attractive sales promotions.

Every community member of our survey spends between 51 and 100 Euros on average per order. Most members (ca. 57%) only order one to four times a year, however, nearly 20 percent of the respondents have indicated to order more than once a month. A comparison of female and male members does not reveal any significant disparities in shopping frequencies.

Figure 11.2 illustrates the results of a multiple item scale and shows that the consumer behavior of club members is, to a high degree, characterized by spontaneous purchases. The business model of the online sales clubs virtually encourages quick orders due to their temporary sales promotions. Furthermore, the respondents do not pre-plan future purchases in the long run either, since sales promotions are announced shortly before. However, it also creates the risk that customers easily become frustrated when articles are sold out after a short time, which is reflected in the ratings of our respondents. Finally, club members of our study highly score regarding multiple purchases to save on shipping costs.

Figure 11.2 Consumer Behavior in Online Shopping Clubs

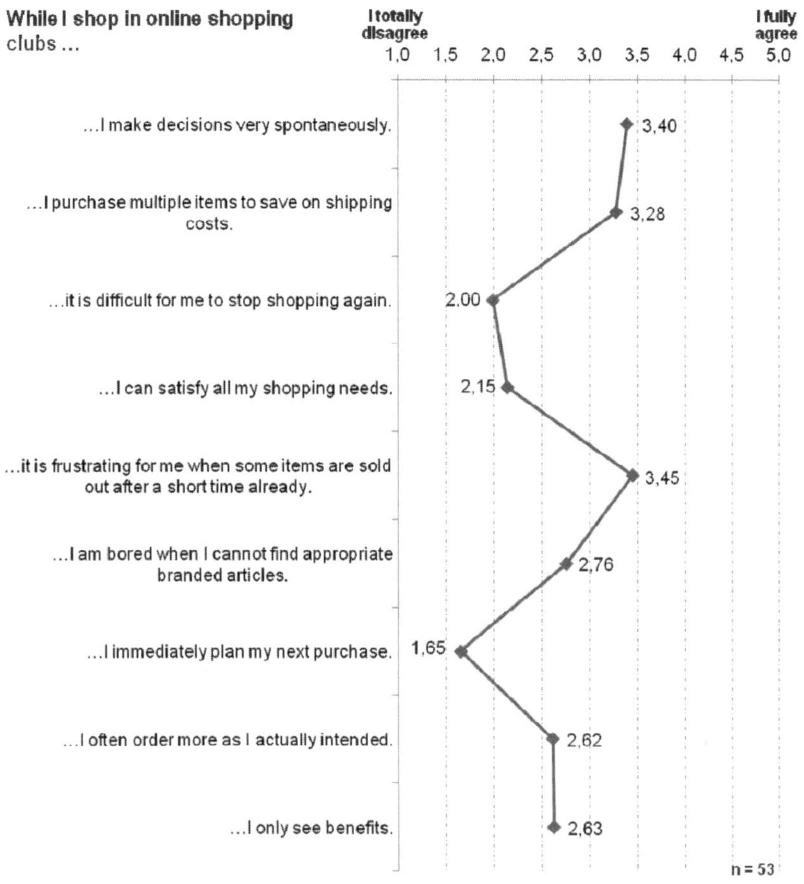

Source: Adapted from Tietjen (2011), p. 72

Motives for Membership

64% have become aware of online shopping clubs through friends, 42% through internet advertisement and ca. 9% through print media. When online shopping club members are asked what the reasons for their membership are, it becomes apparent that practical motives such as convenience, promptness, discounts, and quality prevail (**Figure 11.3**). However, online shopping clubs also appear to be a fun experience for the members. Factors like building self esteem through branded articles, joining an exclusive club, or pursuing a fancy lifestyle seem to be less important.

Figure 11.3 Reasons for Membership

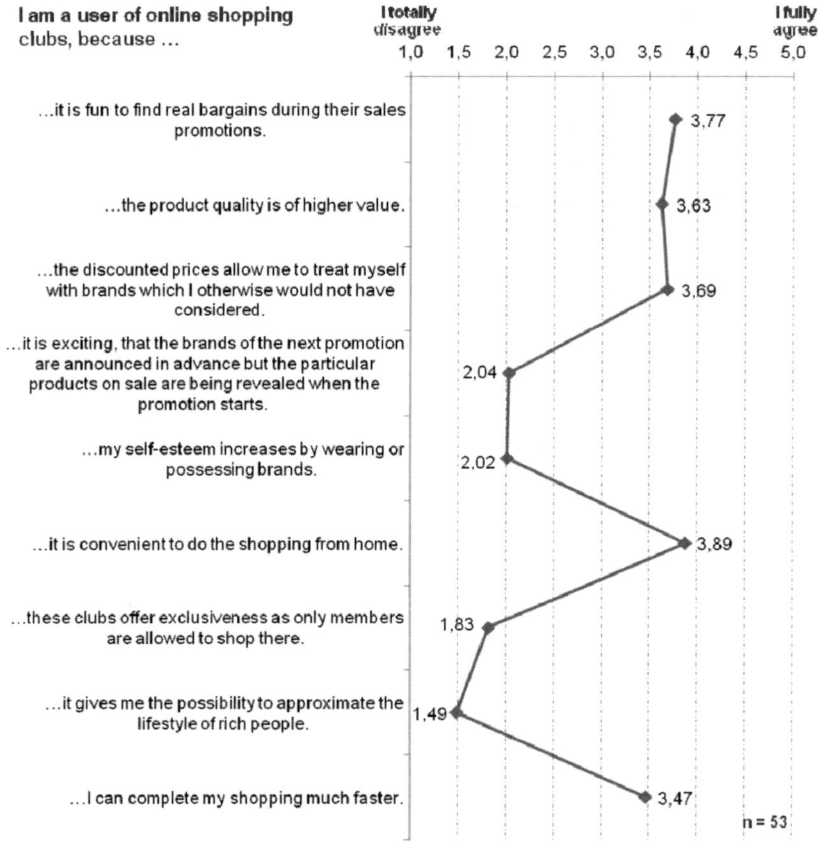

Source: Adapted from Tietjen (2011), p. 71

Furthermore, the respondents were asked to evaluate their favorite shopping club regarding certain aspects. According to the ratings, a well-arranged and appealing design of the website seems to be the most important factor for club members. Beyond that, a good choice of brands, customer-friendly delivery and returning conditions, a regular e-mail newsletter, and a detailed description of the items are highly appreciated.

The reasons why the other respondents of our sample do not join a shopping community are mainly that they generally have no interest in shopping clubs (more than 60%). Further reasons are that the choice of brands does not appeal to them (14%), the assortment is too expensive (13%), or they simply try to avoid e-mail registrations (13%).

Trading Down Categories

In order to allow for the trading down phenomenon, as described in section 1.2, relevant aspects of this trend were integrated in the questionnaire. The chart (**Figure 11.4**) illustrates that besides ordering in online shopping clubs, the respondents of our sales club sample mainly save money by abstaining from information sources that are not free of charge. This result may not be surprising as the internet penetration even further increases due to mobile solutions like smart phones. Additionally, our sample scored fairly high in a third category which is "less expensive furniture".

Figure 11.4 Trading Down Categories

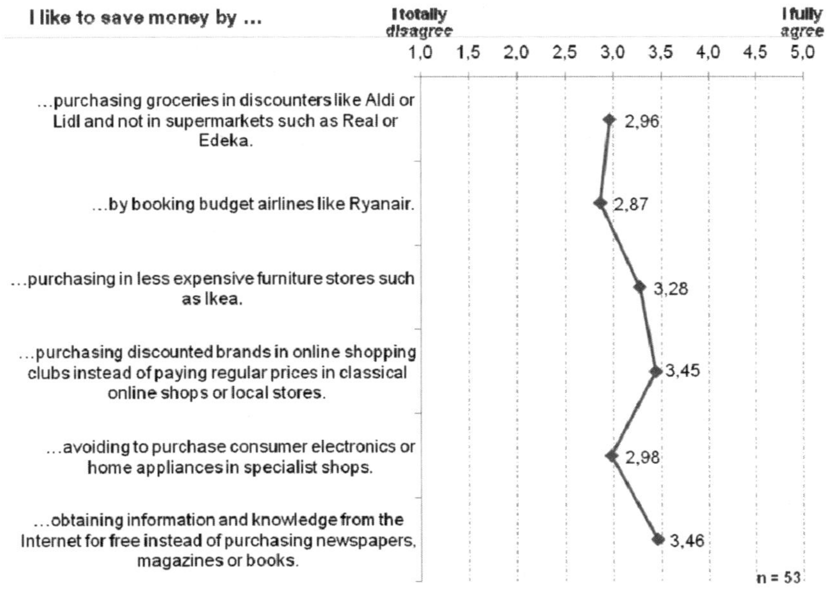

Source: Adapted from Tietjen (2011), p. 74

11.5 Implications

The trading up and down phenomena have had a considerable impact on the consumption behavior in both developed and developing countries over the past decade. Along with the internet, it has changed purchase patterns and created new market potentials [43]. Hence, the luxury industry is in a state of flux. Despite a recession culture, suggested by some authors, the trading up trend and the increasing consumption of new luxury products in the middle-market leads to a general growth of the luxury sector. The business model of "online shopping clubs" clearly benefits from this trend.

11.5.1 Research Implications

The primary goal of this paper was to examine the prospects of luxury marketing with regard to the trading down trend in society which Bosshart (2004) [8] calls the "Age of Cheap". It was shown that consumers nowadays save spending in product categories that are less important to them in order to afford premium and luxury items in other categories. A consequential consumer trend is reflected in the success of online shopping communities in which new luxury goods are sold at greatly reduced prices and, therefore, are affordable for the middle-class segment. An exploratory empirical study was carried out to analyze the behavior, motives and background of online shopping club customers. Based on literature research, an online questionnaire was developed and accessible for a period of four weeks.

As a result of the online survey, it turned out that about half of all respondents know at least one online shopping club and nearly 20% of all respondents already ordered products there. The club members are relatively young consumers with a major proportion of women. Main reasons for joining such a shopping club are, among others, the expectations to get a good bargain, the convenience, and the high quality of products at acceptable prices. In general, apparel, shoes, and accessories are the most important product categories for members of shopping clubs in the internet. A slight tendency towards trading down is visible in categories like groceries, traveling, and consumer electronics. A stronger trading down bias, however, is indicated in the categories knowledge, fashion, and furniture. Given that our study cannot be regarded as representative, follow-up surveys should be carried out on a much broader empirical base.

11.5.2 Implications for New Luxury Marketing

From a managerial point of view, our study may enhance the understanding of motives and consumer behavior of shopping club members. Our results show, for instance, that the motives for purchasing new luxury goods in these shopping communities differentiate to a great extent compared to the motives of traditional luxury consumers. According to the answers of our respondents, branded items are not purchased to raise self esteem, as Gonnermann et al. (2008) [23] suggest, or to pursue a fancy lifestyle. Moreover, the exclusivity, that online shopping clubs want to imply by their registration policies, seems to be less important for their members. The e-mail registration is, on the other hand, a barrier why some consumers do not order in these clubs. However, the exclusive atmosphere may not be of importance for the customers but it avoids a transparency of the brand's sales promotions in the internet and, hence, a dilution of the brand image.

The consulting firm Bain & Company predicts a growth of the luxury sector of six to seven percent until 2014 [24]. Despite an economic crisis, short budgets, and a continued enfolding of the "Age of Cheap", the demand for luxury goods not only remains but even increases. New luxury companies enlarge their business field, for instance, by selling their products on the internet through online shopping clubs and therefore, offer consumers the opportunity to treat themselves with luxury goods at accessible prices. Nevertheless, this business model can only be beneficial for luxury companies if the online sales promotions do not begin to damage their brand image.

Acknowledgements

The authors are grateful to Michael D. Hachar for the care with which he reviewed the manuscript.

References

[1] Alvarez, G./Kemanian, V./Malnight, T. (2004): Keeping The Strategic Balance In The New Luxury Market, http://www.latitudeglobal.com/documents/ceye-sep04-malnight.pdf, accessed Nov. 2011.
[2] Bain & Company (2011): Luxury Goods Worldwide Market Study, May 2011.
[3] Bauer, M./von Wallpach, S./Hemetsberger, A. (2011): ‚My Little Luxury': A Consumer-Centred, Experimental View, in: Marketing ZFP, Vol. 33, 2011 (1), pp. 57-67.
[4] BCG (2011): Luxusmarkt ist knapp eine Billion Euro schwer und wächst weiter, http://www.bcg.de/media/pressreleasedetails.aspx?id=tcm:89-72323, accessed Oct. 2011.
[5] Bellaiche, J.-M./Mei-Pochtler, A./Hanisch, D. (2010): The New World of Luxury – Caught between Momentum and Lasting Change, The Boston Consulting Group, Dec. 2010.
[6] Billharz, M./Belz, F.-M. (2008): Öko als Luxus-Trend: Rosige Zeiten für die Vermarktung „grüner" Produkte?, Marketing Review St. Gallen, 2008, No. 4, pp. 6-10.
[7] Boermsa, T. (2010): Warum Web-Exzellenz Schlüsselthema für erfolgreiche Händler ist: Wie das Internet den Handel revolutioniert, Heinemann, G./Haug, A. (eds.): Web-Exzellenz im E-Commerce: Innovation und Transformation im Handel, Wiesbaden 2010, pp. 21-41.
[8] Bosshart, D. (2004): Billig – Wie die Lust am Discount die Gesellschaft verändert, 2nd ed., Frankfurt 2004.
[9] Brands4Friends (2011a): Homepage, www.brands4friends.de, accessed Nov. 2011.
[10] Brands4Friends (2011b): Fact Sheet – brands4friends, http://presse.brands4friends.de/pressemitteilungen/167-fact-sheet, accessed Nov. 2011.
[11] Bütikofer Schwalbe, R. (2007): Lifestyle of Health and Sustainability – LOHAS, Persönlich, 2007 (8), pp. 42-46.
[12] Büttner, M./ Huber, F./ Rieger, S./ Vollhardt, K. (2008): Phänomen Luxusmarke, 2nd ed., Wiesbaden 2008.
[13] BuyVIP (2010): Amazon unterzeichnet Vertrag zur Übernahme von BuyVIP.com, Pressemitteilung, http://press.buyvip.com/de/news/2010-10-07/amazon_buyvip.pdf, accessed Nov. 2011.
[14] BuyVIP (2011): BuyVIP Presse, http://press.buyvip.com/de/index.htm, accessed Nov. 2011.
[15] Costa, C./Dalens, F./Jacobsen, R./Ramso, J. (2006): The Battle for Europe's Grocery Shoppers, Boston Consulting Group, Aug. 2006.
[16] Dubois, B./Laurent, G. (1995): Luxury Possessions and Practices: An Empirical Scale, European Advances in Consumer Research, Vol. 2, Hansen, F. (ed.): Association for Consumer Research, pp. 69-77.

[17] Focus (2009): Der Markt der Luxusgüter: Daten, Fakten, Trends, http://www.medialine.de/media/uploads/projekt/medialine/docs/bestellung_download/marktanalysen/2009/foc_ma_luxusguet er_200908.pdf, accessed Nov. 2011.
[18] Fritz, W. (2004a) Die Aldisierung der Gesellschaft, bdvb aktuell 85 (2004), pp. 7-8.
[19] Fritz, W. (2004b): Internet-Marketing und Electronic Commerce, 3rd ed., Wiesbaden 2004.
[20] Fritz, W./Lorenz, B./Hauser, U. (2007): Die Discountisierung der Gesellschaft, Gernsbach 2007.
[21] Fritz, W./Lang, F. P. (2008): Der Discountisierungstrend in Wirtschaft und Gesellschaft, Weber, L./Lubk, C./Mayer, A. (eds.): Gesellschaft im Wandel, Wiesbaden 2008.
[22] GfK (2011): Consumer Index – Total Grocery: "Der Osterhase verspätet sich", 2011 (3), http://www.gfkps.com/imperia/md/content/ps_de/consumerindex/ci_03-2011_extern.pdf, accessed Nov. 2011.
[23] Gonnermann, J./Geiss, M./Henning, G. (2008): Luxusmarketing- alles anders?, Marke, 2008, Vol. 41, pp. 60-65.
[24] Handelsblatt (2011): Luxus liegt weiter im Trend – Nachfrage aus China boomt, Handelsblatt, No. 226, Nov. 22, 2011, p.30.
[25] Heitmeyer, C./Naveenthirarajah, S. (2010): Online Customer Segmentation in Shopping-Clubs: Auf dem Weg zur ultimativen Kundenorientierung bei brands4friends, Heinemann, G./Haug, A. (eds.): Web-Exzellenz im E-Commerce: Innovation und Transformation im Handel, Wiesbaden 2010, pp. 71-91.
[26] Hollmann, F. (2011): Paradies für Luxusgüter, absatzwirtschaft Sonderheft 2011, pp. 44-45.
[27] Horx, M. (1995): Trendwörter von Acid bis Zippies, 2nd ed., Düsseldorf et al. 1995.
[28] ILTM (2011): The Future of Luxury Travel: A Global Trends Report, June 2011, http://www.iltm.net/files/the_future_of_luxury_travel_report.pdf, accessed Nov. 2011.
[29] Ko, E./Sung, H. (2007): "Trading up": A Consumption Value Approach, Advances in International Marketing, Vol. 18, pp. 115-137.
[30] Koschel, K./Rademacher, U. (2005): Der Reiz der wahren Werte, Das Magazin von HORIZONT, 2005, Vol. 6, pp. 40-41.
[31] Kotler, P. (2011): Reinventing Marketing to Manage the Environmental Imperative, Journal of Marketing, Vol. 75 (July 2011), pp. 132–135.
[32] Kowalsky, M. (2009): Perspektiven: Die Zukunft des Luxus, Das Schweizer Wirtschaftsmagazin BILANZ, 2009 (19), http://www.bilanz.ch/luxus/perspektiven-die-zukunft-des-luxus, accessed Nov. 2011.
[33] Kull, S. (2004): Aldi und die „Aldisierung". Ein Trendwort verkennt die starke Marke, http://www.jade-hs.de/fileadmin/fb_wirtschaft/downloads/Kull/arbeitspapier1-aldisierung.pdf, accessed Nov. 2011.
[34] Krumrey, K. (2010): Chronischer Schlussverkauf, Der Spiegel, 2010, Vol. 42, p. 117.
[35] Lietz, H. (2010): Neue Perspektiven des Luxus, http://www.weise-kommunikation.de/files/amex_marketing_yearbook.pdf, accessed Nov. 2011.
[36] Merkle, W. (2008): Mango und Zara – Besonderheiten der neuen vertikalen Anbieter im deutschen Textileinzelhandel, Riekhof, H.-C. (ed.): Retail Business in Deutschland, 2nd ed., Wiesbaden 2008, pp. 437-456.
[37] Merten, H.-L. (2009): In Luxus investieren – Wie Anleger vom Konsumrausch der Reichen profitieren, Wiesbaden 2009.
[38] Michael, B.M. (2008): Retail Business: Grenzen der „Geiz ist geil"-Strategien, Riekhof, H.-C. (ed.): Retail Business in Deutschland, 2nd ed., Wiesbaden 2008, pp. 147-178.
[39] Okonkwo, U. (2009): Sustaining the luxury brand on the Internet, Journal of Brand Management, Vol. 16, No. 5/6, pp. 302-310.
[40] Park, J./Ko, E./Kim S. (2010): Consumer Behavior in Green Marketing for Luxury Brands, Journal of Global Academy of Marketing Science, Vol. 20, No. 4, pp. 319-333.
[41] Reischauer, C. (2011): Stresstest für die gute Sache, absatzwirtschaft Sonderheft 2011, pp. 74-77.
[42] Roche, C./Ducasse, P./Liao, C./Grevler, C. (2010): A New World Order of Consumption: Consumers in a Turbulent Recovery, 2010 BCG Global Report on Consumer Sentiment, The Boston Consulting Group, June 2010.

[43] Roche, C./Silverstein, M.J./Charpilo, N. (2008): Trading Up and Down around the world, The Boston Consulting Group, Sept. 2008.
[44] Silverstein, M.J./Fiske, N. (2003): Luxury for the Masses, Harvard Business Review, Vol. 81, No. 4, 48-57.
[45] Silverstein, M.J./Fiske, N. (2004): Trading Up: Trends, brands, and practices – 2004 Research Update, The Boston Consulting Group, May 2004.
[46] Silverstein, M.J./Fiske, N./Buttman, J. (2005): Trading Up –Why consumers want new luxury goods – and how companies create them, New York 2005.
[47] Tietjen, A. (2011): Online Shopping Clubs als aktueller Trend im Luxusgütermarketing – Eine empirische Analyse, Diploma Thesis, Institut für Marketing (TU Braunschweig), Nov. 2011.
[48] Truong, Y./McColl, R./Kitchen, P.J. (2009): New luxury brand positioning and the emergence of Masstige brands, Journal of Brand Management, Vol. 16, No. 5/6, pp. 375-382.
[49] Twitchell, J.B. (2002): Living It Up: Our Love Affair with Luxury, New York 2002.
[50] Vente-Privée (2011a): Presse – Unsere Geschichte, http://pressroom.vente-privee.com/de-DE/History.aspx, accessed Nov. 2011.
[51] Vente-Privée (2011b): Press Room, http://pressroom.vente-privee.com/?sc_lang=de-DE, accessed Nov. 2011.
[52] Wenzel, E./ Kirig, A./ Rauch, C. (2007): Zielgruppe LOHAS – Wie der grüne Lifestyle die Märkte erobert, 1st ed., Zukunftsinstitut 2007.
[53] Wiedmann, K.-P./Klarmann, C./Hennigs, N. (2011): Luxusmarken im Zeitalter des Web 2.0, Wagner, U./ Wiedmann, K.-P./Oelsnitz, D. von der (eds.), Das Internet der Zukunft – Bewährte Erfolgstreiber und neue Chancen, Wiesbaden 2011, p. 77-96.
[54] Wiedmann, K.-P./Siebels, A. (2005): "Geiz ist geil" oder führt er nicht doch zu einem problematischen Wertezerfall? Aktuelle Tendenzen in den Konsumstilen und Skizze zentraler Problemstellungen eines markenwert-zentrierten Risikomanagements, 2nd ed., Schriftenreihe Marketing Management, Universität Hannover, 2005.

12 Luxury Brands in the Digital Age – the Trust Factor

Meng-Shan Wu, Cheng-Hao Chen, Isabella Chaney

12.1 Introduction

Luxury brand marketing has been suggested as one of the fastest growing industries [49] with several research institutions routinely reporting on the market emphasising its considerable value. In a Verdict report, it is predicted that the global market for luxury branded goods will be worth £225bn by 2012 [14]. Until recently luxury brands were the preserve of affluent people from privileged backgrounds but with rising incomes and availability of credit, luxury brands have become more affordable to a wider range of consumers than previously [47], [48].

Luxury brands offer the purchaser a high-quality product coupled with the emotional benefits of prestige and exclusivity. However, when it comes to the purchase situation not all luxury brands are delivering the total product and service experience as there remains a gap between expectations and reality of service encounters for luxury purchases with salespeople being described as "unhelpful, intimidating and rude" [7]. Many young luxury brand purchasers have recently turned their back on the retail malls preferring to purchase via the Internet.

Somewhat belatedly, luxury brands are now acknowledging the benefits of the Internet as a communication media and as a means to increase sale volumes. Okonkwo emphasises the importance of incorporating an e-business into the luxury brand's strategy as this medium is likely to be more effective for the younger upwardly mobile generation [34]. With the global Internet population nearing two billion, having an online presence is one area that cannot be dismissed by retailers including those from the luxury sector [46].

Developing an online presence for any company, but especially with high value luxury products requires the brand manager to ensure there is a focus on establishing trust. Furthermore, for the luxury brand there is the additional complication of the abundance of luxury counterfeit products that could lead to consumers' suspicion of the authenticity of a website and its products. The question is how can trust be established in the online marketplace, especially for the luxury branded product with its high price tag? What strategies do luxury brands have to implement to gain trust from online browsers to turn them into purchasers?

12.2 Luxury Brands and E-Commerce

Seringhaus noted that "luxury brands have a myopic view of the virtual retail environment" [41]. Indeed it was suggested that "for luxury brands, the Internet is very likely to fulfil a communication and information role, but less likely to be used as a customer acquisition channel" [37]. More recently the online luxury market has been referred to as being in its infancy with examples of Versace and Prada waiting until 2005 and 2007 respectively to take advantage of this medium [33], [50]. Furthermore, an article in Business Week stressed that "many luxury brands still treat the Internet with caution, worrying an online presence will dilute a sense of exclusivity" [43].

For some luxury brands, their primary connection with online sales involves the strategy of selling last season's luxury goods over the Internet using authorised third party websites. Discount fashion websites such as the private online shopping club Vente-Privee, sell unsold inventory from luxury retailers, at savings of up to 70 per cent. A tightly controlled membership system results in the creation of an air of exclusivity counteracting any perception of second rate products [51]. For the luxury brand the strategy of using third party websites is an arm's length approach of disposing of last season's products without tarnishing the brand name.

In 2010 Bernstein Research noted a small number of luxury brands were increasing their online efforts with Burberry leading the way by allocating one quarter of their global advertising spend to digital media [4]. Burberry is also cited as one of the most active investors of social media among the luxury brands a strategy of importance when targeting the increasingly affluent 25-44 year old prospective customer with their runway shows streamed live to over 140 countries generating around £8 million of paid-for-media [4], [13], [36]. Other luxury brands, however, are not fully embracing the Internet both as a sales platform and as social media. A search of the UK websites reveal that a minority of the top ten luxury brands sell products on their sites preferring instead to use it as a promotional tool. The use of Facebook, youtube and Twitter is limited with some sites not incorporating a direct link to their social media page.

Over ten years ago in an article titled 'Online Luxury Has Limits', it was predicted that online marketing of luxury goods would meet with failure for two main reasons [32]. Firstly, it was suggested that the store atmosphere and the 'feeling special' element created by well trained sales assistants could not be replicated in the online environment. Luxury brands invest heavily in two main areas of retailing: 1) developing a store's atmosphere to induce a sense of exclusivity and 2) developing and training of sales personnel in the art of making each customer who walks through the door feel extra special. Whether the hedonic experience of the luxury store can be transferred to the Internet by the use of creative art and photography is a major challenge and has resulted in many luxury brands playing a wait and see game before fully committing all their product lines to e-commerce especially after witnessing the failure of the pioneering LVMH-owned e-commerce site eLUXURY after nearly a decade's online presence.

The second reason to limit the applicability of the Internet to luxury brands is the problem of 'trust' both in the form of internet security as well as trust in the process [32]. The development of trust in a website is crucial as it has been shown to be a mediator in the relationship between website attributes (privacy, security, information design and communication) and purchase intention [19]. In the luxury goods sector developing trust is more complicated as the Internet has "spawned a lucrative trade in counterfeit goods" [2]. Even as far back as 2001 companies were being indicted for selling counterfeit luxury goods over the Internet and more recently eBay was fined for failure to prevent sales through its website of counterfeit luxury products purporting to be original Louis Vuitton, Christian Dior and Hermès [35], [45]. Recognising that luxury brand owners require Intellectual Property specialists, Burberry has an active eBay monitoring policy that has resulted in exposing several prosecutions [2]. Other luxury brands such as Louis Vuitton are similarly aggressive in their approach employing large numbers of lawyers full-time to take legal action globally [5]. Even though there have been several headline prosecutions there still remains an abundance of counterfeit luxury products and this "has brought many luxury brands to consider the Internet as a risk" [4]. Moreover, when confronted with various websites selling the same luxury product the consumer may also feel uncertain as to whether a site is in fact the genuine luxury brand's website. Incorporating cues that engender trust in the brand and its website are all important in such a difficult environment.

12.3 Developing Trust in a Luxury Brand Website

Trust plays an important role in many social and economical interactions involving uncertainty and dependency, and is one of the crucial factors that influence a successful e-commerce implementation. Lack of trust has been repeatedly identified as one of the most formidable barriers for people to engage in e-commerce activities [8] and hence building and sustaining consumer trust in the virtual environment is vital for a company to succeed in the digital age. In the case of luxury brands with their high price tag, consumer uncertainty mainly revolves around financial risk. In particular whether the website is owned by the genuine brand and if so, if there is dissatisfaction with the expensive purchase is there an easy and reliable process to return the goods?

It is human nature for consumers to be suspicious while engaging in online shopping. This may be because, in "traditional" business transactions, buyers can interact with sellers and physically touch the products, whereas in most online shops consumers are unable to interact personally with the online retailers and can only rely on the photos and descriptions provided. Therefore, the nature of B2C e-commerce compared with the traditional face-to-face market leads to transaction risks relating to the uncertainty of the online retailer's identity and product quality. It is important to reduce the barriers to purchase, in other words; it is important for the online vendors to develop a trusting relationship in order to maintain consumer loyalty.

In order to create a strong trusting relationship between online luxury seller and consumers, the following factors should be considered [9]:

Table 12.1 Trust Developing Factors

Component	
Usability	Navigation Download speed Ease of use (User friendliness)
Information content	Privacy/Security policies Company information Product Information Company Policies Contact Details
Technological Professionalism	Professionalism Competence
Aesthetics	Overall attractiveness Consistency White space Readability (colour contrast, clear font, etc) Page length Colours Graphics
Safety	Payment system Privacy measures Security measures
e-CRM	Communication Channels After sales services
Order Fulfilment	Delivery time Tracking number

Source: Chen, 2009

12.3.1 Usability

The usability aspect of an online seller's website refers to certain features which make using the site effortless, fast and logical, thus either directly or indirectly creating an impression among the users that the seller is professional, capable and trustworthy [10], [38]. Studies into user Internet behaviour recommend clear, consistent navigation to improve both sales and site traffic [11]. Navigation communicates trustworthiness as it conveys respect for customers by showing that the company is committed to providing a clear, fast service for the "time-deprived" consumer [3]. To promote ease of use certain conventions are suggested such as ways of finding different types of information and accomplishing particular tasks, e.g. groups of related items or in sequences of steps [30]. These conventions make the consumer feel comfortable by making the situation seem normal and familiar. Familiarity, in turn, through situational normality, increases trust [29].

12.3.2 Information Content

Luxury brands should ensure there is sufficient information on the company, products, terms and conditions, privacy and contact details, and that it is clearly written to put the consumer's mind at ease. The provision of comprehensive, current, personalised, and community-specific information, which is sufficient in both depth and width, is crucial in gaining trust [26]. Web design guidelines emphasise the importance of well-written information, and the absence of typographic errors, and more importantly jargon-free content to ensure users are not deterred from the site nor discouraged from bookmarking [17].

Expressed in bureaucratic and legalistic jargon, privacy policies are often buried deep in the website, not to be found and understood by the user [25].Yet, explaining why information is gathered and used, and whether consumers' anonymity is at risk, promotes trust [25], [15]. The importance of this is demonstrated in Keen's description of privacy policies as a "matter of dialog and explanation – shared understanding between your company and its customers" [25]. A complete, comprehensible privacy policy conveys to consumers that the seller conforms to a code of ethics and has nothing to hide, in turn increasing the likelihood of an exchange of personal information.

Whereas detailed, technical explanations of security decrease the level of consumer trust, a clear, concise explanation of the security measures promotes trust [10]. It is important to explain encryption methods and other techniques with which the users are familiar, perhaps through the use of metaphors and textual information, and provide clear visibility of the security techniques employed [15]. The emphasis is on providing convincing, non-technical descriptions of the technical security solutions, thereby showing consumers that security has been "taken care of" and assuring them that the site is trustworthy.

The existence of a "FAQ" section has similarly been claimed to promote trust [6], [27]. Firstly, this section is perceived as evidence that companies are taking an interest in assuring consumers' well-being [6]. Secondly, it is believed that some consumers use it as an indication of size, which, in turn, has been shown to be positively associated with perceived trustworthiness [6], [23].

Providing information about the company, its history, founders and employees has been suggested to promote credibility and trust [15], [27]. Not only does providing company information facilitate familiarisation with the company, but it also assures the consumer that the e-retailer is a real company as opposed to an unreliable firm, thereby enhancing the possibility of relationship building between the consumer and the retailer. A study of luxury brand websites revealed a disregard for the consumer and their information needs. Company information was available in only half of the websites while a help section was included in less than one fifth of sites [37] a situation that needs addressing to fully exploit the benefits of luxury brand e-commerce.

Complete, accurate product information also establishes credibility and trust in an e-retailer [27]. Descriptive, convincing information projects passion about, and interest and knowledge in, the seller's subject area, thereby communicating expertise. However, for product information to be "good" and trustworthy, it must be correct and up to date.

Finally, the provision of contact details encourages trust as it provides a link to the company thus showing the customer that the company is legitimate with a physical location [21]. Providing an e-mail address, toll-free help lines and a physical address may help to establish a relationship with the customer, which, in turn, promotes not only trust but also long-term retention [17].

12.3.3 Technological Professionalism

Professionalism is a compilation of several website features and, as the opposite of amateurism, promotes trust and credibility [17]. The factors that are regarded as having the greatest impact on perceived retailer trustworthiness are technological complexity and the importance of being up to date. Although ignorant of details, consumers are aware of the risks, and thus a trust-promoting website should exude a technical prowess. This is often communicated in two ways: first, through a good overall design, which then shows the consumer that the retailer has sufficient knowledge of and control over the underlying technology involved in the purchase process, and secondly, by assuring the consumer of the retailer's ability to safeguard his/her private information [9] particularly important in the case of wealthy luxury brand consumers.

Maintaining an up to date website has been recognised by retailers, who now advertise their date of last site modification. Simultaneously, the use of "what's new" and "coming soon" sections communicates dynamism and freshness. By encouraging consumers to return to the site, these sections also promote familiarity and, thereby, the formation of trust [21]. Secondly, the home page must be up to date in terms of the technology employed. Although web designers are justifiably wary of using the latest "gadgets" and rightly design pages for the masses, using outdated security techniques, which may/may not have been proven unsafe, indicates an inability to handle the transaction, and privacy and security aspects.

12.3.4 Aesthetics

Trust building is strongly affected by the users' first impressions of a website [15]. As the point of "first look" at the home page plays a pivotal role in gaining and holding the consumers' attention. A convincing website lures customers in, while messy, unorganised pages drive them away [21]. In addition, a sloppy site implies a disregard for potential customers. Following the "KISS" (Keep It Simple and Stupid) principle then the best web pages are "clean", "clear", "relatively simple" and "well laid-out" contributing to the development of trust.

As consumers gain experience in browsing, they become accustomed to certain features. Consistency is important in effective web design and refers to the consistent placing of logos, repeated text, buttons, graphics and navigation tools throughout the website. Consistency should also be employed in the product descriptions and in the chosen colour, and text font and size. It is also a sign of competence and consideration, as it enables users quickly to evaluate the usefulness of the site and match their expectations, while also giving them a sense of control and improving situational normality [10], [19], [39].

The use of colour can also improve page clarity, attract attention, highlight information and facilitate information structure [27]. As well as being part of the aesthetic experience of browsing [28], colours are used to convey the nature of the company behind the website: bright colours usually represent "a fun site", while more serious sites work in shades of blue, grey and brown. In user studies, it has been found that dark colours are unappreciated by users, preferring instead light background colours with high contrast [6].

Since consumers often visit a site after browsing several sites, and since reading from a screen is more difficult and slow than reading from the printed page, web pages should be "spacious". Users get frustrated by, and dislike crammed web pages that offer too much information on one page, or have too many graphics or banner ads [6]. Simultaneously, the use of space enables the user quickly and effortlessly to obtain an overview of the site's content, thereby facilitating navigation.

As the Internet is primarily information based, information should be displayed promptly and clearly. Following the KISS principle, the text should be available immediately, whilst waiting for the graphics to load. It should be clear, strong, and stand out distinctly from the background. For this reason, light colours are ill-suited as text colours. Professionals prefer fonts such as Arial, as these are available on almost all PCs. Although the impact of the text may be greatest in terms of satisfaction, illegible policies do not encourage information sharing or promote trust. Illegibility is also considered to be a sign of amateurism, which, in turn, has been claimed to reduce trustworthiness [21]. Going against the keeping it simple principle a number of luxury brands have employed Flash increasing the time it takes to load a page. This in turn increases the likelihood of frustrating the customer resulting in a quick exit.

12.3.5 Safety

Individual privacy issues have become a critical factor that greatly affects consumers. Privacy and security measures have both been widely adopted by online retailers to ensure the safety of their consumers. There are various tools and techniques which can be used to enhance privacy. In addition to certain techniques, such as passwords, users have the option of deactivating cookies—pieces of information that are sent back and forth between the server and the user and are used to track the consumer's activities. Simultaneously, PGP (US Encryption Software Corporation) is commonly used to encrypt files and e-mail to promote confidentiality, reliability and integrity [40]. Yet, consumers are at the mercy of the retailer in respect to the use and sale of secondary information, and the secure storage of their personal details. As privacy policies are the only way for retailers to express to consumers that details will be kept private, the provision of these policies is vital.

The requirements for security have been recognised by practitioners, and there are already various techniques available that attempt to combat the issue of security threats by providing confidentiality, integrity, availability and non-repudiation. Secure Socket Layer (SSL) can be regarded as the most commonly used security protocol. The security protocol provides data encryption, server authentication, message integrity, and optional client authentication for a TCP/IP connection [18]. 3-D Secure is another common solution used in providing security during transactions. Developed by Visa and adopted by Mastercard, the authentication protocol was designed specifically with e-commerce in mind to provide another layer of security for both the retailer and consumer's peace of mind [1].

12.3.6 Consumer Relationship Management

Consumer relationship management, where users gain enough trust to engage in a commercial relationship with an online retailer, is one of the key areas that most companies are trying to improve. Customer relationship management can also be linked to interface effects by creating a channel for interactive direct communication in the virtual environment or in the real world for consumers. This is important for luxury brands as it is known that timely communication is a factor in generating trust [19]. The use of an online helpline introduces a one-to-one communication vehicle in a similar manner to the personalised experience in the store environment but with the superiority of being available 24 hours per day unlike the limited opening hours of the bricks and mortar store.

Interactivity is crucial in developing a relationship with the customer which in turn creates trust in the company. The CEO of Alexander McQueen was quoted in an article as stating "we've had to become more interactive with customers and have ramped up our activity since the launch of the site" [16]. Kurt Geiger has also invested in developing its interactive features in an attempt to recreate the experience in the retail store [42]. Live podcasts have been undertaken by several luxury companies including Gucci, Christian Dior and Karl Lagerfeld [44].

Developing online brand communities is an intuitive strategy for luxury brands as they have history and tradition that can bring members together and provide a "critical demarcation between users of their brand and users of other brands" [31] and this sense of being special and different to users of other brands can engender loyalty and trust. Cartier was the first in the luxury sector to embrace a brand community strategy by employing a music-focussed community on MySpace to promote brand interaction globally [22]. In 2009 Burberry launched its Art of the Trench social networking site to develop a sense of community within its followers. Drawing like-minded consumers together in such a manner facilitates the development of the community and can be an effective competitive strategy. "A brand related social networking site could be very useful for luxury customers who want recommendations from peers they trust" [24].

The use of social media, however, appears to be a confusing picture with various levels of commitment by luxury brands [36]. For some companies their websites are up front highlighting the social media platforms they employ whereas for others "it is like looking for needle in the hay to find out if they have something" [36].

12.3.7 Order Fulfilment

Order fulfilment is an important factor in developing trust as customers need to have their items delivered on time with complete accuracy in order to be satisfied and loyal. Companies should provide the facility to check the order status as well as informing customers when the purchased product has been dispatched. Should it be necessary to return the product the website should provide an easy to follow returns process as this is important in the development of trust.

12.4 Conclusions

The online purchase of luxury brands by virtue of their high prices might be perceived as a risky undertaking. Developing trust between the company and the customer is an important strategy as it overcomes the key obstacle to making an online purchase. The overall quality of a luxury brand's website and in particular characteristics such as warranties, website security and policies on privacy have been shown to contribute to the consumer's level of trust in purchasing from the site [12], [20]. However, a study of luxury online brands revealed that only a small percentage included term and privacy terms (23 percent and 18 percent respectively) and information on the company was only offered by one quarter of the luxury sites [41].

Having a well designed website that conforms to what the customer expects from such a website i.e. situational normality, contributes to trust [29]. This can be addressed by the inclusion of an uncomplicated, no questions asked product return procedure, overcoming the sense of vulnerability often attached to the online purchase of high value products. Ensuring the entire process of purchasing from the luxury website is easy and straightforward is also important not only for trust-building in the initial purchase stage but also for subsequent purchases as trust in a website has been shown to be a factor of the consumer's satisfaction level with a previous purchase i.e. experience-based trust [29].

Research has highlighted there is a strong association between brand name and trust in the website [39], and this suggests that luxury brands, with their high quality brand image are one step ahead of other brands in the quest for online sales. However, this is not the only factor in trust building and those luxury brands that do not commit to a focus on the four dimensions of usability, information content, technological professionalism and aesthetics might be detering customers. As luxury brands move to change their websites from merely a promotional tool to a sales platform, strategies that create a high level of confidence in purchasing is ever more important.

References

[1] Anon (2011): The Innovate Payments 3-D Secure Hosted MPI is Priced to Minimize Upfront Costs, available www.innovatepayments.com, accessed 10 Nov 2011.
[2] Barraclough, E. (2007): Managing Intellectual Property, April, 168, pp. 42-43.
[3] Bellman, S./Lohse, G.L./Johnson, E.J. (1999): Predictors of online buying behavior, in: Communications of the ACM, Vol. 42, pp. 32-38.
[4] Bernstein Research (2010): Black Book – European General Retail & Luxury Goods: Online – Opportunity or Threat?, Bernstein Global Wealth Management.
[5] Betts, K./Crumley, B./Graff, J./Gibson, H./Gough, N./Israely, J. (2004): The purse-party blues, in: Time, Vol. 164, No. 5, pp. 68-70.
[6] Cardholm, L./Karlsson, S./Karvonen, K. (2000): Trusted E-Business, Userstudy 2000, Helsinki University of Technology.
[7] Cavender, B./Rein, S. (2009): Luxury Goods: Still Strong Sellers, in: China Business Review, March-April, pp. 36-39.
[8] Chen, C. (2006): Identifying significant factors influencing consumer trust in an online travel site, in: Information Technology & Tourism, Vol. 8, pp. 197-214.
[9] Chen, C.H. (2008) Establishment of Trust – The Concepts and Practical Implementations in B2C E-commerce, Ph.D Thesis, Royal Holloway, University of London UK.
[10] Chen, J./Dibbs, S. (2010): Consumer trust in the online retail context: epxloring the antecedents and consequences, in: Psychology & Marketing, Vol. 27, 4, pp. 323-346.
[11] Choi, D./Au, N. (2011): A model of online trust for travel products, in: International Journal of Services Technology & Management, Vol. 15, 1, pp. 132-146.
[12] Christie, J./Kim, S. (2010): Influences of retail brand trust, off-line patronage, clothing involvement and website qaulity on online apparel shopping intention, in: International Journal of Consumer Studies, Vol. 34, 6, pp. 627-637.
[13] Clark, N. (2010): The new luxury market, in: Marketing, 24 Feb, pp. 32-33.
[14] Clark, N. (2011) Luxury: Because you're still worth it, Marketing, http://www.marketingmagazine.co.uk/bullentin/dailynews/article105, accessed 13 June 2011.
[15] Egger, F.N. (2001): Affective Design of E-Commerce User Interfaces: How to Maximise Perceived Trustworthiness, in: Helander, M./Khalid, H.M.Tham (eds.), Proceedings of CAHD2001: Conference on Affective Human Factors Design, Singapore, June 27-29, pp. 317-324.
[16] Farber, A. (2009): Burberry joins push online by luxury brands, in: newmediage, 7 May, pp. 1-2.
[17] Fogg, B.J./Marshall, J./Laraki, O./Osipovich, A./Varma, C./Fang, N./Paul, J./Rangnekar, A./Shon, J./Swani, P./Treinen, M. (2001): What Makes Web Sites Credible? A Report on a Large Quantitative Study, in: CHI2001 Proceedings, Conference on Human Factors in Computing Systems, Seattle, 31 March.
[18] Furnell, S.M./Karweni, T. (1999): Security implications of electronic commerce: a survey of consumers and businesses, in: Internet Research Electronic Networking Applications and Policy, Vol. 9, pp. 372-382.
[19] Ganguly, B./Dash, S.B./Cyr, D. (2009): Website characteristics, trust and purchase intention in online stores: – an empirical study in the Indian context, in: Journal of Information Science and Technology, Vol. 6, 2, pp. 22-44.
[20] Gefen, D./Karahanna, E./Straub, D.W. (2003): Trust and TAM in online shopping: an integrated model, in: MIS Quarterly, Vol. 27, No. 1, pp. 51-90.
[21] Geissler, G.L. (2001): Building customer relationships online: the Web site designers' perspective, in: Journal of Consumer Marketing, Vol. 18, pp. 488-502.
[22] Goldie, L. (2008): Cartier becomes first luxury brand to set up community on MySpace, in: newmediaage, 5 June, p. 3.
[23] Jarvenpaa, S.L./Tractinsky, N./Vitale, M. (1999): Consumer trust in an Internet store, in: Information Technology and Management, Vol. 1, 12, pp. 45-71.

[24] Karimzadeh, M./Corcoran, C.T. (2008): Luxe brands follow the money – to the Internet, in: Women's Wear Daily, Vol. 195, No. 71, p. 12.
[25] Keen, P. (2000): Designing Privacy for your e-business, in: PC Magazine, available www.ZDNet.com.au, accessed 10 Nov 2011.
[26] Kim, S.Y./Lim, Y.J. (2001): consumers' perceived importance of and satisfaction with Internet shopping, in: Electronic Markets, Vol. 11, pp. 148-154.
[27] Lohse, G.L./Spiller, P. (1999): Internet store retail design: how the user interface influences store traffic and sales, in: Journal of Computer-Mediated Communication, Vol. No. 5, 2, p. 0.
[28] Mandel, T. (1997): The Elements of Effective Web Design, John Wiley, Toronto.
[29] Martin, S.S./Camarero, C. (2008): Consumer trust to a web site: moderating effect of attitudes toward online shopping, in: CyberPsychology & Behavior, Vol. 11, No. 5, pp. 549-554.
[30] McGraw, G./Viega, J. (2000): Software Security Principles, IBM developerWorks.
[31] Muniz, A.M./O'Guinn, T.C. (2001): Brand community, in: Journal of Consumer Research, Vol. 27, No. 4, pp. 412-432.
[32] Neuborne, E. (2000): Online luxury has limits, in: Business Week, 18 Sept, p. 1.
[33] Okonkwo, U. (2009): Sustaining the luxury brand on the Internet, in: Brand Management, Vol. 16, No. 5/6, pp. 302-310.
[34] Okonkwo, U. (2010): Luxury Online, Palgrave Macmillan, New York, NY, USA.
[35] Passariello, C./Mangalindan, M. (2008): Ebay fined over selling counterfeits, in: Wall Street Journal – Eastern Edition, Vol. 252, No. 1, pp. 2-5.
[36] Phan, M. (2011): Do social media enhance consumer's perception and purchase intentions of luxury fashion brands?, in: Vikalpa: The Journal for Decision Makers, Vol. 36, No. 1, pp. 81-84.
[37] Riley, F.D./Lacroix, C. (2003): Luxury branding on the Internet: lost opportunity or impossibility? In: Marketing Intelligence & Planning, Vol. 21, No. 2, pp. 96-104.
[38] Roy, M./Dewitt, O./Aubert, B.A. (2001): The impact of interface usability on trust in Web retailers, in: Internet Research, Vol. 11, No. 5, pp. 388 – 398.
[39] Ruparelia, N./White, L./Hughes, K. (2010): Drivers of brand trust in internet retailing, in: Journal of Product & Brand Management, Vol. 19, No. 4, pp. 250-260.
[40] Schneier, B. (1995): Applied Cryptolography: Protocols, Algorithmsm and Source Code in C, John Wiley, New York.
[41] Seringhaus, F.H.R. (2005): Selling luxury brands online, in: Journal of Internet Commerce, Vol. 4, No. 1, pp. 1-25.
[42] Shearman, S. (2011): Kurt Geiger steps up its online retail experience, in: Marketing, 13 April, p. 9.
[43] Sherman, L. (2009): Online luxury retail remains elusive, in: Business Week Online, 20 Nov., p. 9.
[44] Socha, M. (2007): Shooting at the Net: luxury brands boost their online profiles, in: Women's Wear Daily, Vol. 193, No. 39, pp. 1-18.
[45] The Economist, (2008): Handbagged, Vol. 387, No. 8585, p. 76.
[46] The Economist (2010): The World in 2010, 24th edition
[47] Tsai, S. P. (2005): Impact of Personal Orientation on Luxury-Brand Purchase Value, in: International Journal of Market Research, Vol. 47, No. 4, pp. 429-454.
[48] Unity Marketing (2004): "Luxury Report, 2004", [online] Available at: http://www.unitymarketingonline.com, accessed: 27 Feb 2009.
[49] Vigneron, F./Johnson, L. W. (1999): A review and a conceptual framework of prestige-seeking consumer behaviour, in: Academy of Marketing Science Review, Vol.99, No. 1, pp.1-15.
[50] Walmsley, A. (2008): Web should be in the lap of luxury, in: Marketing, 20 Aug, p. 16.
[51] Warren, M. (2010): By invitation only, in: Profit, Vol. 29, No. 1, pp. 11-12.

Part 6

Counterfeits Challenging the Luxury Industry, Consumers and Society

13 Luxury Goods vs. Counterfeits: An Intercultural Study

Udo Wagner, Seung-Hee Lee, Sabine Kleinsasser, Jutatip Jamsawang

13.1 Statement of problem

Branding has been around for centuries as a means to distinguish one producer's goods from those of another. The earliest signs of branding in Europe can be traced back to medieval guilds, which required craftspeople to mark their products as a means of protection against inferior quality ([29], p. 276). Branding remains one of the most important concepts for marketing management, but its scope now extends to a vast variety of marketing activities.

Building brand equity is a long-term undertaking that requires substantial financial investments over extended periods of time. Competitors looking for an easier, faster way to succeed in a market often offer imitations or counterfeits. An imitator copies a few features of an established brand but maintains differentiation in terms of other characteristics (e.g., packaging); a counterfeiter instead works to duplicate key elements of the established brand and sell it on the black market or through disreputable dealers. In this sense, luxury brands are especially attractive victims for counterfeiters, because of their strong brand equity and customers' willingness to pay a price premium for luxury. These customers might purchase a counterfeit in two types of buying decisions: (1) they *consciously* prefer a counterfeit product; often because it is cheaper, or (2) they are *unaware* they have purchased it, usually because they cannot judge the quality of the product accurately and have been cheated by the seller.

In accordance with these decision processes, there are two main sources of damage to legitimate luxury producers: an immediate loss of sales to brand pirates, and a long-run negative consequence when customers confront defective (e.g., drugs, software, food items) or inappropriate product quality. Their resulting dissatisfaction with what they perceive to be the legitimate brand can cause them to refrain from further purchases. Their word-of-mouth communication might further devalue corporate reputations.

The Counterfeiting Intelligence Bureau, a specialized division of the International Chamber of Commerce, estimates that counterfeiting accounts for between 5 and 7% of world trade, worth an estimated $600 billion a year [14]. Overwhelming evidence suggests that this volume is increasing too, with severe macroeconomic consequences, including lost tax revenues and the cost of thousands of people's livelihoods.

Although such brand piracy occurs in virtually every industry, it is predominant in product categories that require long-term investments, whether in product development (e.g., software, pharmaceuticals) or marketing (e.g., luxury brands). This paper focuses on the latter, because counterfeits are frequent in luxury product categories. The European Commission estimates that in 2010, clothing, shoes, and accessories accounted for 56% of brand piracy, watches for 6%, and other electrical equipment (e.g., hair irons) for 14% [18]. Because buying behavior is highly involved for such products, it is also relatively easy to investigate. In this sense, the problem of brand piracy may be partly the fault of the industries themselves, which turn luxury items into "credence goods," such that their value depends on the credence given to them by others (e.g., designer's or distributor's reputation, use by a distinctive set of fashionable consumers) [26]. Furthermore, consumers' perception of luxury and their purchase behavior depend substantially on social, individual, functional, and financial value dimensions [57], [58].

This article uses this setting and describes an intercultural project on consumer attitudes toward fake luxury goods, ethical issues surrounding counterfeits, and the resulting purchase intentions. The next section provides a brief overview of research into the factors that influence attitude and intention, as well as the research questions that this study analyzes empirically. The third section describes the study design and scales used. The results of this investigation appear in the fourth section. Finally, a discussion section concludes.

13.2 Literature review and research questions

The pirated products considered herein are high involvement goods, such that many variables influence the relevant consumer purchase decisions. A common practice [29] classifies these factors into six categories: cultural, social, psychological, demographic, marketing-mix, and situational factors. By applying this schema to a search for research papers on counterfeiting and its impacts, in computer databases available at the University of Vienna (ABI/Inform, ProQuest, JSTOR), a total of 48 articles emerged, as listed in the Appendix. **Table 13.1** summarizes some of the key results.

Starting in the 1990s, approximately three pertinent papers were published in scholarly journals every year; the rate has increased slightly in recent years. Although in the empirical investigations, no consistent pattern emerges with respect to sample sizes, students have been the primary respondents (i.e., 21 studies employ student samples; **Table 13.1**). Most of the studies have been conducted in (East) Asian, European or American countries, usually in relation to clothing and accessories or electronic items. The primary means of communication were data collection forms, especially questionnaires. Finally, the table presents the most commonly investigated determinants of purchases of pirated brands, including demographic variables like income, gender, and age; marketing-mix variables, especially price and product attributes; cultural characteristics; and ethical concerns. In the past five years, other determinants have attracted more attention, including attitude toward counterfeits, social concerns, and previous buying experience.

Despite the presence of several relationships that were analyzed repeatedly, consistent findings remain rare, with one notable exception. Residents of East Asian countries appear more likely to have positive attitudes toward counterfeiting than residents of Western countries (e.g., [6], [32], [44]). Otherwise, the results of different studies are not directly comparable, because they entail dissimilar environmental conditions. For example, though women seem to have weaker intentions to buy counterfeit products (e.g., [30]), this tendency might not apply for fashion items [26].

Table 13.1 Summary of characteristics of prior literature

Respondents	Analyzed countries	Product categories	Study design	Factors influencing buying decision
Students: 21	Asia: 20	Clothing & accessories: 31	Questionnaire: 38	Demographic: 39
Consumers: 16	Europe: 19	Electronic devices: 16	Focus group: 10	Marketing-mix: 33
Pedestrians: 6	North America: 16	General counterfeits: 12	Experiment: 7	Psychological: 21
Quota sample: 4	South America: 5	Software: 9	Interview: 7	Cultural/ethical: 20
Others: 4	Others: 5	Food & drug: 9	Online survey: 3	Social: 16
		Luxury good: 2		Situational: 7
		Others: 3		

Table 13.1 also reveals the repeated consideration of the link between cultural and ethical antecedents of attitude formation with respect to counterfeiting. This study therefore adopts an intercultural approach to contribute to theory, beginning with the impact of culture on ethical concerns related to brand piracy. Several authors suggest that demographics moderate this relation, so this study also accounts for gender and age. Therefore, the first research question asks:

> **(RQ-A):** Does culture influence ethical concerns related to counterfeiting, and do gender or age moderate this relation?

Focusing on demand for counterfeit goods, Tom et al. [55] develop a typology of consumer orientation toward such products. They distinguish between consumers who consider ethical questions, whether they prefer legitimate goods (e.g., for ethical reasons) or counterfeit goods (e.g., because they object to the dominance of multinational companies), and those who focus on the product, whether they perceive high product parity (i.e., similar quality of counterfeit and legitimate goods) or low product parity (i.e., counterfeit is inferior to legitimate goods on quality but superior in price). As empirical evidence, they confirm the validity of their typology among U.S. consumers and for products such as software, T-shirts, and purses. This article replicates their study in an intercultural context for luxury goods and thereby analyzes the suitability of their segmentation scheme to predict intentions to purchase faked and original luxury goods, while also addressing the potential impacts of demographics. Specifically,

> **(RQ-B):** Can an existing consumer typology extend to different cultural contexts and luxury goods? Should gender or age also be considered?
>
> **(RQ-C):** Is the existing consumer typology a suitable segmentation tool to predict consumer intentions to buy fake versus legitimate luxury goods? Should gender or age also be considered?

13.3 Method

13.3.1 Cultural context

The preceding literature review demonstrates the significant number of studies that focus on Asian and European countries. However, even within the European Union, substantial differences mark the economic, cultural, and legal setting, especially as it continues to expand. The adoption of cultural norms also is a long-term process, such that consumers in the new member states (e.g., Rumania, Bulgaria) likely behave differently in relation to the focal issues than do those in "older" member states such as Austria. In particular, the markets for counterfeits are substantial in Rumania, especially for luxury goods, fashion, perfume, and alcohol. Their presence is likely facilitated by the current legal environment; counterfeits flourish particularly in countries with weak legal infrastructures and corruptible public officials [23].

To address the potential cultural effects, this joint intercultural research project spans Austria, South Korea, and Rumania. The considerable geographic distances among these countries imply that their unique cultures can be measured with sufficient accuracy according to nationality. Korea provides approximately 70% of counterfeit products worldwide, together with China, Thailand, and Taiwan, and counterfeits made in Korea are often rated as the best in quality [21]. In addition, a recent study showed that 67% of Korean college students had purchased pirated brands of handbags or shoes at least once [33]. Austria and Korea thus provide benchmark comparisons with previous studies and represent Western and Asian industrialized economies, respectively; Rumania represents a new emerging economy in the European Union.

13.3.2 Measures

As outlined in **Section 13.2**, this project entails a replication study that relies on a descriptive design. In accordance with extant literature and to minimize instrument variation, a written questionnaire served to collect the primary data in the three countries. International marketing research projects face several major difficulties when measuring latent constructs such as attitude or ethical concerns, especially if cultural biases are possible. Therefore, this study employs previously validated scales and exhibits great care in the translation and cultural adoption of the items to other languages. **Table 13.2** presents the measures in the questionnaire. Added items came from general screening procedures and focus group interviews and were subjected to extensive pretests in all three countries.

Counterfeit purchase behavior may depend on the product category or specific brands within categories. Product category often plays an important role in luxury markets too [58]. In addition, brand awareness and familiarity with product categories may depend on gender, age, and nationality. Therefore, this study features two versions of the questionnaire: for female and for male respondents. Both start with an introductory question about awareness of specific luxury brands in three product categories: handbags for women, watches for men, and clothes and sunglasses for all respondents. To account for national differences, the language of the questionnaires changed, as did the brands listed, though the lists always featured top-ranked brands (according to *BusinessWeek* [8]), such as Armani, Chanel, Gucci, Hugo Boss, and Rolex. However, these lists did not include labels that only targeted a specific age group.

Table 13.2 Summary of scales

Variable	Source	Number of items	Type of scale
Awareness of luxury brands	This study	1	{Armani, Chanel, Gucci, Hugo Boss, Rolex, ...}
Ethical concerns	Muncy and Vitell (1992)	20	7-point Likert scale
Attitude toward counterfeits	Tom et al. (1998)	12	7-point Likert scale
Purchase intention for specific luxury goods (originals or fakes) (clothes, sunglasses, handbags/watches)	This study	2 each	7-point rating scale
Age category Gender Country		3	{<30, 30-50, >50} {female, male} {Austria, Korea, Rumania}

The first question mainly aimed to establish cooperation and orient respondents to the theme of luxury brands. Then they evaluated ethical concern items pertaining to counterfeiting and indicated their attitudes toward counterfeits (**Table 13.2**). The questionnaire proceeded with questions about pirated and original luxury brands. To aid recall, the questionnaire included a list of prominent brand labels in each product category; the items requested specific information related to intentions to buy such products. Finally, the questionnaire ended with a selection of demographic items.

13.3.3 Sampling

A disproportionate quota sampling procedure attempted to interview 50 respondents from each country, of each age group, and of both genders. Thus, the targeted overall sample size was approximately 900. Unlike most studies that exhibit a bias toward younger samples (i.e., students), the sampling procedure in this study therefore offers additional benefits. Despite the challenge of identifying older respondents that were willing to complete the questionnaire and establishing cooperation with them, significant effort supported the achievement of the sample goal. The interviews averaged between 20 and 30 minutes.

Finally, 1,032 questionnaires were useable for the study analysis. The sample consisted of 309 Austrian, 371 South Korean, and 352 Rumanian respondents. As **Table 13.3** reveals, 507 female (49%) and 525 male (51%) respondents participated, and in terms of age, approximately 37% were younger than 30 years, 33% were between 30 and 50 years, and 30% were older than 50 years. About three-quarters (76%) of the sample had purchased counterfeits in the past.

Table 13.3 Overall sample characteristics

Percent	Austria (n = 309)	South Korea (n = 371)	Romania (n = 352)
Female	51	46	50
Male	49	54	50
Younger than 30	36	41	33
30 - 50	32	33	35
Older than 50	32	26	32

13.4 Results

In terms of common knowledge, luxury brands and counterfeit products were well known in all investigated countries; international luxury brands were identified more frequently by female respondents. Calvin Klein, Chanel, Christian Dior, and Gucci achieved recognition by more than 95% of the Austrian, 85% of the Korean, and 80% of the Rumanian participants.

13.4.1 Ethical concerns about counterfeiting in general

A measurement scale from a previous study of ethical consumer behavior issues [35] consisted of 20 items, each describing a certain consumer behavior; respondents indicated whether they regard each behavior as ethical. Four items pertained to situations related to counterfeiting. **Table 13.4** presents the results from these items; an added fifth item referred to compliance with the fiscal system. The first three columns of **Table 13.4** reveal the mean evaluation per country (higher values imply greater assessments of ethical correctness). Austrians are the least strict with respect to their moral standards; the Korean respondents are the most conservative (cf. item five, for which Rumanians indicate the most stringent fiscal standards).

Table 13.4 Cultural differences in ethical concerns

Mean evaluation			MANOVA (*p*-level)					
Austria	South-Korea	Romania	Total	Country	Gender	Age	Country × gender	Country × age
Buying a counterfeit CD instead of the real thing								
4.0	3.1	3.7	<0.01	<0.01	n. s.	<0.01	0.09	n. s.
Taping a movie off the television								
6.5	4.6	5.3	<0.01	<0.01	n. s.	0.02	n. s.	<0.01
Recording a CD instead of buying it								
5.1	4.2	4.9	<0.01	<0.01	0.02	0.01	n. s.	n. s.
Using computer software or games that you did not buy								
4.5	2.9	4.0	<0.01	<0.01	n. s.	<0.01	n. s.	0.07
Stretching the truth on income tax return								
3.2	2.6	1.5	<0.01	<0.01	n. s.	0.07	n. s.	n. s.

Notes: 7 = strongly agree, 1 = strongly disagree.
n. s. = not significant

Consistent with the visual inspection of the mean evaluations, the MANOVAs confirm a significant main effect of country (column five, **Table 13.4**), and post-hoc tests further strengthen the significant differences that emerge from the comparisons of just two cultures at a time. Gender does not moderate the mean evaluations though (column six, **Table 13.4**), with the exception of the third item; women think more freely about recording a CD instead of buying it. Age has a significant impact: Younger respondents are less concerned about ethical correctness. Second-order effects seem negligible (cf. older Koreans, who are more concerned about taping a movie off television than younger ones). In summary, the results offer face validity.

13.4.2 Typology of consumer orientations toward fake luxury goods

Segmentation is key to marketing; it is particularly relevant for luxury goods. Because including counterfeits at the same time complicates the situation, it makes sense to adopt the previously mentioned, existing segmentation proposal [55]. This scheme notes four segments, reflecting unique preferences for counterfeit/legitimate products and high/low product parity between counterfeit and legitimate products. Therefore, they include

- **Sly shoppers** who evaluate fake products as comparable to originals in terms of style, quality, and function and superior in price; they feel smart when buying counterfeits for various reasons (e.g., circumvent or beat the system).
- **Economically concerned shoppers** who assess fake products as inferior to the originals in terms of product attributes but consider price as a sufficient inducement to buy counterfeits. The superior price makes the fake product affordable or worthwhile, but these shoppers do not prefer designer products, because they perceive the prices as unfair.
- **Ethical shoppers** who evaluate fake products as comparable to the originals but object to counterfeits for various ethical reasons or concerns that brand piracy has negative macro-economic effects.
- **Risk-averse shoppers** who share reservations about the quality of fake products and do not believe their superior prices compensate for such deficiencies, so they prefer to spend more money on superior products to minimize the potential for post-purchase frustration.

As a starting point for a priori segmentation, this study adopted selected items from the attitude toward counterfeits scale [55] and ran a k-means cluster analysis (with k = 4). Considerable cultural influences on ethical concerns about counterfeiting have been established (**Table 13.4**), which makes it necessary to run a separate analysis for each country. Assuming the general consumer typology should hold across cultures, the intensities of attitudes toward counterfeiting (as expressed by responses to the different items) might differ, such that Austrians might be less strict about ethical concerns than Koreans on average, but their dispersion patterns should be similar. Attitude toward counterfeits is measured by an interval, rather than a ratio, scale, which provides further justification for this procedure.

The results in **Table 13.5** offer several insights. First, different segment sizes appear in each country. In Korea, there are more economically concerned shoppers (31%) than in Austria (18%) or Rumania (10%). Risk-averse shoppers represent the largest segment for all cultures, particularly Rumania (67%), which appears plausible because the highly priced luxury goods under consideration have achieved relatively less distribution in emerging economies such as Rumania. Second, the results include the coordinates of the cluster centroids. On the horizontal dimension, higher values indicate greater product parity between fake and original products; on the vertical dimension, higher values imply a higher preference for counterfeits. The locations of the centroids vary for the three countries, but they are consistent with regard to the interpretative content and the segment patterns, especially if the classification is reduced to a categorical level (i.e., high vs. low): Austrian sly shoppers possess the highest product parity (5.0) and highest preference for counterfeits (4.9). The results also are satisfactory from a statistical point of view, in that the cluster solutions explain 50%, 60%, and 52% of the variation in the data for Austria, Korea, and Rumania, respectively. According to Franke et al. [19], this evidence indicates acceptable magnitudes.

With regard to the influences of gender and age, the results indicate:

- In Austria, ethical shoppers are likely to be women and older; risk-averse shoppers are likely to be men and younger.

- For Korean respondents, gender does not vary by segment membership, though sly shoppers tend to be older, and economically concerned shoppers tend to be younger.
- In Rumania, gender does not vary by segment membership; ethical shoppers are older, and risk-averse shoppers are younger in general.

Table 13.5 Consumer typology [55]

Austria / South Korea / Rumania	High product parity			Low product parity		
	Sly shoppers			Economically concerned shoppers		
Preference for counterfeit goods	21 (5.0/4.9)	22 (4.6/4.6)	9 (2.7/3.7)	18 (2.2/4.6)	31 (2.7/3.3)	10 (2.3/4.1)
	Ethical shoppers			Risk-averse shoppers		
Preference for legitimate goods	17 (4.8/3.5)	14 (4.5/2.2)	14 (3.9/1.5)	44 (2.1/2.5)	33 (1.9/1.7)	67 (1.3/1.3)

Notes: The cells indicate the segment sizes as percentages and the centroid coordinates.

13.4.3 Intention to buy fake luxury goods

The third research question relates to whether the previously developed segmentation can better predict the buying behavior of members than a non-segmented approach. To this end, this study employs purchase intention data that has not been used previously for segmentation purposes. That is, the respondents to this study provided their intentions to buy fake luxury clothes, sunglasses, and handbags/watches (**Section 13.2**), as well as branded luxury goods in the same product categories. Averaging the data across the three product categories produces the results in **Figure 13.1** and **Table 13.6**.

Panel 5 at the bottom of **Figure 13.1** clearly shows that overall, intentions to buy original luxury goods are greater than the propensity to purchase fake luxury goods (3.8 vs. 2.4). Koreans are more willing to buy either type of goods than Rumanians and Austrians; the latter are the most conservative in this respect. The bar charts in Panels 1–4 of **Figure 13.1** also indicate little difference between intentions to buy fake or original goods among sly shoppers, in accordance with extant expectations. In addition, the (relative) preference for counterfeits among economically concerned shoppers (cf. ethical and risk-averse shoppers) is reflected by the comparatively smaller differences between intentions to buy each type of luxury good. In other words, intentions to purchase fakes are much smaller for ethical and risk-averse shoppers, whereas intentions to buy originals do not differ extensively between these three segments. This discrepancy is particularly pronounced among Koreans.

Figure 13.1 Purchase intentions for fake and original luxury goods per segment

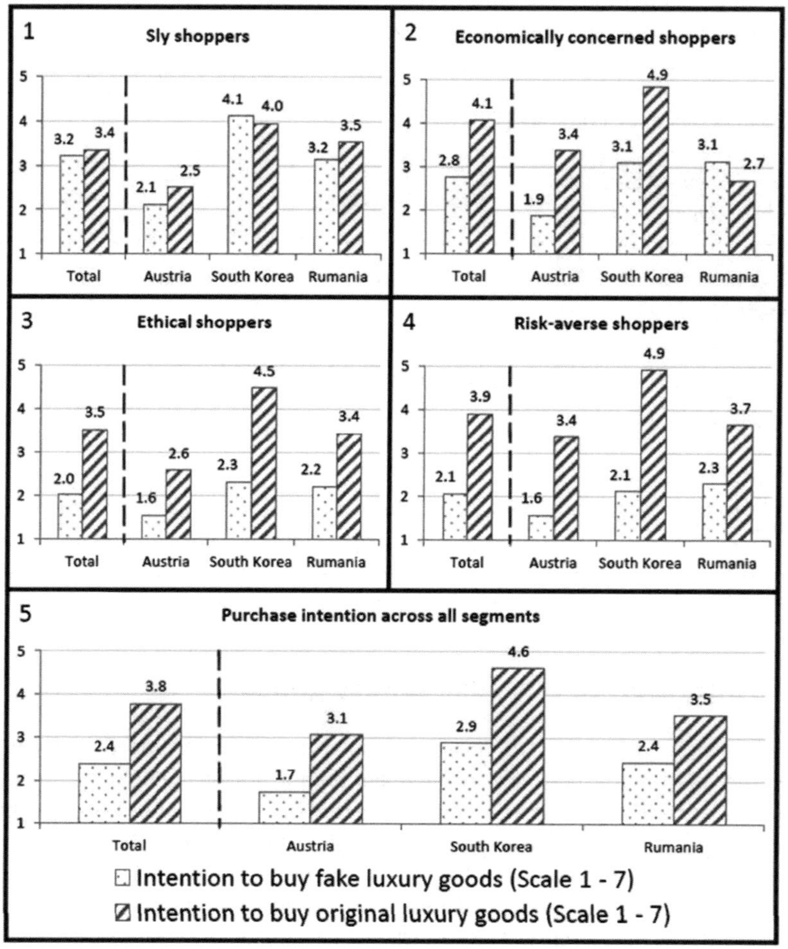

The bar charts per country in Panels 1–4 in **Figure 13.1** also reveal two anomalies: Korean sly shoppers' tendencies to buy fake goods are substantial and even greater than tendencies to buy original goods. The same outcome emerges from Rumanian economically concerned shoppers. Their intentions to buy fakes equal those of Koreans, but their intentions to buy originals are smaller than those of Austrians.

In **Table 13.6**, the results of two MANOVA analyses reveal the outcomes when intention to buy fake/original luxury goods is the dependent variable, and country, segment membership, gender, and age represent independent variables. In accordance with the visual depiction of the data in **Figure 13.1**, the main effects of country and segment and their interaction effect are highly significant. A more detailed analysis confirms that the other previously obtained results are statistically significant.

Table 13.6 Cultural differences in intention to purchase specific goods

MANOVA (*p*-level)							
Total	Country	Segment	Gender	Age	Country × segment	Country × gender	Country × age
Purchase intentions *fake* luxury goods (i.e., clothes, sunglasses, handbags/watches)							
<0.01	<0.01	<0.01	n. s.	n. s.	<0.01	<0.01	n. s.
Purchase intentions for *original* luxury goods (i.e., clothes, sunglasses, handbags/watches)							
<0.01	<0.01	<0.01	0.05	<0.01	<0.01	<0.01	n. s.

n. s. = not significant

The impacts of gender and age differ by purchase intention. For fake luxury goods, the significant interaction effect between country and gender reflects the greater purchase intentions of Korean men. For original luxury goods, the main effects of gender and age are significant; post hoc tests show that men's purchase intentions are greater than those of women, and younger respondents are more willing to buy than are older respondents. In addition, the country × gender interaction is highly significant in this case, because Korean men are more reluctant to spend money for luxuries.

13.5 Summary

13.5.1 Discussion

Building on an established concept, this research proposes a basis for segmenting intercultural markets for fake and original luxury goods. This behavioral segmentation approach uses attitudinal data pertaining to quality and preference for counterfeits versus original products. As further validation of the procedure, this study includes data related to purchase intentions. Overall, these analyses suggest the following results.

- **Impact of culture on the purchase of luxuries**
  Austrians are the least stringent with respect to ethical concerns about consumer behavior. Moreover, they hold the most positive views about their preferences for and the

quality of counterfeits; nevertheless, their purchase intentions to buy luxury goods (fakes and originals) are the lowest.

Koreans adopt the most conservative positions in relation to the ethical issues of buying behavior, though their opinions vary considerably across the well-defined segments and for the rather substantial cluster of ethical shoppers. In addition, their purchase intentions for both fake and legitimate luxuries are very high.

Rumanians worry about consumer ethics but perceive the inferior quality of counterfeits, which is not very detrimental to demand. Although currently, there seems to be little difference between sly and economically concerned shoppers (i.e., both segments are small and located close to each other), Rumanians seem eager to buy luxury goods (both fake and legitimates), as indicated by their purchase intentions.

- **Impact of gender on the purchase of luxuries**
  Gender does not seem much of an issue, with a few exceptions. Men are more prepared to buy legitimate luxuries, probably because of the inequality in pay between men and women. In Austria, more women are clustered in the ethical shoppers group, but more men are risk-averse shoppers. Korean men are more inclined toward fakes but not toward original luxury items.
- **Impact of age on the purchase of luxuries**
  On average, younger respondents are less concerned with ethical issues. In Korea, economically concerned shoppers are younger, and sly shoppers are older. Ethical shoppers are older, whereas risk-averse shoppers are younger in both Austria and Rumania.

The results achieve face validity, are statistically significant, and appear consistent with extant literature (see **Section 13.1**).

On a more general level, two lessons emerge. First, there are considerable differences in ethical concerns, attitudes, and purchase intentions for counterfeits between consumers in the three countries analyzed herein. Second, the segmentation typology proposed by Tom et al. [55] applies to luxury goods and on a cross-cultural level. That is, despite cultural differences, patterns of attitude and intention seem to follow similar principles.

These findings are relevant for both managers and public policy makers trying to constrain counterfeiting. Managers of luxury brands might find it useful to target a certain market segment by issuing high quality/high reliability arguments (to appeal to risk-averse shoppers), emphasizing good value for money or launching flanker brands (to economically concerned shoppers), or adopting moral reasoning and accentuating the corporate social responsibility of the brand (ethical shoppers). Efforts to limit counterfeiting may be more effective if they target consumer accomplices whose attitudes are more supportive of counterfeiting and who appear satisfied with fake goods (sly shoppers). For example, a public service campaign might counter conceptions that people who purchase counterfeits are being clever or committing only a small peccadillo. Although in no countries is the sly shoppers' segment dominant, it is larger in more advanced economies such as Austria and Korea.

13.5.2 Limitations and further research

The segmentation model comes from Tom et al. [55], who provided some theoretical reasoning but based their approach on data. The results in this study support their findings; however, it remains unknown whether the results confirm an even more general pattern. Empirical studies in other countries, for other products, and with a broader scope would provide further necessary evidence. Moreover, the theoretical foundation still demands refinement, which could be a compelling topic for conceptual research.

This theoretical shortcoming motivates another limitation of this study, which used a priori segmentation and fixed the number of segments in advance. A more precisely validated measurement tool is required to determine segment membership. Statisticians would probably prefer to decide on the number of segments, using indices designed for the specific clustering procedure, rather than a priori.

As is common for behavioral segmentation, issues of effectiveness also may arise: How can marketers reach these segments without excessive waste? How should firms develop products to attract and serve these segments? Unfortunately, this analysis of the influence of demographic information does not provide implementable guidelines. More research on the psychographic characteristics of different types of consumers could help make the proposed segmentation procedure more actionable.

Overall though, this study provides insights into consumer behavior toward luxury goods, both real and fake, on an international level. Further research should continue to advance this effort.

References

[1] Albers-Miller, N.D. (1999): Consumer misbehaviour: Why people buy illicit goods, in: Journal of Consumer Marketing, Vol. 16, No. 3, pp. 273-287.

[2] Ang, S.H./Cheng, P.S./Lim, E.A.C./Tambyah, S.K. (2001): Spot the difference: Consumer responses towards counterfeiting, in: Journal of Consumer Marketing, Vol. 18, No. 3, pp. 219-235.

[3] Bian, X./Moutinho, L. (2009): An investigation of determinants of counterfeit purchase consideration, in: Journal of Business Research, Vol. 62, No. 3, pp. 368-378.

[4] Bian, X./Moutinho, L. (2011a): The role of brand image, product involvement, and knowledge in explaining consumer purchase behaviour of counterfeits: Direct and indirect effects, in: European Journal of Marketing, Vol. 45, No. 1/2, pp. 191-216.

[5] Bian, X./Moutinho, L. (2011b): Counterfeits and branded products: Effects of counterfeit ownership, in: Journal of Product & Brand Management, Vol. 20, No. 5, pp. 379-393.

[6] Bian, X./Veloutsou C. (2007): Consumers' attitudes regarding non-deceptive counterfeit brands in the UK and China, in: Journal of Brand Management, Vol. 14, No. 3, pp. 211-222.

[7] Bloch, P.H./Bush, R.F./Campbell, L. (1993): Consumer accomplices in product counterfeiting – A demand side investigation, in: Journal of Consumer Marketing, Vol. 10, No. 4, pp. 27-36.

[8] BusinessWeek (2005): The 100 top brands, in: BusinessWeek, August 1, 2005, pp. 90-94.

[9] Casola, L./Kemp, S./Mackenzie, A. (2009): Consumer decisions in the black market for stolen or counterfeit goods, in: Journal of Economic Psychology, Vol. 30, No. 2, pp. 162-171.

[10] Chan, A./Wong, S./Leung, P. (1998): Ethical beliefs of Chinese consumers in Hong Kong, in: Journal of Business Ethics, Vol. 17, No. 11, pp. 1163-1170.

[11] Chaudhry, P./Stumpf, S. (2011): Consumer complicity with counterfeit products, in: Journal of Consumer Marketing, Vol. 28, No. 2, pp. 139-151.
[12] Cheng, H.K./Sims, R.R./Teegen, H. (1996): Toward a profile of student software pirates, in: Journal of Business Ethics, Vol. 15, No. 8, pp. 839-849.
[13] Cheung, W./Prendergast, G. (2006): Buyers' perceptions of pirated products in China, in: Marketing Intelligence & Planning, Vol. 24, No. 5, pp. 446-462.
[14] Counterfeiting Intelligence Bureau (2011): online resource: http://www.icc-ccs.org/home/cib, accessed on October 2011.
[15] Cordell, V.V./Wongtada, N./Kieschnick, R.L. Jr. (1996): Counterfeit purchase intention: Role of lawfulness attitudes and product traits as determinants, in: Journal of Business Research, Vol. 35, No. 1, pp. 41-53.
[16] De Matos, C./Ituassu, C./Rossi, C. (2007): Consumer attitudes toward counterfeits: A review and extension, in: Journal of Consumer Marketing, Vol. 24, No. 1, pp. 36-47.
[17] Eisend, M./Schuchert-Güler, P. (2006): Explaining counterfeit purchases : A review and preview, in: Academy of Marketing Science Review, Vol. 12, No. 6, pp. 1-26.
[18] European Commission (2010): online resource: http://ec.europa.eu/taxation_customs/resources/documents/customs/customs_controls/counterfeit_piracy/statistics/statistics_2010.pdf, accessed on October 2011.
[19] Franke, N./Reisinger, H./Hoppe, D. (2009): Remaining within-cluster heterogeneity: A meta-analysis of the "dark side" of clustering methods, in: Journal of Marketing Management, Vol. 25, No. 3-4, pp. 273-293.
[20] Furnham, A./Valgeirsson, H. (2007): The effect of life values and materialism on buying counterfeit products, in: Journal of Socio-Economics, Vol. 36, No. 5, pp. 677-685.
[21] Gentry, J.W./Putrevu, S./Shultz, C.J./Commuri, S. (2001): How now Ralph Lauren? The separation of brand and product in a counterfeit culture, in: Advances in Consumer Research, Vol. 28, No. 1, pp. 258-265.
[22] Gentry, J.W./Putrevu, S./Schultz, C.J. (2006): The effects of counterfeiting on consumer search, in: Journal of Consumer Behaviour, Vol. 5, No. 3, pp. 245-256.
[23] Green, R.T./Smith, T. (2002): Executive insights: Countering brand counterfeiters, in: Journal of International Marketing, Vol. 10, No. 4, pp. 89-106.
[24] Gupta, P./Gould, S./Pola, B. (2004): To pirate or not to pirate: A comparative study of the ethical versus other influences on consumer's software acquisition-mode decision, in: Journal of Business Ethics, Vol. 55, No. 3, pp. 255-274.
[25] Harvey, P.J./Walls, W.D. (2003): The revealed demand for pirate goods: Probit analysis of experimental data, in: International Journal of Management, Vol. 20, No. 2, pp. 194-202.
[26] Hilton, B./Choi, C.J./Chen, S. (2004): The ethics of counterfeiting in the fashion industry: Quality, credence and profit issues, in: Journal of Business Ethics, Vol. 55, No. 4, pp. 345-354.
[27] Kini, R.B./Rominger, A./Vijayaraman, B.S. (2000): An empirical study of software piracy and moral intensity among university students, in: Journal of Computer Information Systems, Vol. 40, No. 3, pp. 62-72.
[28] Koklic, M. (2011): Non-deceptive counterfeiting purchase behavior: Antecedents of attitudes and purchase intentions, in: Journal of Applied Business Research, Vol. 27, No. 2, pp. 127-137.
[29] Kotler, P./Keller, K.L. (2009): Marketing management, 13th edition, Pearson Education, New Jersey, NJ.
[30] Kwong, K.K./Yau, O.H.M./Lee, J.S.Y./Sin, L.Y.M./Tse, A.C.B. (2003): The effects of attitudinal and demographic factors on intention to buy pirated CDs: The case of Chinese consumers, in: Journal of Business Ethics, Vol. 47, No. 3, pp. 223-235.
[31] Kwong, K.K./Yu, W./Leung, J./Wang, K. (2009): Attitude toward counterfeits and ethnic groups: Comparing Chinese and Western consumers purchasing counterfeits, in: Journal of Euromarketing, Vol. 18, No. 3, pp. 157-168.
[32] Lai, K.K./Zaichkowsky, J.L. (1999): Brand imitation: Do the Chinese have different views?, in: Asia Pacific Journal of Management, Vol. 16, No. 2, pp. 179-192.

[33] Lee, S.-H. (2006): The effects of consumer-brand relationship on attitudes of counterfeits, in: Journal of the Korean Society of Clothing & Textiles, Vol. 30, No. 9/10, pp. 1445-1454.
[34] Mavlanova, T./Benbunan-Fich, R. (2010-11): Counterfeit products on the internet: The role of seller-level and product-level information, in: International Journal of Electronic Commerce, Vol. 15, No. 2, pp. 79-104.
[35] Muncy, J.A./Vitell, S.J. (1992): Consumer ethics: An investigation of the ethical beliefs of the final consumer, in: Journal of Business Research, Vol. 24, No. 4, pp. 297-311.
[36] Nia, A./Zaichkowsky, J.L. (2000): Do counterfeits devalue the ownership of luxury brands?, in: Journal of Product & Brand Management, Vol. 9, No. 7, pp. 485-497.
[37] Norum, P.S./Cuno, A. (2011): Analysis of the demand for counterfeit goods, in: Journal of Fashion Marketing and Management, Vol. 15, No. 1, pp. 27-40.
[38] Penz, E./Schlegelmilch, B.B./Stöttinger, B. (2008): Voluntary purchase of counterfeit products: Empirical evidence from four countries, in: Journal of Internation Consumer Marketing, Vol. 21, No. 1, pp. 67-84.
[39] Penz, E./Stöttinger, B. (2005): Forget the "real" thing – take the copy! An explanatory model for the volitional purchase of counterfeit products, in: Advances in Consumer Response, Vol. 32, No. 1, pp. 568-575.
[40] Penz, E./Stöttinger, B. (2008): Corporate image and product similarity — Assessing major demand drivers for counterfeits in a multi-country study, in: Psychology & Marketing, Vol. 25, No. 4, pp. 352-381.
[41] Perez, M.E./Castano, R./Quintanilla, C. (2010): Constructing identity through the consumption of counterfeit luxury goods, in: Qualitative Market Research: An International Journal, Vol. 13, No. 3, pp. 219-235.
[42] Phau, I./Sequeira, M./Dix, S. (2009a): To buy or not to buy a „counterfeit" Ralph Lauren polo shirt: The role of lawfulness and legality toward purchasing counterfeits, in: Asia-Pacific Journal of Business Administration, Vol. 1, No. 1, pp. 68-80.
[43] Phau, I./Sequeira, M./Dix, S. (2009b): Consumers' willingness to knowingly purchase counterfeit products, in: Direct Marketing: An International Journal, Vol. 3, No. 4, pp. 262-281.
[44] Phau, I/Teah, M. (2009): Devil wears (counterfeit) Prada: A study of antecedents and outcomes of attitudes towards counterfeits of luxury brands, in: Journal of Consumer Marketing, Vol. 26, No. 1, pp. 15-27.
[45] Phau, I./Teah, M./Lee, A. (2009): Targeting buyers of counterfeits of luxury brands: A study on attitudes of Singaporean consumers, in: Journal of Targeting, Measurement and Analysis for Marketing, Vol. 17, No. 1, pp. 3-15.
[46] Prendergast, G./Chuen, L.H./Phau, I. (2002): Understanding consumer demand for non-deceptive pirated brands, in: Marketing Intelligence & Planning, Vol. 20, No. 7, pp. 405-416.
[47] Rutter, J./Bryce, J. (2008): The consumption of counterfeit goods: Here be pirates?, in: Sociology, Vol. 42, No. 6, pp. 1146-1164.
[48] Sahin, A./Atilgan, K.O. (2011): Analyzing factors that drive consumers to purchase counterfeits of luxury branded products, in: Journal of American Academy of Business, Vol. 17, No. 1, pp. 283-292.
[49] Schlegelmilch, B.B./Stöttinger, B. (1999): Der Kauf gefälschter Markenprodukte, [The purchase of counterfeit products], in: Marketing ZFP, Vol. 3, No. 2, pp. 196-208.
[50] Sharma, P./Chan, R.Y.K. (2011): Counterfeit proneness: Conceptualisation and scale development, in: Journal of Marketing Management, Vol. 27, No. 5-6, pp. 602-626.
[51] Simmons, L.C. (2004): An exploratory analysis of software piracy using cross-cultural data, in: International Journal of Technology Management, Vol. 28, No. 1, pp. 139-148.
[52] Swami, V./Chamorro-Premuzic, T./Furnham, A. (2009): Fake it: Personality and individual difference predictors of willingness to buy counterfeit goods, in: Journal of Socio-Economics, Vol. 38, No. 5, pp. 820-825.
[53] Swinyard, W.R./Rinne, H./Kau, A.K. (1990): The morality of software piracy: A cross cultural analysis, in: Journal of Business Ethics, Vol. 9, No. 8, pp. 655-664.

[54] Taormina, R.J./Chong, K.Y. (2010): Purchasing counterfeit and legitimate products in China: Social and psychological correlates and predictors, in: Journal of East-West Business, Vol. 16, No. 2, pp. 118-145.
[55] Tom, G./Garibaldi, B./Zeng, Y./Pilcher, J. (1998): Consumer demand for counterfeit goods, in: Psychology & Marketing, Vol. 15, No. 5, pp. 405-421.
[56] Wee, C.-H./Tan, S.-J./Cheok, K.-H. (1995): Non-price determinants of intention to purchase counterfeit goods: An exploratory study, in: International Marketing Review, Vol. 12, No. 6, pp. 19-46.
[57] Wiedmann, K.-P./Hennigs, N./Siebels, A. (2007): Measuring consumers' luxury value perception: A cross-cultural framework, in: Academy of Marketing Science Review, Vol. 2007, No. 7, pp. 1-21.
[58] Wiedmann, K.-P./Hennigs, N./Siebels, A. (2009): Value-based segmentation of luxury consumption behavior, in: Psychology & Marketing, Vol. 26, No. 7, pp. 625-651.
[59] Wilcox, K./Kim, H.M./Sen, S. (2009): Why do consumers buy counterfeit luxury brands?, in: Journal of Marketing Research, Vol. 46, No. 2, pp. 247-259.
[60] Yelkur, R. (2000): Consumer perceptions of generic products: A Mexican study, in: Journal of Product & Brand Management, Vol. 9, No. 7, pp. 446-456.

Appendix: Literature Review

Author	Date	Sample size	Respondents	Country	Product categories	Study design	Determinants of buying decision
Albers-Miller	1999	153	Graduates	USA	Color TVs	Questionnaire	Price
Ang, Cheng, Lim, Tambyah	2001	3,621	Diverse	Singapore	Music CDs	Questionnaire, face-to-face	Gender, income, perceived social status
Bian, Moutinho	2009	321	Shoppers in supermarket	UK	Watches	Focus group, questionnaire	Perceived brand personality, perceived product attributes
Bian, Moutinho	2011a	321	Shoppers at supermarket entrances	UK	Watches	Focus group, questionnaire	Perceived brand personality, perceived product attributes, perceived benefits
Bian, Moutinho	2011b	321	Shoppers at supermarket entrances	UK	Watches	Focus group, questionnaire	Perceived competence personality, satisfaction benefit, functional attribute, social risk
Bian, Veloutsou	2007	2 samples: 295, 230	Pedestrians in flea market and high street	China, UK	Sunglasses	Focus group, questionnaire	Gender, age (only for UK)
Bloch, Bush, Campbell	1993	100	Consumers in shopping malls	USA	T-shirts	Field experiment	Price, perceived social status
Casola, Kemp, Mackenzie	2009	3 samples: 80, 53, 64	Students, consumers	New Zealand	LCD TV, leather jacket, DVD, watches	Questionnaire, experiments	Price
Chan, Wong Leung	1998	242	Students	China, Hong Kong	Counterfeits in general	Questionnaire	Ethical concerns
Chaudhry, Stumpf	2011	254	Students	USA	Movie, pharmaceuticals	Online questionnaire	Attitudes toward counterfeits, hedonic shopping experience
Cheng, Sims, Teegen	1996	340	Students	USA	Software	Questionnaire	Gender, risk aversion, product familiarity
Cheung, Prendergast	2006	1,152	Quota sample of consumers	China	VCDs, clothing, accessories	Focus groups, questionnaire	Gender, income, education, age, marital status, occupation
Cordell, Wongtada, Kieschnik	1996	221	Students	Asia	Cameras, T-shirts	Experiment, questionnaire	Product quality, price, legal concerns
De Matos, Ituassu, Rossi	2007	400	Consumers	Brazil	Counterfeits in general	Interviews	Attitudes toward counterfeits, perceived risk, buying experience, subjective norm, integrity, price-quality inference, personal gratification

Author	Date	Sample size	Respondents	Country	Product categories	Study design	Determinants of buying decision
Eisend, Schuchert-Güler	2006	12	Students, non-students	Germany	Counterfeits in general	Focus group, in-depth interviews	Attitudes towards counterfeiting, purchase situation, mood, social and cultural context, product, person, coping strategies for cognitive dissonance
Furnham, Valgeirsson	2007	102	Consumers	UK	Counterfeits in general	Questionnaire	Marital status, materialism, beliefs, traditional value
Gentry, Putrevu, Schultz	2006	102	International students	Australia	Counterfeits in general	Interviews	Brand decision
Gupta, Gould, Pola	2004	689	Internet users	USA	Software	Online survey	Ethical concerns, age
Harvey, Walls	2003	60	Students	Hong Kong	Software, audiovisual products	Experimental market	Legal concerns
Kini, Rominger, Vijayaraman	2000	800	Students	USA	Software	Questionnaire	Moral concerns, age, education
Koklic	2011	1,211	Quota sample of consumers	Slovenia	Counterfeits in general	Questionnaire	Attitudes toward counterfeiting, perceived risk
Kwong, Yau, Lee, Sin, Tse	2003	306	Students	Hong Kong	Music CDs	Questionnaire, face-to-face	Ethical concerns, age, gender
Kwong, Yu, Leung, Wang	2009	500	Consumers	Hong Kong, UK	Handbags	Questionnaire	Attitude toward counterfeits, ethnic groups, ethical belief, social benefit of reproduction
Lai, Zaichkowsky	1999	137	Local immigrants	Taiwan, Hong Kong, North America	Foods	Questionnaire	Buying experience, country of origin
Mavlanova, Benbunan-Fich	2010	174	Students	USA	Watches	Experiment	Trust in on-line sellers
Nia, Zaichkowsky	2000	74	Owners of luxury goods	Canada	Luxury goods	Questionnaire	Income, brand image
Norum, Cuno	2011	437	Students	USA	Handbags, medicine, clothing, jewels, cards, sporting goods, toiletries, car parts, music	Questionnaire	Income, education level, attitudes about counterfeit buyers and sellers (being criminals)
Penz, Schlegelmilch, Stöttinger	2008	4 samples: 385, 118, 195, 242	Consumers	Austria, Mexico Slovenia, Czech republic	Watches, T-shirts	Expert interviews, questionnaire	Price-value relationship
Penz, Stöttinger	2005	1,040	Quota sample of consumers	Austria	Different branded products	Questionnaire	Price, ethical concerns, consumer characteristics

Author	Date	Sample size	Respondents	Country	Product categories	Study design	Determinants of buying decision
Penz, Stöttinger	2008	6 samples: 1,040, 125, 130, 213, 211, 128	Quota sample of consumers	Austria, Mexico Slovakia, Sweden, Ukraine, USA	Counterfeits in general	Questionnaire	Corporate image, perceived product similarity
Perez, Castano, Quintanilla	2010	37	Female consumers	Mexico	Fashion	In-depth interviews	Inner goals (memorable experience filled with fun, risk and adventure), positive identity
Phau, Sequeira, Dix	2009a	202	Students	Australia	Polo shirts	Focus group, questionnaire	Integrity
Phau, Sequeira, Dix	2009b	202	Students	Australia	Watches	Questionnaire	Product durability
Phau, Teah	2009	202	Shoppers in a major shopping complex	China	Counterfeits in general	Questionnaire	Status consumption, integrity, normative susceptibility, information susceptibility, personal gratification, value consciousness, novelty seeking, attitude toward counterfeits of luxury brands
Phau, Teah, Lee	2009	204	Postgraduate students	Singapore	Counterfeits in general	Online questionnaire	Social influence, price quality inference
Prendergast, Chuen, Phao	2002	200	Pedestrians in subway stations	Hong Kong	VCDs, T-shirts	Focus group, questionnaire	Price, income, age, education
Rutter, Bryce	2008	2,370	Diverse	Great Britain, Northern Ireland	Fashion clothes, music, film, games, software	Focus group, postal and online questionnaire	Price, age, gender, income, socio-economic status
Sahin, Atilgan	2011	404	Consumers	Turkey	Counterfeits in general	Questionnaire	Social impacts, price-quality inference, ethical issues
Schlegelmilch Stöttinger	1999	230	Students	USA	Shirts	Questionnaire	Price, GNP of the country
Sharma, Chan	2011	348	Shoppers in a shopping mall	Hong Kong	Counterfeits in general	Questionnaire	Gender, age, income, education, buying experience
Simmons	2004	386	Students	USA, Taiwan, Australia, Hong Kong, Singapur	Software	Focus group, questionnaire	Cultural and social influences
Swami, Chamorro-Premuzic, Furnham	2009	237	Adults at a large train station	UK	Diverse	Questionnaire	Attitudes towards counterfeiting, buying experience, material values

Author	Date	Sample size	Respondents	Country	Product categories	Study design	Determinants of buying decision
Swinyard, Rinne, Kau	1990	371	Students	USA, Singapore	Software	Questionnaire	Cultural influences, legal concerns, individual benefit
Taormina, Chong	2010	226	Adults in shopping malls	Macau	Clothing, accessories, electronic devices	Questionnaire	Risk taking, worry about inflation, need for dominance, employment status
Tom, Garibaldi, Zeng, Pilcher	1998	3 samples: 208, 434, 268	Consumers in shopping malls & flea markets	USA	CDs, perfumes, software, purses & wallets, T-shirts	Experiment, questionnaire	Price, product quality, age, income
Wee, Tan, Cheok	1995	949	Students, professionals	Southeast Africa	Technical literature, software, purses & wallets, watches	Questionnaire	Product quality, brand image, novelty seeking, income, age, education
Wilcox, Kim, Sen	2009	3 samples: 79, 138, 176	Under-graduates	USA	Counterfeits in general, handbags	Questionnaire, experiment	Social motivations
Yelkur	2000	200	Consumers in supermarkets	Mexico	Foods, cosmetics, sanitary products	Questionnaire	Price, income, age

14 Brazil: Luxury and Counterfeits

Suzane Strehlau

14.1 The Brazilian context

Luxury companies operating in Brazil have to deal with a specific context, one that results from the country's history, from it economic development, its cultural and fiscal aspects. A salient factor in consumption in Brazil is the crime rate: armored cars, security systems, high gates and walls are part of everyday life – it's important to know where and how to walk.

It is customary to give more prestige to imported brands comparing to national ones. One reason for this originates from the colonization process, supported by the exploitation of natural resources and large rural plantations. Manufactured goods were imported by the elite, whose children were sent abroad to study. Moreover, in the process of acculturation, immigrants gave products from their country of origin more value than product produced locally.

The complex system of taxes is based on flows such as the movement of goods or the importation of goods. There is a cumulative tax effect, one being calculated upon the other. The Brazilian Association of Luxury (ABRAEL) asks for tax refund for foreigners who buy products in Brazil [55]. Depending on the nature of the product, the free-on-board (FOB) price may increase from 100 to 120%, before the importer or retailer margin.

Depending on the exchange rate the primary competition for a luxury brand in Brazil is the same brand in another country, usually in the United States (Miami, New York) and or France (Paris). Five out of ten *Montblanc* pens sent for technical assistance in Brazil were bought abroad. Cartier in Brazil lowered prices in order to achieve a price difference of only 15% in comparison to the price practiced in overseas stores [26]. For many Brazilian it has become a habit to shop during trips abroad, especially when the exchange rate is advantageous. Travel agencies offer itineraries developed according to shopping opportunities.

The primary advantage for Brazilians of buying in Brazil is then the possibility of paying in installments. This helps consumers with lower purchasing power, but is also a strategy for hiding expenses from a spouse. Installments used to be without interest when inflation was controlled by the government; however, as inflation is growing many retailers have begun to charge interest on installments.

The consumption of luxury services is not associated with this behavior because other people, like friends and family, are present and because there is a different rhythm of consumption. High end restaurants, hotels, spas and resorts will sustain the sector's growth over the coming years together with the beauty, automobile and real estate industries [45].

14.1.1 Figures for the Brazilian luxury market

The Boston Consulting Group (BCG) estimates the global market for luxury goods at 1 trillion euros, significantly above expectations which were 150 to 180 billion euros [5]. However, despite a regular growth rate of over 10% for the last decade, luxury sales in the Brazilian market in 2010 (6424 billion euros – exchange rate of 12 October 2011) still represent around 3% of the global movement. The Datamonitor Report [36] gives Brazil a slightly larger share: 7% of world annual sales.

According to the World Wealth Report 2011, the number of High Net Worth Individuals (HNWIs) in Brazil is 155,000. Although only 1% of the country's total population, these approximately 10.9 million people are quite resilient to crisis. Latin America still has the highest percentage of Ultra-HNWIs relative to the overall HNWI population—2.4%, compared with the global average of 0.9% [54]. Another aspect that makes the country interesting is the wide base of aspiring customers.

The individual wealth, stocks, real estate, art, aircrafts, cash and other assets officially reported suggest that the wealth of 4720 people in Brazil is superior to USD 30 million, and that there 50 billionaires. The ranking of the North American magazine, Forbes, gives a smaller number of billionaires for the country, at only 30. That said, its latest survey shows that the total of Brazilians with assets of more than 1 billion US dollars increased from 18 to 30 in one year [43].

Figure 14.1 Number of new registred helicopters

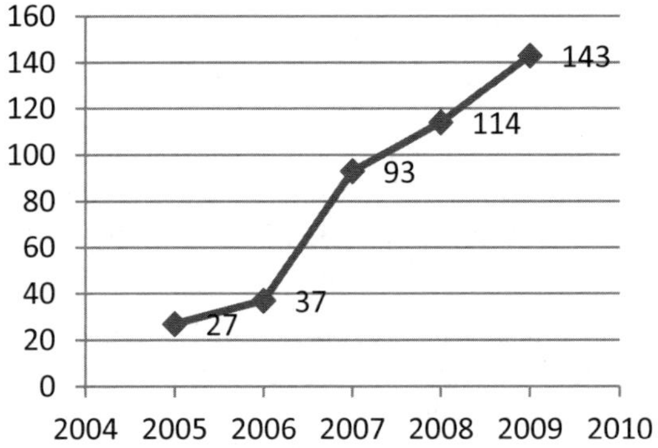

Source: Civil Aviation Department – DAC

This is increase is reflected in the reports of the Civil Aviation Department (DAC) of the Brazilian government (See **Figure 14.1**). The figures presented concerns new registered helicopters in Brazil. By 2010 the total amount of helicopters was 1325, of which 541 were in the state of Sao Paulo, 285 in Rio de Janeiro, 139 in Minas Gerais, 50 in Brasilia, 46 in Parana.

Brazil is the third largest market for business jets in the world with 626 units, trailing behind the USA with 11,000 and Mexico with 713 [13].Between 2005 and 2009, the Brazilian private jet fleet increased by 8.6%, significantly above the world average of 6% [41].

The Government Census of 2010 recorded a total population of 190,755,799 in Brazil. Only 5.1% of Brazilian households have a monthly income above 110 euros (the equivalent of five minimum wages) [29]. 307,000 families have an income superior to 8,000 euros [44].

The "Brazil Criterion" (Brazil Economic Classification Criterion – CCEB) is a measurement instrument based upon the presence and quantity of some items in the household as well as the educational level of the head of the family. The points are assigned according to the characteristics of each household; there is a correspondence between the score ranges and income levels which are defined using the categories A1 (highest) A2, B1, B2, C1, C2, D, E (lowest) [12]. In Brazil, 42% of Mastercard's revenues come from people earning more than R$ 4.5 thousand, in other words form class A1 and A2 [54].

Table 14.1 Brazilian Market trends Title

Class	A1	A2	B1	B2
% of population	0.5%	4%	9.1%	19.3%
Montly earnings	€ 4691 R$11,480	€3389 R$8295	€1942 R$4754	€ 1085 R$ 2656

Source: ABEP http://www.abep.org/novo/Content.aspx?ContentID=301

Companies show positive earnings in the Brazilian Market. In 2009, the Brazilian branch was BMW's fastest growing branch, up 83% in terms of volume and market share: for the company headquarters in Germany, Brazil is the hub of this region [37]. Hermès ended 2009 with sales growth of 8.5% [22]. Swarovki cristals annouced that Latin American market growth was 40% and between 2000 [20].

Wealth is largely concentrated in the Southeast region (approximately 70%) and to some extent in the South (10%). São Paulo Metropolitan area, Rio de Janeiro city, Brasilia, Belo Horizonte, Porto Alegre, Curitiba, Salvador present the greatest concentration of higher income[42].

Brazil does not have department stores like Harrods or Selfridges. Luxury brand stores are concentrated on certain streets and malls (called "shopping centers" in Portuguese) like Fashion Mall, Village Mall in Rio de Janeiro and Iguatemi, Cidade Jardim in São Paulo. More malls will open in 2014 [45]; security and car parking facilities are among the main factors behind the popularity of shopping malls. However, stores in SP concentrate 75% of the market volume for luxury shopping [27].

There are some Brazilian luxury brands for jewelry (H. Stern, Amsterdam Sauer), hotels (Copacabana Palace, Fasano, Emiliano, Estrela D'agua), and fashion goods (Osklen, Victor Hugo, Ricardo Almeida, Carlos Miele, Alexandre Herchcovitch).

14.2 Different perspectives of Luxury

There are several theories in different research fields which propose different frameworks to analyze this sort of consumption. Here, we are looking at Luxury from the point of view of a postmodernism trend, where contextuality, native perspectives and outliers are significant postmodern considerations [47]. The basis of postmodernism, the thesis from which much else follows, comes under the heading of the linguistic turn. This means that reality can only be apprehended through the intervention of language because all perceptions, concepts and claims to truth are constructed in language.

14.2.1 Marketing theory and Luxury

Here are a few ideas from authors within a field of marketing; their thoughts on the concept of luxury are briefly presented, as the objective is not to provide a literature review.

French authors emphasize tradition and history with relation to the concept of luxury [18], [1], [31]. Other, non-French authors [19], [39], [53], connect the concept of luxury to that of status. Giving value to tradition will give a competitive edge to companies owning ancestral French brands and will create obstacles to new brands, especially those from other countries. The traditional luxury management is a pyramid system in which prestige is provided from the top, with *haute couture,* and financial support from the bottom, with licenses and mass produced products. During the 1970's however, a new generation of luxury brands appeared using a new business style: the Galaxy model. It is not hierarchical and products gravitate around the world view of the designer or *créateur* [48]. Brand identity is strategically built and communicated to consumers via a story telling process.

The definition of a luxury product includes its nature (built with rare and good raw materials, good reputation or special technology); origin (source brings social recognition); role (what the product represents in terms of interpersonal relations and in terms of the system of social influence [46]). These products are symbols of social hierarchy.

Luxury products are associated with a brand that will give prestige to its owners [17]. The luxury concept is increasingly present inside the brand and less present in the object itself. For example, Chanel is without doubt a luxury brand but sometimes being carried in a simple product like a t-shirt or cotton pads to clean off make-up.

Prestige is linked to the evaluation of others during social interactions, although a luxury good can be kept for private consumption. One can talk about having luxury for oneself but not prestige for one's own [53]. The concept of luxury is then associated with prestige when there is an evaluation by others.

In the 19th century prestige goods confirmed and strengthened a preexisting status; in consumer societies these goods have themselves become status markers regardless of the social background of owner [15].

In the 21st century, new luxury arose as a business strategy but this approach to luxury is controversial. Kapferer and Bastien state that premium products or "masstige" are an extension of a classical marketing strategy [30]. Going further, the authors pointed out that brand extension activity often involves trading down, targeting the middle class with products that are not luxury products. The luxury brand might be used for a "normal" product in order to enhance market acceptance and to increase profit margin. The concept of luxury is sustained only through brand associations. And experiences became central to luxury consumption activity [3].

14.2.2 Theory of consumption and Luxury

The objective in his section is to present a few theories of consumption linked to luxury brands, developed by authors with a social sciences orientation. Again, the intention here is not to make a complete review of consumption literature.

The consumer society system has a logic that organizes social life around the consumption of mass-produced commodities. Several researchers devote attention to this subject, and while the concept of luxury goods appears, the concept of brand is not really researched.

Consumption is a ritual process: its primary function is to make sense of events and goods communicate categories of culture, they are artifacts of rituals [16]. Rituals are the process whereby meaning is transferred from the product to the consumer. Objects turn into symbols, building a coherent system with meaning; the symbolic side prevails over the utilitarian side [4]. Not only does the object, but the brand also has a role in the consumption ritual. Since brands are used to create and maintain social distinction [21], it is expected that they play an important role in the consumption ritual.

The meaning of Luxury is specific to a society. While the categories of luxury and necessity remain stable, a good representing one or the other might migrate from one to the other as time goes by [6]. Luxury is based on the possession of a certain good, in the past this could have been a porcelain tea set. When the object became more popular, Luxury was linked to the brand of the tea set, like Limoges. The origin of many luxury brands is linked to a manufacturer of a particular good: the manufacturer's name served to identify the producer'.

The concept of Luxury is transitory. It is dynamic not static [6], in constant movement. First was on the product, then on the brand and moving nowadays to experience. So, not surprisingly Berry [6] states that luxury is not linked to a product's attributes of price, rarity, and exclusivity. It is not a question of what the producer defines as attributes, but rather what consumer experiences.

Consumers bring their subjective experiences of the world to bear on their consumption practices. This agrees well with Fiske's idea that producers can inscribe preferred meanings into artifacts but that these meanings might not be activated by consumers in the way producers want them to be [33].

The meaning of consumption, according to Veblen, is almost exclusively concentrated in social status. Conspicuous consumption is seen as a way of gaining honor through displaying waste. Indeed, dominant roles in society are often associated with the possession, and sometimes with the ostentatious consumption, but primarily with the redistribution of goods difficult to obtain [14]. Leibenstein [34] added two effects to Veblen's theory: the bandwagon and the snob. The first one concerns imitation, which is the manifestation of the need to look like the others. In this case, the demand for a product may increase because certain individuals already have this good An example is, a neighbor purchasing a particular car because he wants be seen in a similar car to the ones in his neighborhood. The bandwagon effect indicates a desire for conformity, a product is purchased mainly because it is owned by others. The consumer shows his membership to particular group, improving his self-concept by the use of well-known brands which signal affiliation to the desired group. The origin of snob effect is the need for differentiation. The consumer's desire is to be unique and exclusive. It is quite the opposite of the bandwagon effect: the demand for a particular product decreases because many people already have it.

Another interesting point regarding conspicuous consumption is that there is a difference between compulsory and voluntary behavior. In the case of voluntary behavior, the person generally wants to be showcase of their wealth or to demonstrate a higher social status. In the case of compulsory behavior, the person consume in order to protect their achieved social status from another social group or to protect their status within the same group.

Additionally, Vigneron and Johnson [53] propose two effects based on individual factors, factors that do not depend on the consumption of others: hedonism and perfectionism. In the first one, consumer will use the prestige brand to stimulate feelings and affective states, targeting his own pleasure. In the second, the perfectionist effect, the consumer demand for better quality and prestige brand serves as an indicator.

During the 1980 decade the yuppie movement promoted brands to a supreme position in consumption. Specific knowledge about brands' differences and meanings became important not only in the choice of the brand but also the knowledge of how to use it correctly.

Even more important than the know-how required for choosing and combining brands with occasions, is the requirement for a cultural capital. Bourdieu proposed a framework to analyze the logic behind consumption, explaining how taste is built. The classification moves within contemporary society [7]. Habitus generates a schema that creates cultural competency practices: it is "a system of durable and transposable dispositions which expresses systematic preferences" [8].

The elite dominate a set of information and master a level of education that requires money as well as time to acquire and also depends on social insertion. Fashion is a justification for the intentional search for distinction through appearance; pure ostentation is seen as pretentious and decreases the amount of distinction [7], [8]. The family trajectory is based on the amount of social, economic, cultural and symbolic capital that its members have had for several generation. Bourdieu made groups based on differences on two axes: ancestry in the bourgeoisie and possession of economic and cultural capital.

The *nouveaux riches* have a strong need to validate a newly acquired social position and so are very conscious of the necessity to consume conspicuously but with taste if they aspire to a higher status [32]. As Brazil is a country built by immigrants and social mobility is very common, to obtain status by consumption acquiring possessions is easier than getting it by birth (royalty) or professional achievement. The attitudes of many Brazilians towards luxury tend to be associated with pleasure and status exhibition.

In Brazil, consumers called "emergentes" traditionally try to appropriate the elite's pattern of consumption. They have imported cars, wear designer clothes and accessories. They live inside condominiums with large security systems and have domestic workers who live in private rooms near the kitchen. These consumers like to point out that their upward movement in the social hierarchy is a result of labor, discipline and entrepreneurship. The media celebrates their success but also mocks their efforts to respect the protocol copied from the elite. These women are described as "peruas"(female turkey in Portuguese) with a connotation of frivolity and bad taste; their consumption tends to be classified as excessive and ostentatious [35]. Negative opinions about the "emergentes" appear in magazines and talks.

The proliferation of cultural meanings and the fragmentation of identity weaken the link between social categories and consumption patterns [28]. Class distinctions and ideological differences between social classes are less visible, but as inequality in income grows with a knock-on effect on spending patterns, class perceptions are affected. As the importance of social class declines, some authors claim that other social differentiations, like authenticity, become of greater concern to consumers [24].

The main focus is on how to deliver authenticity – a real experience – during the ritual of consumption in such a way that luxury will pass from the brand/object to the consumer. The issue of real versus fake is not the same as the issue of true versus false; being true to oneself does not define truth itself [24].

When there is a discontinuity of knowledge with regards to the movement of the commodity from producer to consumer, expertise and authenticity will arise. Once the distance between luxury good and consumer shrinks, the issue of exclusivity provides a platform for authenticity. Under pre-modern conditions, the geographical distance between producer and consumers made costs higher, as a consequence, the acquisition cost characterized exclusivity and underlined the sumptuary distinction. Nowdays technology enables the mass reproduction of luxury objects, and the dialogue between the consumer and the producer is more direct. Moreover, middle class consumers are now capable economically of vying for these objects. Therefore, the way to preserve the distinctive function of these objects is by increasing the criterion of authenticity [2].

Luxury must remain inaccessible for majority of the population to maintain the social distinction it produces, but the concept goes beyond the object or the brand.

14.2.3 Modes of consumption of fake Luxury

The objective here is to analyze the purchase and the use of counterfeit luxury brands from the consumer's point of view, investigating this phenomenon through their speech and language (for more details regarding methodological aspects, see Strehlau [50]). There are different theories to explain the luxury and counterfeit consumption; here the emphasis will be placed on the marketing and sociological approach background. Below is a brief presentation of these approaches.

One of the major problems plaguing the luxury industry is the lack of respect for intellectual property; the counterfeit products that result from this can be high or low quality [50], [23]. Many researchers have attempted to define profiles for counterfeit luxury brand buyers, however, in Brazil we find hybrid consumers, consumer that possess genuine items as well as forgeries (usually another model or product type).

The purchase of a fake luxury brand does not occur as a result of deception [25] – the consumer often realizes through a hints at the point of sale, the price and the style, that the products offered are not originals. The purchase then takes place because the consumer perceives greater value for money or because he perceives a value in the label and / or design itself.

Two factors help a consumer make a conscious choice when acquiring a counterfeit product. The first is the increasing acceptance of these products by members of a social group and the second is the emergence of new criteria for evaluation. As the fake represents to some degree the brand being copied, there is a desire for the brand, but not necessarily the product [23].

Tom at al sought to identify the consumers most likely to buy fakes and to determine what product attributes would attract them. Here we focus on consumers that prefer fake products, no matter what their perception of the quality [52].

Counterfeit brands affect the prestige of the original. Bourdieu claimed that when there is a significant spread across different strata of society, the effect is a decrease of rarity and threat to the distinction of the owner [8].

14.2.3.1 Method

The qualitative field research was conducted through in-depth interviews with consumers of luxury counterfeited products. The need to find individuals who were accessible and willing to provide information led to the use of a snowball sample. The selection of the initial group was made by e-mail asking for contact details of people who had bought counterfeit products. Interviews were conducted by the researcher, recorded and later transcribed. The saturation criterion for the responses showed that the number of interviews, 15 in total, was sufficient; content was repetitive and similar.

This consumer is particularly difficult to access due to illegal nature of the fake luxury trade. Approaching the consumer at the point of sale is not recommended for two reasons. Firstly, the researcher is seen as a supervisory or a "dangerous person" by both seller and buyer. Secondly, speaking to these consumers on the street is not recommended as the sensitive content of the discussion can cause distrust, and also danger.

The interview script covered three areas. In the first one, the focus was on the family, its origin and social trajectory, with the aim of identifying the evolution of cultural capital and the development of *habitus*.

The second was focused on *Habitus* and interviewees reference of brands. Here the interviewees were stimulated to talk about themselves as consumers, their relationship with brands and also to show their knowledge about luxury brands.

The trajectory of the family was used to evaluate the *habitus* of the respondent, particularly through the occupations of their ancestors: paternal and maternal grandparents, parents and the respondent. The occupations cited by respondents were classified into two categories prepared according to the socio-professional categories used by Bourdieu [8]. The original classification used by Bourdieu was however adapted because new professions have emerged since the survey was carried out in France in 1963 (complemented in 1968 and 1969). In addition, France has more stable social strata than Brazil, where the immigration movement and social mobility is much stronger. The two categories are: a) popular strata: artisans, small traders, small and middle managers, secretaries, office workers, technicians, instructors; b) elite stratum: higher education teachers and secondary producers of art, professionals, industrialists, managers, engineers. Many respondents were unsure of the schooling of their ancestors because Brazil has gone through five major reforms in the school system in the last 100 years.

The first classification of the occupations of the respondents was made by the author, a second by a doctoral student reading the interviews, a third academic person then provided an opinion to resolve the divergences.

Habitus develops signs of respect. The *etiquette* and *savoir vivre* that distinguish the acts of one person, therefore, can be evaluated through their pattern of consumption. During the interview the researcher therefore tried to determine the framework for the brands with which the consumer interacts actively. In other words, what the consumer's knowledge was with regards to brands in general and with regards to luxury brands in particular.

The topic of the consumption of a fake luxury brand was introduced: how the interviewee bought the product, where it was the purchased, where and how it was used, and their overall evaluation of it.

The final area explored through the script looked at the forms of consumption of a fake luxury brand, focusing on how the person buys and uses the product.

14.2.3.2 Findings

The consumption of counterfeit brands is related to the knowledge necessary for their "correct" consumption. This means knowing where to shop to get better quality (appearance and functionality) and how to use it (at what occasions, how to proceed in social interactions in order to obtain social benefits from the fake). These consumers have specific knowledge and share common interests, building a kind of complicity.

The concept of family trajectory explains the purchase and use of counterfeits. Three recurring patterns came to light in the interviews, with the result that three groups were formed.

The first group of consumers belongs to the traditional elite, i.e., those with stability at the top of the social hierarchy; they admitted having purchased a counterfeit and presented a specific pattern of consumption. They have several luxury brands that are present in everyday life such as perfumes or cars, but when they were asked about luxury, they spontaneously mentioned travels and moments of life, not luxury brands they possess or like. They position themselves as shapers of taste, arbiters of elegance whose transgressions are not errors but provocations or announcements of a new fashion, forms of expression or actions with the objective of imposing themselves a model [8]. In their speech there is a clear use of the strategy of distinction, although it is not experienced as such [9]. It means they also like to demonstrate their freedom of choice and of consumption.

Under these rules, noted throughout their speech, the brand elements "name" and "logo" are devaluated, and knowledge is enhanced as a definer of a luxury product. Two forms of ostentation of freedom appear in their speech: the addition of requirements, imposing more elements for appropriate use, and the addition of new rules of consumption. For example, one consumer had a watch collection with approximately 100 pieces from the Omega brand, both originals and counterfeits. This expresses a transgression of the limits of necessity and, at the same time, increases the level of requirements for consumption; it implies economic and cultural capital. Bourdieu [9] further stated that a competence for consumption only has value if there is a market for it. This stable member of the elite group of consumers expresses his freedom in this regard, clearing showing that luxury is carried by the individuals and not by the brands they posses.

> "We have a friend who provides it: perfect copies. This friend, for example, today [he] brought to Tony a copy of Vacheron Constantin's that I ordered immediately. I have a real VacheronConstantin [...]" (Interview number 12).

This group of consumers buys counterfeit luxury goods and they display them in a specific way. Firstly, they provide the circumstances of the purchase followed a judgment of the fake's similarity to the original. The purchase of the counterfeit product occurs outside of Brazil, mainly in the USA (New York) or in China; or it takes place through a "representative" who sells the best quality fakes. They then share cosmopolitan references, similar to what Holt [28] pointed out as being a characteristic of American high cultural capital. Secondly, social interactions are an opportunity to share rare and extravagant experiences in picturesque places where the fake was bought. These are authentic experiences [24]. Talking about an experience at an official brand store can be seen as pretentious behavior or as uninteresting, in their circle all or almost all have already been inside a flagship store. These users master the rules and norms of consumption and "legitimize" the fake. They mix original brands with the fake brands to compose an aesthetic visual, with consideration of the quality limitations of each forgery. This group has close contact with genuine products and brands.

> „[...]I learned this from her: she wears a Issey Myake suit , Ferragamo shoes and then she has a fake Fendi that she bought around the corner from a big black guy holding many, many bags. [...] No one thinks it is fake Fendi. Because the rest is not. [...] it is a posture"(interview no. 7)

Interviewees moving between social levels depend more on the brands to express their position. This group attempts to follow elite's pattern of consumption, however they sometimes present characteristic elements of the popular stratum. These consumers possess a formal education (the sign of cultural competence) and seek to demonstrate their knowledge during informal meetings, for example, by discussing brands. Thus, they demonstrate awareness of the taste system promoted by the original brand. The use of counterfeit brands provides a platform to show that the individual both recognizes and is aware of the brand values and its symbolism. As luxury brands are "tangible parameters of taste" [31], it is important to demonstrate cultural capital, in this case knowledge of luxury brands and of their counterfeits as well as of how to compare them. Since those with high institutionalized cultural capital are able to synthesize and manipulate information [28], they use their cultural competence in order to balance out their economic capital constraints. For example, a consumer who uses a counterfeit product of good quality to make success visually apparent and to leverage social capital in professional situations.

> "In my consulting activity: there comes a wretched man, disheveled, with a shoe stuck: they will not believe the person [...] So for me the matter of clock, pen, suit, tie, it's just professional image. Just that. Nothing more. I do not like walking in a suit over the weekend or at night. Things I do not have to put. An image you have to give".(interview number10)

In this kind of social interaction there is no claim that the product passes for the original; reporting that it is false offers an opportunity to demonstrate the knowledge embedded in the act of consumption (by choosing the best forgery) and finally they emphasize a cosmopolitan view and proof of self-realization through having traveled abroad. The "sin" of pretending to have distinction through the use of a fake product is thus avoided.

The concept of brand is emptied or reduced in discussions with interviewees situated in the popular stratum, without any ascending social movement. A luxury brand means no more than a name and a logo to them; they judge quality upon the presence of the logo. A parallel with the analysis of Bourdieu [8] on works of art can be drawn. In the former, the viewer ends up being pushed to reproduce the artist's original operation in a naive exhibitionism. In the latter, the consumer is looking for the distinction in an primary exhibition; there is a poor mastery of the concept of luxury. These respondents are essentially restricted to the position of spectator: looking at luxury brands from outside the store's glass window.

> "I arrived at the stall and said – Wow! Perfect. With label and everything! It's just like it... put the label, quality is the same."

They display the label with the brand name with the conscious or unconscious intention of achieving distinction; it is a component of their aesthetic disposition. That is, the look of popular aesthetic is based on the continuity of art and life, and it is explicit in the subordination of form to function. One interviewee said „jeans are jeans". Another asked himself the question: what is the difference between an expensive purse and a cheap one. They compare the fake to the original based upon the performance of the proposed function and the presence of the brand's logo. The relationship of this group with the consumers of these brands is contentious.

14.2.4 Final remarks

A large part of the Brazilian market is still learning about luxury brands from high luxury to *masstige* items. The consumption of luxury products will increase in the next years, due not only to economic growth but also to cultural reasons. It is worth remembering that in the XVIII century, because of their need not to appear "provincial", settlers in the United States were manipulated by fashion more easily than they would have been in a large city [10].

The growth of aspirants and socially emerging consumers coincides with the so-called strategy of new luxury. The taste of the elite tends to be redefined; it will manifest itself in another way to create social distinction. Referring back to history again, the American colonial elite adopted three strategies to respond to competitive consumption and fashion: sticking to old, traditional behaviors; position themselves in favor of simplicity and restraint in consumption; and competing consuming more [51]. Drawing a parallel, we can expect the elite to ignore new brands and new luxury strategies, to stick to traditional and historical brands, to adopt the "no logo" movement and to buy more and more brands. The last approach is highly compatible with the consumption of fake brands, the consumer having a great quantity of goods (genuine or not). For consumers with social mobility using counterfeit of luxury brands gives them an opportunity to demonstrate taste.

These elite consumers then impose additional requirements for the classification of a brand as a luxury which need to be fulfilled by companies. One approach so far has been launching new and more expensive models, giving the opportunity for the consumer to show their habitus. Another approach is binding a luxury brand with art.

The widespread and trivialization of luxury brands is leading them inexorably to the ordinary world. "The perpetual re-creation of the difference lies in luxury" [11]. The authenticity criterion of luxury brands is in checkmate. The brand itself doesn't have an inherent value: the value comes from consumers'judgment of its value. In future, luxury will be represented in another way, like through authentic experiences.

References

[1] Allérès, D. (1997): Luxe: Stratégies marketing, 2ed., Paris: Economica.

[2] Appadurai, A. (1986): Introduction: commodities and the politics of value, in: (Ed.) The Social Life of Things Commodities in cultural perspective. Cambridge University Press.

[3] Atwal, G./Williams, A. (2009): Luxury brand marketing — The experience is everything! Journal of Brand Management 16, 338–346

[4] Baudrillard, J. (1995): A sociedade de consumo. Lisboa. Edições 70.

[5] Bellaiche, J./Mei-Pochtler, A./Nanisch (2010): The new World of Luxury. Cought between growing momentum and lasting change. The Boston Consulting Group – BCG.

[6] Berry, C. (1994): The Idea of Luxury. A conceptual and historical investigation. New York: Cambridge University Press.

[7] Bourdieu, P. (1974): A economia das trocas simbólicas, São Paulo: Perspectiva.

[8] Bourdieu, P. (1979): La Distinction. Critique sociale du jugement. Paris: les Éditions de Minuit. 1979

[9] Bourdieu, P. (1983): Gostos de classe e estilos de vida, in: Ortiz, R. (org.) Pierre Bourdieu: sociologia. São Paulo: Ática.

[10] Carr, L. G./Walsh, L.S. (1994) Changing Lifestyles and Consumer Behavior in Colonial Chesapeake, in: Hoffman, R./Albert, P. (eds.), Of Consuming Interests. The Style of Life in the Eighteenth Century. Charlottesville: University Press of Virginia.

[11] Conquery, N. (1998): L'Hotel Aristocratique. Le marché du luxe à Paris au XVIIIe siècle, Paris: Publications de la Sorbonne.

[12] Critério Brasil http://www.abep.org/novo/Content.aspx?ContentID=301

[13] Dal-Bó, G. (2011): Clube do Milhão. América Economia, São Paulo, p. 26-32, agosto

[14] Daloz, J.-P. (2010): The Sociology of Elite Distinction. From theoretical to comparative perspectives. Palgrave macmilan:Houndmills.220p

[15] Daloz, J.-P. (2010): Ostentation in comparative in comparative perspective: Culture and elite legitimating Comparative Studies of Culture and Power. Comparative Social Research, v. 21, pp.29–62

[16] Douglas, M/Isherwood, B. (1979): The uses of goods from the world of good. London: Routledge.

[17] Dubois, B./Czellar, S. (2002): Prestige brands or luxury brands? An exploratory Inquiry on Consumer Perceptions, in: 31st European Academy Conference Preeciddings, Braga, Portugal. European Marketing Academy.

[18] Dubois, B./Laurent, G./Czellar, S. (2001): Consumer rapport to luxury. Analyzing complex and ambivalent attitudes. Cahier de Recherche.

[19] Eastman, J.K. et al. (1991): Status consumption in consumer behavior: Scale development and validation, in: Journal of Marketing Theory.

[20] Farah, A.(2011): Swarovski planeja abrir 76 unidades na América Latina, in: Brasil Econômico, São Paulo, 23 de agosto.

[21] Featherstone, M. (1995): A cultura do consumo e pós-modernidade, São Paulo: Studio Nobel.

[22] Gaspar, P. (2010): Hermès fecha o ano de 2009 com crescimento de 8,5 % nas vendas, in: Gestão do Luxo, 9 abril. Disponivel em: <http://www.gestaodoluxo.com.br/>. Acesso em: 12 abril 2010.

[23] Gentry, J./Putrevu, S./Shutz II, C./Commuri, S. (2001): How now Ralf Lauren? The separation of brand and product in a counterfeit culture, in: Advances in Consumer Research, v.28, pp.258-265.

[24] Gilmore, J. H./Pine II, B. J. (2007): Authenticity. Boston: Harvard Business School Press.

[25] Grossman, G./Shapiro, C. (1988): Foreign counterfeiting of status goods, in: The Quarterly Journal Of Economics, Feb. pp.79-100.

[26] Guerra, C. (2009): Fogo amigo no mercado de luxo, in: Isto é Dinheiro, São Paulo, 27 dec. 2009.

[27] Holanda, I. (2011): Design de interiores cresce por meio de polos de luxo, in: DCI, São Paulo, p. A8, 11 out.

[28] Holt, D.B. (1998): Does cultural capital structure American consumption, in: Journal of Consumer Research, v.25, pp. 1-25.

[29] IBGE Instituto Brasileiro de Geografia e Estatística
[30] Kapferer, J.-N./Bastien, V. (2009): The specificity of luxury management: Turning marketing upside down. Journal of Brand Management. 16, 311 – 322.
[31] Kapferer, J.-N. (2003): As Marcas Capital da Empresa: criar e desenvolver marcas fortes. Porto Alegre: Bookman. 459 p.
[32] Lauman, E. O./House, J. S. (1973): Living room styles and social attributes: the patterning of material artifacts in a modern urban community, in: Kassarjian, H. H./Robertson, T. S. Perspectives in consumer behavior. Glenview, Ill.: Scott Foresman.
[33] Lee, M. J. The Consumer Society Reader. Malden: Blackwell, 2000. 325 p
[34] Leibenstein, H. (1950): Bandwagon, snob and Veblen effects in the theory of consumers' demand. The Quarterly Journal of Economics, 1950
[35] Lima, D. N. D. O. (2005): Sujeitos e Objetos do "Sucesso". Uma etnografia do brasil emergente. Rio de janeiro: Universidade Federal do Rio deJaneiro. Museu Nacional.
[36] LVMH Moet Hennessy Louis Vuitton SA Company Profile. (2010) DATAMONITOR. 30 Sep
[37] Marcelo, P.(2010): Entrevista de Alan Parker Crean: O prazer é o conceito da marca BMW, in: Propmark, São Paulo, p.7, 22 março.
[38] McCracken, G. (1988): Culture and consumption: new approaches to the symbolic character of consumer goods and activities. Bloomington: Indiana Press University.
[39] Nia, A./Zaichkowsky, J. (2000): Do counterfeits devalue the ownership of luxury brands? Journal of Product and Brand Management, v.9, 7, pp.485-497.
[40] O'Shaughnessy, J./ O'Shaughnessy, N. J. (2008): The dominance of the adaptive unconscious: The undermining of beliefs in the autonomy and rationality of consumers. New York: Routledge.
[41] Oscar, N. (2010): Mercado de jatinhos avança no pais. O Estado S. Paulo, São Paulo, 12 de agosto de 2010 p.B14
[42] Pochman, M. et al (2004): Atlas da exclusão social no Brasil. Os ricos no Brasil. v.3 São Paulo: Cortez.
[43] Ribas, S. (2011): Brasil tem quase um terço dos milionários da AL, in: Jornal do Comercio, Rio de Janeiro, p. A-3, 14 jun.
[44] Ricciardi, A. (2011): Brasileiros devem gastar R$ 20,94 bi em itens de luxo, in: DCI, São Paulo, 28, 29 e 30 maio, Primeiro Caderno, p. A9
[45] Rolli, (2010): Mercado de luxo espera crescer o triplo do PIB, in: Folha de S.Paulo, São Paulo, Mercado, p. B4.
[46] Serraf, G. (1991): Le produit de luxe: somptuaire ou ostentatoire? Revue Française de Marketing, n.2-3, p.7-15.
[47] Sherry JR, J. F. (1991): Postmodern alternatives: the interpretive turn in consumer research, in: Robertson, T. S./Kassarjian, H. H. (1991): Handbook of Consumer Behavior. Englewood Cliffs, New Jersey: Prentice-Hall, pp. 548-591
[48] Sicard, M.-C. (2008): Luxe, Mensonges & Marketing. 2. ed. Paris: pearson education.
[49] Silverstein, M./Firske,N. (2005): Trading Up: why consumers want new luxury goods – and how companies create them. London: Penguin Group. 290p.
[50] Strehlau, S. (2008): O Marketing do luxo. São Paulo: Cengage.
[51] Sweeney, K. (1994): High–Style Vernacular: Lifestyles of the Colonial Elite, in: Hoffman, R./Albert, P. (eds.), Of Consumig Interests. The Style of Life in the Eighteenth Century. Chalottesville: University Press of Virgin.
[52] Tom,G./Garibaldi,B./Zeng, Y./Pilcher,J. (1998): Consumer demand for counterfeit goods, Psychology and Marketing, v.15 (5), Aug. pp. 405-421.
[53] Vigneron, F./Johnson, L. W. (1999): A Review and a Conceptual Framework of Prestige. Academy of Marketing Science Review Available: http://www.amsreview.org/articles/vigneron01-1999.pdf
[54] Weiling, J. (2010): Público de alta renda corresponde a 3% da população brasileira, in: Propmark, São Paulo, p. 4, Agosto.
[55] World Wealth Report (2011): Capgemini and Merryll Lynch.
[56] http://www.abrael.org.br

15 Luxury Longing and Counterfeit Complicity: A Consumer Typology based on the Perception of Luxury Value and Counterfeit Risk

Christiane Klarmann, Klaus-Peter Wiedmann, Nadine Hennigs

15.1 Introduction

In last decades, there has been an enormous theoretical as well as practical debate on concepts of consumer misbehavior, which also included the purchase of counterfeit goods, due to the increasing economic importance of this illicit market. Counterfeits constitute an important economic, political and social issue [12], [73]. The market for counterfeit goods has spread to almost all product categories, is estimated to account for as much as ten percent of world trade [66], and is often seen as one of the fastest growing industries in the world [40]. More precisely, according to the International Chamber of Commerce (ICC) based on 2008 data, the total global economic value of counterfeit and pirated products accounts for as much as $650 billion every year and estimated to be $1.77 trillion in 2015 [36].

Poddar et al. [58] argue that *"...counterfeit purchases are one of the most serious problems facing luxury brand marketers."* Counterfeiting harms the legitimate producers, as substandard imitations may affect consumers' confidence in the genuine brand, destroy brand equity and companies' reputation, cause loss of revenues, and threaten consumer health and safety [12], [26], [28]. Especially in a luxury product context, the prevalence of low-cost counterfeits may also reduce the perceived exclusiveness [89], while threaten revenues and intangible values of the genuine brands [14]. Although counterfeits encompass various negative effects consumers disregard these issues [57].

Even though the purchase of counterfeits may include risks in getting caught by authorities [2], the demand for counterfeit products still increases [36]. Counterfeiting constitutes an international phenomenon; nevertheless, punishment for participating in this illegal market at both the supply and the demand side is internationally inconsistent [15]. Against this backdrop, a better understanding of the antecedents of counterfeit demand will enable marketers to implement effective customized countermeasures. Therefore, focusing on consumers' motives for counterfeit consumption as well as drivers for counterfeit non-consumption becomes more worthwhile than ever before. Given that the market for counterfeit luxury brands relies on consumers' desire for real luxury brands [34] [54], it is critical for researchers and marketers to investigate both, consumers' perceived luxury value and perceived counterfeit risks as drivers affecting buying behavior in the trade-off between authentic or counterfeit products. Reasoning this, the aim of the present study is to begin filling this research gap by identifying groups of consumers who differ in their value and risk perception related to the trade-off between genuine and counterfeit luxury goods.

15.2 Theoretical Background

15.2.1 Luxury Consumption

Although the concept of luxury becomes more often scientifically considered, still, the term "luxury" elicits no clear understanding. This might be due to its subjective concept that depends on each consumer's perception of indulgent value [55], [9], Tyan et al. [77] state that luxury starts where the ordinary ends, whereby boundary is a matter of degree as judged by consumers. Generally, luxury can be seen as goods for which the simple use or display of a particular branded product brings esteem for its owner, as apart from functional needs luxury goods also enable consumers to satisfy psychological needs [83]. As a subjective and multidimensional construct, a definition of the concept of luxury should follow an integrative understanding [87], [88]. Following Vigneron and Johnson [82] luxury brands can be seen as the highest level of prestigious brands encompassing several physical and psychological values. Thus, to evoke exclusivity, brand identity, brand awareness, as well as perceived quality from the consumer's perspective becomes essential for luxury brand management [55]. Although the term "buying to impress others" has often been used in order to explain luxury consumption behavior, with reference to Wiedmann, Hennigs and Siebels [87] interpersonal aspects like snobbery and conspicuousness [39], [42] should be enhanced through personal factors such as hedonism and perfectionism [21] as well as situational conditions (e.g., economic and societal factors) [82], [83]. Therefore, the perceived luxury value can be explained through four dimensions: financial, functional, individual and social value [87], [88].

15.2.2 Counterfeit Consumption

Counterfeits are products which are bearing a trademark that is identical to, or indistinguishable from, a trademark registered to another party, and thus are infringing the rights of the holder of the trademark [10]. More specific, counterfeiting can be defined as *"...any manufacturing of a product which so closely imitates the appearance of the product of another to mislead a consumer that it is the product of another or deliberately offer a fake substitute to seek potential purchase from non-deceptive consumers"* [51]. According to this it has to be distinguished between deceptive and non-deceptive counterfeit consumption [29]. In case of deceptive counterfeiting, consumers can be regarded as a victim that unknowingly purchases a counterfeit product [56]. Markets for automotive parts, pharmaceuticals and medical devices are often threatened by deceptive counterfeiting [29]. Non-deceptive counterfeiting is related to the situation when the consumer knows or strongly suspects that the purchased product is not an original [29]. Customer complicity is often mentioned related to this buying behavior, as counterfeits are actively demanded [15]. This article focuses on non-deceptive counterfeiting, which is prevailing in the luxury market [48], and enables the identification of consumers' perceptions which influence their choice processes in the trade-off between genuine and counterfeit luxury goods [11]. The most promising way to defeat counterfeiting is to reduce the demand, which underlines the importance of investigating the antecedents of the counterfeit buying intention.

Existing studies have investigated the consumer's attitudes to as well as underlying assumptions for counterfeit consumption. According to Wiedmann, Hennigs and Klarmann [86], it can be suggested that the trade-off between genuine and counterfeit is determined by the value of the original. Furthermore, we assume that the choice-process is significantly influenced by the risk perception in view of the faked product.

15.3 Conceptualization

15.3.1 The Conceptual Model

In order to identify groups of consumers who differ in their buying decision in the trade-off between genuine and counterfeit luxury goods, both perceived values and risk should be integrated into a single model.

As illustrated in **Figure 15.1**, the study presented here considers a combination of four value dimensions (financial, functional, individual, and social value specified as price, usability, quality, uniqueness, self-identity, hedonic, materialistic, conspicuousness, and prestige value) as well as four risk dimensions (financial, functional, individual, and social risk) related to the trade-off between luxury and counterfeit consumption.

Figure 15.1 The Conceptual Model

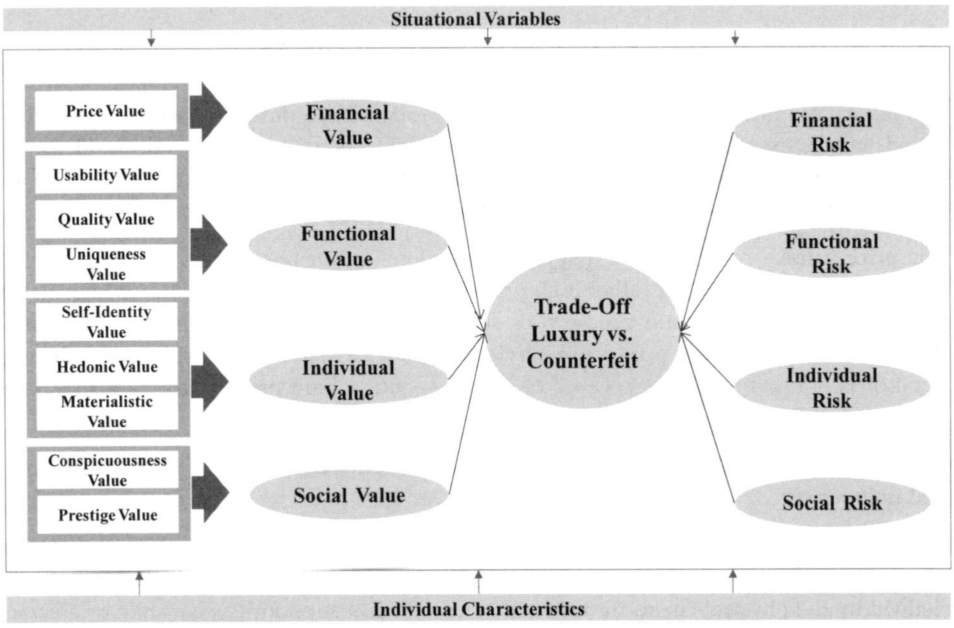

Assuming that the market for counterfeit luxury brands relies on consumers' desire for real luxury brands [34], [54], it is critical for researchers and marketers to investigate consumers perception of luxury value. Generally, values can be regarded as beliefs that guide the selection or evaluation of desirable behavior [65]. With regard to consumption values which directly explain why consumers choose to buy or avoid particular products [67], different types of values influence consumers' purchase choices. In relation to luxury consumption, Wiedmann, Hennigs and Siebels [87] revealed four dimensions – financial, functional, individual and social value – in order to enhance the current understanding of consumer motives and value perception. These four dimensions remain the basis for the presented model.

Besides any potential values and benefits consumers may also experience risks during their purchase decision process [31]. Generally, consumers try to reduce negative consequences and uncertainty before purchasing products [16], [18], and therefore it is assumed, that perceived risk primarily relates to potentially negative outcomes within the context of consumer behavior [72]. Basic approaches which analyzed the 'if' and 'how' perceived risk influences consumer choices in terms of non-deceptive counterfeits already exist (e.g. [10], [81], [31], [3], [54], [75]. Nevertheless, past research is lacking in contrasting consumer's perceived values and risks in the trade-off between genuine and counterfeit goods.

In this context, there is a research gap related to the segmentation of counterfeit and luxury consumers related to their perceived values and risks in terms of counterfeit and genuine luxury consumption.

15.3.2 Dimensions of Luxury Value Perception

In a luxury product context, the evaluation and propensity to purchase or consume luxury brands can be explained by four dimensions [87], [88]. As the consumer's demand for the counterfeit good is related to their desire for the original [34], [54], the trade-off between authentic and faked products will be affected by the four value dimensions [86].

The **price value** – related to the financial dimension – addresses direct monetary aspects. Prestige pricing, as setting a rather high price to suggest i.e. high quality and/or high status [43], may even make certain products or services more desirable [30]. Nevertheless, it is important to realize that a product or service does not have to be expensive to be a luxury good or is not luxurious just because of its price – but a high price might be a signal for exclusivity, uniqueness and high quality.

The functional dimension addresses product related attributes, such as usability, quality, and uniqueness:

The **usability value** is related to the performance and particular function of a product. The concept of usability has been examined and understood in terms of ease of use and can be defined by the physical-chemical-technical (e.g., technical superiority), concrete or abstract product/service dimensions (e.g., [53]). Consumers expect the item they buy to work right, look good, last a long time, and perform as expected and as promised (e.g., [24]). It has to be stated that usability needs to be differentiated between an objective and subjective judgment of usability which depends on individual evaluation and the specific purpose of use [87].

The **quality value** in terms of luxury is often referred to the fact that products are not being mass-produced, but often hand-made and thus, offer excellent product quality and performance as compared to non-luxury brands (e.g., [21], [22], [27], [63], [60], [48], [49], [83]). Consumers may associate luxury products with a superior brand quality and reassurance so that they perceive more value from it [1]. In accordance to this, the literature on luxury consumption often underlines the importance of quality to ensure the perception of and therefore the value of luxury [60], [61], [27], [30], [63].

The **uniqueness value** is related to the assumption that the perceived exclusivity and rareness of a limited product enhances the consumer's desire or preference for a brand [80], [41], [52]. Therefore, the more unique a brand is perceived, the more valuable the brand becomes [79]. The wish of consumers for differentiation and exclusivity can only be fulfilled when the consumption and use of a certain brand is only given to an exclusive clientele [39], [82], [83].

The individual dimension addresses personal matters such as materialism, hedonism, and self-identity:

The **self-identity value** refers to the internal (private) facet of one's self in terms of the way the individual perceives him or herself [45], [69], [37]. Concerning luxury brands, [59] confirmed the significant impact of self-congruity on luxury-brand purchase. From this point of view, consumers may use luxury items to integrate the symbolic meaning into their own identity [35], [83] or they use a luxury brand to support and develop their own identity [20], [33], [19].

The **hedonic value** is based on the fact, that certain products and services carry an emotional value and provide intrinsic enjoyment in addition to their functional utility [32], [67], [85]. Research concerning the concept of luxury has repeatedly identified the emotional responses associated with luxury consumption, such as sensory pleasure and gratification, aesthetic beauty, or excitement [8], [23], [64], [83].

The **materialistic value** is connected to the degree to which individuals principally find possessions to play a central role in one's life. The more materialistic a consumer is, the more likely he is to acquire possessions, to have positive attitudes related to acquisition, and to assign a high priority to material possessions [87]. Additionally, research has found that materialistic oriented consumers rely heavily on external cues, favoring those possessions that are worn or consumed in public places [62], [50], whereby consumers' possessions serve as a signal or source of communication to others for portraying and managing impressions of who they are and what their status or position is [20], [6].

The social dimension refers to aspects of status consumption and prestige orientation:

The **conspicuousness value** is linked to individuals' search of social status and representation. Conspicuous consumption means in particular that the ranking in a society is associated with a brand [87]. For example, [4] concluded that luxury goods consumed in public were more likely to be conspicuous goods than privately consumed luxury goods, and therefore, conspicuous consumption plays a significant part in shaping preferences.

The **prestige value** assumes that products often enclose prestigious values, and thus, social referencing and the construction of one's self appear to be determents of luxury consumption [87]. People's desire to possess luxury brands will serve as a symbolic sign of group membership. This bandwagon effect influences an individual to conform to affluent lifestyles and/or to distinguish themselves from non-affluent lifestyles [25], [68], [47], [70], [46], [44], [7], [19].

With reference to counterfeit luxury goods, it is expected that consumers who have a high value perception of genuine luxury goods are less willing to purchase counterfeits because he or she is worried about the buying decision and has a higher risk perception of the counterfeit alternative. On the other hand it might be that the positive perception of the luxury brand is transferred to the counterfeited alternative. Then, the faked good would fulfill e.g. the social value and the consumer might look in particular for the counterfeit product.

15.3.3 Dimensions of Counterfeit Risk Perception

Besides any potential values and benefits consumers may experience feelings of risk during the purchase decision process [31]. Within the context of consumer behavior the concept of perceived risk primarily relates to potentially negative outcomes [72]. The level of perceived risk of counterfeit luxury consumption might conclusively influence the trade-off purchase decision [86]. As in case of non-deceptive counterfeits, where the consumer is well aware or strongly suspects that the product is faked, the risk associated with its purchase might be higher [17]. Some scholars have already attempted to analyze how perceived risk influences consumer choices in terms of counterfeit goods (e.g. [78], [81], [31]). Evidence was found that consumers are aware of potential consequences of purchasing these illicit products, e.g. loss of money or dissatisfaction with the performance (see [78], [3]). Further research indicates perceived risk to have a significant negative effect on the attitude towards counterfeits [18].

In accordance to Stone and Grønhaug [72], the construct of perceived risk can be divided into six dimensions: financial, performance, physical, psychological, social and time-related risk. Nevertheless, to match the key luxury value dimensions, the present study focuses on the perceived financial, functional, individual and social risk:

The **financial risk** in terms of counterfeit consumption might be associated to the lower quality compared to genuine goods or a lack of performance which are often ascribed to counterfeit products (e.g. [54]). Therefore, the chance of a monetary loss can be regarded as being higher. Additionally, in the case of poor performance or usability, consumers may not have the opportunity to return or exchange the product because the illegal manufacturer remains largely unknown and no warranties are given [18], [16].

The **functional risk** is related to consumers' perception of counterfeits being inferior in quality, usability, as well as uniqueness as they are easily affordable (e.g. [84], [76], [16]). In their early work about product counterfeiting, Bamossy and Scammon [3] describe poor quality materials, poor performance, and manufacturing defects as 'trademarks' of counterfeit products. Their interviewees indicated these aspects as the three top reasons for dissatisfaction with the product.

The **individual risk** dimension is related to the consumers self-construct. Veloutsou and Bian [78] suggest that consumers may fear that consuming counterfeits will destroy their self-concept or self-esteem captures. Moreover, consumers' might fear that buying a fake product is not be perceived as very wise and therefore, consumers may feel uncomfortable with the purchase decision in hindsight or may even experience feelings of guilt [38]. This would be contrary to a hedonically motivated shopping experience.

The **social risk** refers to the consumer's fear, that the peer group may evaluate the possession of counterfeit goods in a negative way. This is especially important for publicly consumed goods [18] or when peers have expert knowledge and are able to distinguish authentic luxury from fake [56]. Especially when the impression on others is important, buyers of counterfeits face the risk of being detected wearing fake and "have to reckon with social sanctions" [54].

15.4 Methodology

In order to identify types of genuine and counterfeit luxury consumers in the context of our conceptual model, we used existing and tested measures, as shown in **Table 15.1**.

Table 15.1 The Questionnaire Scale

Scale	Author (s), year
Luxury Values	
Price Value	Yoo et al. [90] Sweeney and Soutar [74] Wiedmann et al., 2009 [88]
Functional Value	Wiedmann et al. [88]
Uniqueness Value	Wiedmann et al. [88]
Quality Value	Sweeney and Soutar [74]
Self-Identity Value	Wiedmann et al. [88]
Hedonic Value	Spangenberg et al. [71]
Materialistic Value	Richins and Dawson, [62]
Conspicuousness Value	Sweeney and Soutar [74]
Prestige Value	Bearden et al. [5], Wiedmann et al. [88]
Counterfeit Risks	
Financial Risk	Stone and Grønhaug [72]; Ha and Lennon [31]
Functional Risk	Stone and Grønhaug [72], Tan [75] , Ha and Lennon [31]
Individual Risk	Ha and Lennon [31]
Social Risk	Stone and Grønhaug [72]

All items were rated on five-point Likert scales (1 = strongly disagree to 5 = strongly agree). After examining the length and layout of the questionnaire and the quality of the items used, to investigate the research model, personal interviews were conducted among consumers in Germany in summer 2011. A total of 123 questionnaires were received. Regarding sample characteristics' gender distribution, 60.2% of the respondents were female. The mean age of the respondents amounts 26.2 years. Regarding the study context of luxury and counterfeit goods, 82.9% of the respondents have already bought a genuine luxury product at least once, 56.9% have already bought a counterfeit luxury product. Although this is not a representative one, with reference to the given research focus, the convenience sample used in this study offers a balanced set of data.

15.5 Results and Discussion

SPSS 19.0 was used to analyze the data. A principal component factor analysis with varimax rotation revealed a 13-factor structure that summarized 44 items with medium to high factor loadings. To conduct the cluster analysis, the factor scores for each respondent were saved.

Table 15.2 Cluster Results: Factor Mean Scores

	Cluster 1	Cluster 2	Cluster 3	Cluster 4	F	Sig.
% of the sample	37.4	19.5	22.0	21.0		
Luxury Value Dimension						
Price Value	-.17005	-.47456	.50480	.21470	5.482	.001
Usability Value	.22205	.80954	-.46752	-.65463	15.983	.000
Uniqueness Value	-.29069	.78251	.24641	-.46391	10.648	.000
Quality Value	-.26049	.87982	.14690	-.50383	12.303	.000
SelfID Value	-.17832	.16200	.42281	-.27312	3.106	.029
Hedonic Value	.26363	.92492	-.69453	-.59896	24.076	.000
Materialistic Value	.38578	.80714	-.68889	-.71220	26.159	.000
Social Value	.40416	.78406	-.46186	-.95917	29.415	.000
Prestige Value	.43730	.62574	-.56742	-.76204	20.814	.000
Counterfeit Risk Dimension						
Financial Risk	-.53597	.98141	.63609	-.61822	36.381	.000
Functional Risk	-.34462	.81240	.74904	-.91805	33.543	.000
Social Risk	-.15048	1.05835	.27681	-.99816	43.380	.000
Individual Risk	-.46819	1.06233	.61609	-.79206	5.482	.000

In our analysis, we used a combination of Ward's method of minimum variance and non-hierarchical k-means clustering. The results strongly suggested the presence of four clusters. With regard to classification accuracy, we also used discriminant analysis to check the cluster groupings once the clusters were identified; 95.1% of the cases were assigned to their correct groups, validating the results of cluster analysis for the useful classification of consumer subgroups based on the factors included in the model. To develop a profile of each market segment, more detailed information was obtained by examining the factor scores cross-tabulated by cluster segment, as presented in **Table 15.2** and **Figure 15.2**.

Figure 15.2 Cluster Segments

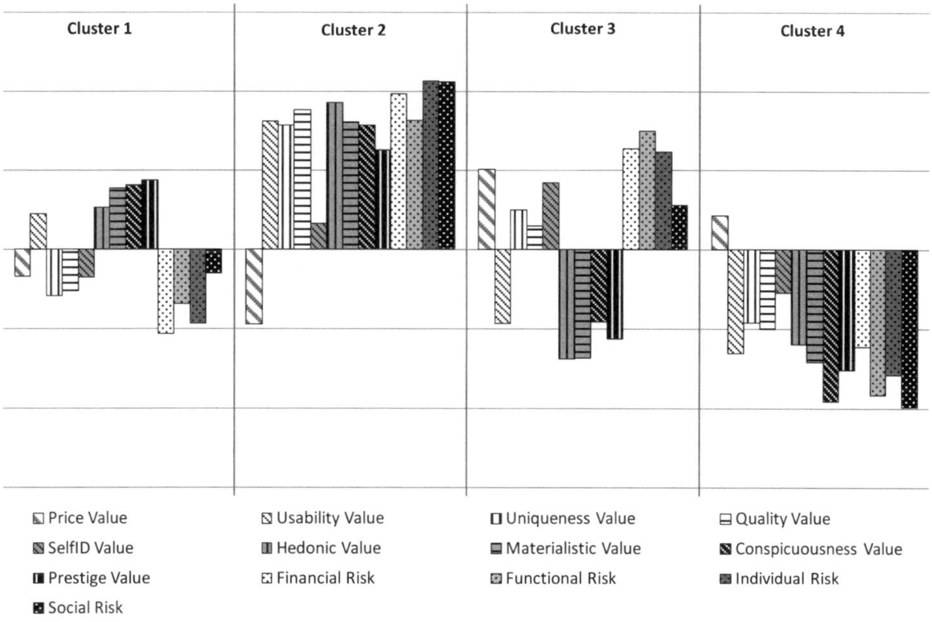

Based on the variables from which they were derived, the four clusters were labeled as follows:

Cluster 1: **The Prestige-Seeking Counterfeit Consumers** with the lowest mean age of 25.2 years and 50% female respondents form 37.4% of the sample. Referring to our study context, 96% of the respondents state that they purchased a genuine luxury good at least once. 73% have also already bought a counterfeit luxury product once, of whom 43% have already bought a counterfeit luxury product several times – this is the highest percentage of all groups. In terms of future purchase, with 52% (at least considering counterfeit consumption) this group has the highest intention to buy counterfeits (again). Typical consumers in this cluster can be considered as frequent consumers of fake luxuries, as evidenced by a low counterfeit risk perception combined with high counterfeit experience. Significantly more than cluster 3 and 4, they value the social-oriented aspects of luxury and are not likely to fear social sanctions, as compared to cluster 2 and 3.

Cluster 2: **The Risk-Averse Luxury Shoppers** with a mean age of 25.9 years and 67% female respondents form 19.5% of the sample. Overall, 88% of the respondents in this group state that they have already bought a genuine luxury product at least once, 54% have never bought a counterfeit luxury good. Except the price and self-identity value, this cluster possesses the highest factor means, for both, perceived luxury value and perceived counterfeit risk. Especially the low price value is remarkable, as this consumer group possesses the lowest income level. Even on this backdrop, referring to their future behavior, 83% consider buying authentic luxury goods in the future, and with the lowest percentage of the identified clusters, only 4% consider buying a counterfeit luxury product.

Cluster 3: **The Identity-Oriented Luxury Buyers** with the highest mean age of 27.4 years and 56% female respondents comprise 22% of the sample. The Identity-oriented Luxury Buyers perceive the highest financial value. Compared to cluster 2, the self-identity value is higher, whereas, the other individual values are lower. The social-oriented values are even below average. 74% of these respondents state that they have bought a genuine luxury product at least once. In accordance with the high risk perception, less than half of the respondents of this cluster already bought a counterfeit luxury product. In the future, 67% of the Identity Oriented Luxury Buyers consider buying genuine luxury goods, whereas 15% consider buying counterfeit luxury goods.

Cluster 4: **The Indifferent Risk-Takers** with a mean age of 26.7 years and 65% female respondents comprise 21% of the sample. In this cluster, overall, the perceived luxury value – except the financial value – is very low. Characteristically is the low counterfeit risk perception. 35% of this cluster have never bought a genuine luxury product, which is compared to the other clusters the highest percentage. Besides, 54% have never bought a counterfeit luxury product. Even if the risk perception is low; referring to future behavior, none of the respondents is sure about buying genuine or counterfeit luxuries. This might be due to the underrepresented luxury value perception. Nevertheless, 77% at least consider buying genuine luxuries and 31% consider buying counterfeit luxury goods.

Our results show that the four perceived value and risk dimensions are able to segment consumers referring to their decision in the trade-off between luxury and counterfeit consumption. In view of developing effective countermeasures, our results might be a motivational basis for ongoing research regarding the concept of perceived counterfeit risk. According to this, in the next section based upon a short conclusion, implications for further research as well as managerial implications will be presented.

15.6 Conclusions and Implications

The topic of counterfeiting has become more important for marketers and researchers in last decades. This might be due to the increasing global impact of counterfeiting, which effects are perceptible at both macro- and microeconomic level. Even though governments, supranational organizations, and industry associations have undertaken various attempts – as e.g. IPR protection and law enforcement – the business of counterfeiting will be a prosperous business as long as there exists a high demand for counterfeited brands. Focusing on the demand side – while investigating the specific consumer motivation for purchasing faked goods as well as their reasons for counterfeit non-consumption – builds the basis for the development of strategies that aim to reduce the global appetite for counterfeits. Therefore, this paper aimed to fill the existing research gap by identifying different consumer segments related to their perceived luxury value and counterfeit risk.

Interestingly, the Prestige-Seeking Counterfeit Consumers have both, the highest luxury as well as counterfeit experience. This confirms the assumption that both shopping behaviors are not independent from each other. Even though price is often believed to be the main reason that causes counterfeit purchases, this study reveals that there are multifaceted motives that affect consumer attitudes and behavior. Cluster 2, the consumer group with the lowest price value and the highest counterfeit risk perception, has the highest intention to buy genuine goods and the lowest intention to buy counterfeit goods in the future. Similarly, the Identity-Oriented Luxury Buyers with the second highest risk perception are not interested in counterfeit products in the future. In this context, the results confirm that counterfeit risk perception negatively affects counterfeit shopping behavior.

Understanding the motives of counterfeit consumption and the determinants of counterfeit resistance enables luxury brand managers to implement customized countermeasures. Based upon our results, the threat of a high counterfeit demand can be avoided by communicating the risks associated with faked products as well as highlighting the values of the genuine goods.

As our results revealed, the trade-off between authentic and counterfeit consumption is affected through four value and risk dimensions. With regard to possible directions for future research, to advance current research focusing on counterfeit shopping behavior, our first approach should be pursued in view of situational contingencies as well as emotional effects, probably in a combination of explicit and implicit measurement methods. Additionally, the effectiveness of countermeasures directed to the consumer's perceived counterfeit risk would constitute a promising field of research.

References

[1] Aaker, D. A. (1991): Managing brand equity: Capitalizing on the value of a brand name. New York: Free Press.

[2] Ang, S. H./Cheng, P. S./Lim, E.A.C./Tambyah, S. K. (2001): Spot the difference: consumer responses towards counterfeits, in: Journal of Consumer Marketing, Vol. 18, No. 3, pp. 219 – 235.

[3] Bamossy, G./Scammon D. L. (1985): Product Counterfeiting: Consumers and Manufacturers Beware, in: Advances in Consumer Research, Vol. 12, No. 1, pp. 334 – 340.

[4] Bearden, W. O./Etzel, M. J. (1982): Reference group influence on product and brand purchase decisions, in: Journal of Consumer Research, Vol. 9, pp. 183–194.

[5] Bearden, W. O./Netemeyer, R. G./Teel, J. E. (1989): Measurement of Consumer Susceptibility to Interpersonal Influence, in: The Journal of Consumer Research, Vol. 15, No. 4, pp. 473 – 481.

[6] Belk, R.W. (1985): Materialism: Traits aspects of living in the material world, in: Journal of Consumer Research, Vol. 12, pp. 265–280.

[7] Belk, R.W. (1988): Possessions and the Extended Self, in: Journal of Consumer Research, Vol. 15, No. 2, pp. 139-168.

[8] Benarrosh-Dahan, E. (1991): Le contexte lexicologique du luxe, in: Revue Française du Marketing, Vol. 132/133, No. 2-3, pp. 45-54.

[9] Bian, Q. & Forsythe, S. Purchase intention for luxury brands: A cross cultural comparison. Journal of Business Research, Available online 27 December 2011.

[10] Bian, X./Mouthino, L. (2009): An investigation of determinants of counterfeit purchase consideration, in: Journal of Business Research, Vol. 62, pp. 368–378.

[11] Bian, X./Mouthino, L. (2011): The role of brand image, product involvement, and knowledge in explaining consumer purchase behaviour of counterfeits: Direct and indirect effects, in: European Journal of Marketing, Vol. 45, No. 1/2, pp. 191-216.

[12] Bian, X./Veloutsou, C. (2007): Consumers' attitudes regarding non-deceptive counterfeit brands in the UK and China, in: Journal of Brand Management, Vol. 14, No. 3, pp. 211–222.

[13] Bian, X./Veloutsou, C: (2008): A cross-national examination of consumer perceived risk in the context of non-deceptive counterfeit brands, in: Journal of Consumer Behaviour, Vol. 7, pp. 3-20.

[14] Bush, R. F./Bloch, P. H./Dawson, S. (1989): Remedies for product counterfeiting, in: Business Horizons, January-February, pp. 59-65.

[15] Chaudhry, P./Stumpf, S. (2011): Consumer Complicity with Counterfeit Products, in: Journal of Consumer Marketing, Vol. 28, No. 2, pp. 139 – 151.

[16] Cordell, V. V./Wongtada, N./Kieschnick, R.L. (1996): Counterfeit Purchase Intentions: Role of Lawfulness Attitudes and Poduct Traits as Determinants, in: Journal of Business Research, Vol. 35, pp. 41 – 53.

[17] d'Astous, A./ Gargouri, E. (2001): Consumer evaluations of brand imitations, in: European Journal of Marketing, Vol. 35, No. 1/2, pp. 153 – 167.

[18] de Matos, C./Augusto, C./Ituassu, T./Rossi, C. A. V. (2007): Consumer attitudes toward counterfeits: a review and extension, in: Journal of Consumer Marketing, Vol. 24, No. 1, pp. 36 – 47.

[19] Dittmar, H. (1994): Material possessions as stereotypes: Material images of different socioeconomic groups, in: Journal of Economic Psychology, Vol. 15, No. 4, pp. 561-585.

[20] Douglas, M./Isherwood, B. (1979): The world of goods. New York: Basic Books.

[21] Dubois, B./Laurent, G. (1994): Attitudes toward the concept of luxury: An exploratory analysis, in: Asia Pacific advances in consumer research, Vol. 1, pp. 273–278.

[22] Dubois, B./Laurent, G. (1996): The Functions of Luxury : A Situational Approach to Excursionism, in: Advances in Consumer Research, Vol. 23, No. 1, pp. 470-477.

[23] Fauchois, A./Krieg. A. (1991): Le discours du luxe, in: Revue Française du Marketing, Vol. 132/133, No. 2-3, pp. 23-39.

[24] Fennell, G. G. (1978): Perceptions of the product-in-use situation, in: Journal of Marketing, Vol. 42, No. 2, pp. 39-47.

[25] French, J. R. Jr./Raven, B. H. (1959): Bases of social power, in: Cartwright, D. (Ed.): Studies in social power (pp. 150-167). Ann Arbor, MI: Institute for Social Research.
[26] Furnham, A./Valgeirsson, H. (2007): The effect of life values and materialism on buying counterfeit products, in: The Journal of Socio-Economics, Vol. 36, pp. 677-85.
[27] Garfein, R. T. (1989): Cross-cultural perspectives on the dynamics of prestige, in: Journal of Services Marketing, Vol. 3, No. 3, pp. 17-24.
[28] Green, R T./Smith, T. (2002): Countering brand counterfeiters, in: Journal of International Marketing, Vol. 10, No. 4, pp. 89-106.
[29] Grossman, G.M./Shapiro, C. (1988): Foreign Counterfeiting of Status Goods, in: The Quarterly Journal of Economics, Vol. 103, No. 1, pp. 79 – 100.
[30] Groth, J./McDaniel, S.W. (1993): The exclusive value principle: The basis for prestige pricing, in: Journal of Consumer Marketing, Vol. 10, pp. 10–16.
[31] Ha, S./Lennon, S. (2006): Purchase intent for fashion counterfeit products: Ethical ideologies, ethical judgments, and perceived risks, in: Clothing and Textiles Research Journal, Vol. 4, No. 4, pp. 297-315.
[32] Hirschman, E. C./Holbrook, M. B. (1982): Hedonic consumption: Emerging concepts, methods and propositions, in: Journal of Marketing, Vol. 46, pp. 92–101.
[33] Hirschman, E. C. (1988): Upper class WASPs as consumers: A humanistic inquiry, in: Hirschman, E. (Ed.): Research in Consumer Behavior, Vol. 3, pp. 115-148. Greenwich, CT: JAI Press.
[34] Hoe, L./Hogg, G./Hart S. (2003): Fakin' it: Counterfeiting and consumer contradictions, in: European Advances in Consumer Research, Vol. 6, pp. 60-67.
[35] Holt, D. B. (1995): How consumers consume: A typology of consumption practices, in: Journal of Consumer Research, Vol. 22, pp. 1–16.
[36] ICC Counterfeiting Intelligence Bureau (2005): The International Anti-Counterfeiting Directory 2005 [accessed on June 9th 2011].
[37] Jamal, A./Goode, M. (2003): A study of the impact of self-image congruence on brand preference and satisfaction, in: Marketing Intelligence & Planning, Vol. 19, No. 6/7, pp. 482-492.
[38] Kim, J.-E./Cho, H. J./Johnson, K. P. (2009): Influence of moral affect, judgment, and intensity on decision making concerning counterfeit, gray-market, and imitation products, in: Clothing and Textiles Research Journal, Vol. 27, No. 3, pp. 211-226.
[39] Leibenstein, H. (1950): Bandwagon, snob, and Veblen effects in the theory of consumers' demand, in: Quarterly Journal of Economics, Vol. 64, pp. 183-207.
[40] London, E. (2003): Criminal remedies against counterfeiting, in: Journal of Brand Management, Vol. 11, No. 2, pp. 137–140.
[41] Lynn, M. (1991): Scarcity effects on value: A quantitative review of the commodity theory literature, in: Psychology & Marketing, Vol. 8, pp. 45–57.
[42] Mason, R. S. (1992): Modeling the demand for status goods. Working paper, Department of Business and Management Studies, University of Salford, U.K. New York: St Martin's.
[43] McCarthy, E. J./Perreault,W. D., Jr. (1987): Basic marketing: A managerial approach. 9th ed. Homewood, IL: Irwin.
[44] McCracken, G. (1986): Culture and consumption: A theoretical account of the structure and movement of the cultural meaning of consumer goods, in: Journal of Consumer Research, Vol. 13, No. 1, pp. 71-84.
[45] Mehta, A. (1999): Using self-concept to assess advertising effectiveness, in: Journal of Advertising Research, Vol. 39, No. 1, pp. 81-89.
[46] Mick, D. G. (1986): Consumer research and semiotics: Exploring the morphology of signs, symbols, and significance, in: Journal of Consumer Research, Vol. 13, No. 2, pp. 196-213.
[47] Midgley, D. F. (1983): Patterns of interpersonal information seeking for the purchase of a symbolic product, in: Journal of Marketing Research, Vol. 20, No. 1, pp. 74-83.
[48] Nia, A. & Zaichkowsky, J. L. (2000): Do counterfeits devalue the ownership of luxury brands?,in: Journal of Product & Brand Management, Vol. 9, No. 7, pp. 485 – 497.

[49] O'Cass, A./Frost, H. (2002): Status brands: Examining the effects of non-product brand associations on status and conspicuous consumption, in: Journal of Product & Brand Management, Vol. 11, No. 2, pp. 7-88.
[50] O'Cass, A./Muller, T. E. (1999): A study of Australian materialistic values, product involvement and self-image/product-image congruency relationships for fashion clothing. Proceedings of the Ninth Biennial World Marketing Congress, Academy of Marketing Science.
[51] OECD (1998): The economic impact of counterfeiting, available: http://www.oecd.org/dataoecd/11/11/2090589.pdf, Retrieve date: 23rd July 2010.
[52] Pantzalis, I. (1995): Exclusivity strategies in pricing and brand extension. Unpublished doctoral dissertation, University of Arizona, Tucson, AZ.
[53] Park, C. W./Jaworski, B. J./McInnis, D. J. (1986): Strategic brand concept-image management, in: Journal of Marketing, Vol. 50, No. 3, pp. 135-145.
[54] Penz, E./Stöttinger, B. (2005): Forget the real thing-take the copy! An explanatory model for the volitional purchase of counterfeit products, in: Advances in Consumer Research, Vol. 32, pp. 568-75.
[55] Phau, I./Prendergast, G. (2000): Consuming luxury brands: The relevance of the "rarity principle.", in: Journal of Brand Management, Vol. 8, pp. 122–138.
[56] Phau, I./Teah, M. (2009): Devil wears (counterfeit) Prada: a study of antecedents and outcomes of attitudes towards counterfeits of luxury brands, in: Journal of Consumer Marketing, Vol. 26, No. 1, pp. 15-27.
[57] Phau, I./Sequeira, M./Dix, S. (2009a): Consumers' willingness to knowingly purchase counterfeit products, in: Direct Marketing: An International Journal, Vol. 3, No. 4, pp. 262 – 281.
[58] Poddar,A./Foreman, J./Banerjee S./Scholder Ellen, P. (2011): Exploring the Robin Hood effect: Moral profiteering motives for purchasing counterfeit products, in: Journal of Business Research, Available online 29 October 2011.
[59] Puntoni, S. (2001): Self-identity and purchase intention: An extension of the theory of planned behavior, in: European Advances in Consumer Research, Vol. 5, pp. 130-134.
[60] Quelch, J. A. (1987): Marketing the premium product, in: Business Horizons, Vol. 30, No. 3, pp. 38-45.
[61] Rao, A. R./Monroe, K. B. (1989): The effect of price, brand name, and store name on buyers' perceptions of product quality: An integrative review, in: Journal of Marketing Research, Vol. 26, No. 3, pp. 351-357.
[62] Richins, M./Dawson, S. (1992): A consumer values orientation for materialism and its measurement: Scale development and validation, in: Journal of Consumer Research, Vol. 19, No. 2, pp. 303-316.
[63] Roux, E. (1995): Consumer evaluation of luxury brand extensions. EMAC Conference, May, ESSEC, Paris, France.
[64] Roux, E./Floch, J-M. (1996): Gérer l'ingérable: la contradiction interne de toute maison de luxe. Décisions Marketing, Vol. 9, No. 2-3, pp. 15-23.
[65] Schultz, P. W./Zelezny, L. C. (1999): Values as predictors of environmental attitudes: Evidence for consistency across 14 countries, in: Journal of Environmental Psychology, Vol. 19, No. 3, pp. 255-265.
[66] Sharma, P./Chan, R.Y.K. (2011): Counterfeit proneness: Conceptualisation and scale development, in: Journal of Marketing Management, Vol. 27, No. 5–6, pp. 602–626.
[67] Sheth, J. N./Newman, B. I./Gross, B. L. (1991): Why we buy what we buy: A theory of consumption values, in: Journal of Business Research, Vol. 22, pp. 159–170.
[68] Sirgy, M. J. (1982): Self-concept in consumer behavior: A critical review, in: Journal of Consumer Research, Vol. 9, No. 3, pp. 287-300.
[69] Sirgy, M. J./Johar, J. S. (1999): Toward an integrated model of self-congruity and functional congruity, in: European Advances in Consumer Research, Vol. 4, pp. 252-256.
[70] Solomon, M. R. (1983): The role of products as social stimuli: A symbolic interactionism perspective, in: Journal of Consumer Research, Vol. 10, No. 3, pp. 319-329.

[71] Spangenberg, E. R./Voss, K.E./Crowley, A.E. (1997): Measuring the He-donic and Utilitarian Dimensions of Attitude: A Generally Applicable Scale, in: Advances in Consumer Research, Vol. 24, No. 1, pp. 235 – 241.
[72] Stone, Robert N./Grønhaug, Kjell (1993): Perceived Risk: Further Considerations for the Marketing Discipline, in: European Journal of Marketing, Vol. 27, No. 3, pp. 39 – 50.
[73] Swami, V./Chamorro-Premuzic, T./Furnham, A. (2009): Faking it: Personality and individual difference predictors of willingness to buy counterfeit goods, in: The Journal of Socio-Economics, Vol. 38, pp. 820-825.
[74] Sweeney, J. C./Soutar, G. N. (2001): Consumer Perceived Value: The Development of a Multiple Item Scale, in: Journal of Retailing, Vol. 77, No. 2, pp. 203 – 220.
[75] Tan, Benjamin (2002): Understanding Consumer Ethical Decision Making with Re-spect to Purchase of Pirated Software, in: Journal of Consumer Marketing, Vol. 19, No. 2, pp. 96 – 111.
[76] Tsai, S. (2005): Impact of personal orientation on luxury-brand purchase value, in: International Journal of Market Research, Vol. 47, No. 4, pp. 429-454.
[77] Tynan, C./McKechnie, S./Chhuon, C. (2010): Co-creating value for luxury brands, in: Journal of Business Research, Vol. 63, pp. 1156–1163.
[78] Veloutsou, Cleopatra/Bian, Xuemei (2008): A Cross-National Examination of Con-sumer Preceived Risk in the Context of Non-Deceptive Counterfeit Brands, in: Journal of Consumer Behaviour, Vol. 7, No. 1, pp. 3 – 20.
[79] Verhallen, T. M./Robben, H. S. (1994): Scarcity and preference: An experiment on unavailability and product evaluation, in: Journal of Economic Psychology, Vol. 15 (June), pp. 315-331.
[80] Verhallen,T. M. (1982): Scarcity and consumer choice behavior, in: Journal of Economic Psychology, Vol. 2, pp. 299–321.
[81] Vida, I. (2007): Determinants of consumer willingness to purchase non-deceptive counterfeit products, in: Managing Global Transitions, Vol. 5, pp. 253-270.
[82] Vigneron, F./Johnson, L.W. (1999): A review and a conceptual framework of prestige-seeking consumer behavior, in: Academy of Marketing Science Review, Vol. 1, pp. 1-15.
[83] Vigneron, F./Johnson, L. W. (2004): Measuring perceptions of brand luxury, in: Journal of Brand Management, Vol. 11, No. 6, 484-506.
[84] Wall, D. S./Large, J. (2010): Jailhouse Frocks – Locating the Public Interest in Policing Counterfeit Luxury Fashion Goods, in: British Journal of Criminology, Vol. 50, No. 6, pp. 1094 – 1116.
[85] Westbrook, R. A./Oliver, R. L. (1991: The dimensionality of consumption emotion patterns and consumer satisfaction, in: Journal of Consumer Research, Vol. 18, pp. 84–91.
[86] Wiedmann, K.-P./Hennigs, N./Klarmann, C. (2012): Luxury consumption in the trade-off between genuine and counterfeit goods: What are the consumers' underlying motives and value-based drivers?, in: Journal of Brand Management, Advance online publication 2 March 2012.
[87] Wiedmann, K.-P./Hennigs, N./Siebels, A. (2007): Measuring consumers' luxury value perception: a cross-cultural framework, in: Academy of Marketing Science Review, Vol. 7, pp. 1-21.
[88] Wiedmann, K.-P./Hennigs, N./ Siebels, A. (2009): Value-Based Segmentation of Luxury Consumption Behavior, in: Psychology & Marketing, Vol. 26, No. 7, pp. 625–651.
[89] Wilke, Ricky/Zaichkowsky, J. L. (1999): Brand imitation and its effects on innovation, competition, and brand equity, in: Business Horizons, Vol. 42, No. 6, pp. 9-18.
[90] Yoo, B./Donthu, N./Lee, S. (2000): An Examination of Selected Marketing Mix Elements and Brand Equity, in: Journal of the Academy of Marketing Science, Vol. 28, No. 2, pp. 195 – 211.

16 What is the Harm in Fake Luxury Brands? Moving Beyond the Conventional Wisdom

Aaron Ahuvia, Giacomo Gistri, Simona Romani, Stefano Pace

16.1 Introduction

When reading the literature on counterfeit branded luxury goods (henceforth counterfeit BLGs) in both the scientific and popular press, one quickly comes across what we will summarize as the conventional wisdom. We do not mean the phrase "conventional wisdom" to have disparaging connotations. On the contrary, this view is widely popular because it makes sense and fits with common observation. The conventional wisdom holds that:

a. counterfeit BLGs are a prevalent phenomenon;
b. the people who buy counterfeit BLGs generally know they are purchasing fakes, and hence are not defrauded when purchasing them;
c. the consumer of a fake BLG is highly unlikely to have purchased the genuine article had the fake not been available, and hence luxury goods firms do not lose a significant amount of sales from direct cannibalization by counterfeits;
d. nonetheless the legitimate producers of the faked BLGs are significantly victimized by counterfeiting because the existence of fakes reduces the desirability (i.e. brand equity) of the real product among consumers who might have purchased the genuine article.

This paper reviews published data and opinion as it applies to all four elements of the conventional view. While academic research is often criticized for confirming the obvious, we find that research on counterfeits has significantly challenged elements B, C and D of the conventional wisdom.

16.2 The Scope of the Problem

The conventional wisdom holds that the counterfeiting of BLGs is a highly prevalent phenomenon. This is the only element of the conventional wisdom which is consistently confirmed by the available research, although even here some estimates of the extent of the problem have been insupportably high. To understand the scope of the problem we start by looking at counterfeiting across all product categories, which in some analyses includes the illegal digital copies of music, movies, software, etc. In the popular imagination, counterfeit physical products (i.e. excluding digital piracy) are mostly thought to be the fake brand name luxury goods sold by sidewalk vendors. Today however, counterfeiting physical products extends into many segments of the economy, from auto parts to alcoholic beverages to pharmaceuticals [47]. It is estimated that counterfeiting amounts to an astonishing $ 600 billion a year globally and that between 5% and 7% of all goods traded are counterfeit [1]. A skeptic might point out that the International Anti-Counterfeiting Coalition has an interest in sounding the alarm about counterfeiting, and claiming that it is a huge business helps them do just that. Staake, Thiesse and Fleisch (2009) [47] provide an insightful critique of the assumptions and methods used that used to arrive at these staggering figures. But even if we approach these staggering estimates with a dose of skepticism, counterfeiting across all product categories is no doubt one of the planet's more widespread criminal activities.

In this paper we are concerned with counterfeit BLGs, rather that all counterfeit products. Finding estimates specifically on the volume of counterfeit BLGs is difficult. In the US fake BLGs make up about 33% of all counterfeit goods seized by customs, in Europe the figure is around 60% [10]. A re-analysis of data provided by Gallup (in Stewart 2005a [48]), suggests that around 27% of fake purchases in the US are fake luxury brands. But this 27% figure is based on self report data from consumers who admit to having purchased these goods. Therefore this figure is probably too low because it misses instances where the consumer withheld information about their illegal purchases from the pollster, and it misses instances in which the consumer unknowingly purchased a counterfeit and never discovered the deception. In the UK, almost half of all consumers (44%) have knowingly purchased fake designer clothing or footwear (Alliance Against IP Theft 2007), and this does not even include handbags, watches and other often counterfeited luxury goods. So while counterfeit BLGs are only a segment of the total counterfeit trade, they still constitute a large industry.

16.3 Consumers as Co-Conspirators

Grossman and Shapiro (1988) [27] introduced the terminology of deceptive and non-deceptive counterfeits, in which non-deceptive counterfeits are products whose consumers know they are buying fakes. Conventional wisdom dictates that most fake BLGs are sold as non-deceptive counterfeits [19], [44]. Buyers are presumed to know they are fakes because they are very low priced, purchased through a non-authorized outlet, and of noticeably low quality. The belief that fake BLGs are sold as non-deceptive counterfeits has led most of the ethical discussion around counterfeit luxury brands to focus on the harm they may create for producers rather than defrauded consumers, since the latter are seen as few and far between. For example, they discuss health and safety problems caused by deceptive counterfeits of pharmaceuticals and replacement parts; but they assume that most counterfeit BLGs are non-deceptive counterfeits and thus they focus mainly on how counterfeit BLGs damages legitimate producers, rather than how these counterfeit BLGs might harm consumers.

Contrary to this conventional wisdom, the available data indicate that large numbers of fake buyers are in fact deceived at the time of purchase [22]. Just over half (51%) of US consumers who reported having purchased fake brand name clothing say that they did so without knowing the product was a fake, and the same is true for 39% of purchases of fake brand name watches [48]. In the UK 31% of fake buyers report buying a counterfeit which they thought was real [15]. Moreover, these statistics likely understate the extent of the problem since they exclude people who bought fakes thinking they were real, and still mistakenly think they are real. While many consumers do purchase fakes knowingly, these figures show that consumer fraud should be a significant aspect of the debate around the ethics and impact of counterfeit luxury goods.

How could this many consumers could be deceived into thinking that a fake luxury good was genuine? This question can partly be answered by noting the frequent purchase of goods over the internet. Almost half (49%) of UK fake buyers reported buying a fake on line, mostly through eBay [15]. Matlack and Mullaney, (2006) [34] found that 90% of "Vuitton" items on eBay were counterfeit, and Tiffany conducted a study which found that 95% of the "Tiffany" items for sale on eBay were counterfeits [9]. It is possible that many consumers shopping on-line think they are getting used or grey market genuine merchandise when in fact they are getting fakes. However, the internet may not remain such an easy distribution channel for deceptive counterfeit BLGs, as legitimate firms continue to mount legal challenges to eBay, Google, and other websites [32]. This trend began in 2006 when LVMH, the luxury goods provider, won a legal challenge against Google. The Paris Court of Appeals ordered Google to pay LVMH €300.000 for misleading advertising, unfair competition and trademark infringement. Then in 2008, a Paris court ordered eBay to pay LVMH €35.5 million in damages for its sale of counterfeit items [25].

Another reason that fake BLGs can sometimes be sold deceptively, is that their quality has in some cases become extremely good [22], allowing them to be sold into legitimate distribution channels by dishonest distributors. In fact, the appearance of the counterfeit version has so closely matched the genuine article, that major retailers such as Wal-Mart and Tuesday Morning have unknowingly purchased counterfeits and stocked them in their stores [3].

Finally, further complexity is added by the prevalence of knockoffs, which are products which imitate well known brands, yet sometimes stay within the law by not using the exact brand name or logo of the original producer. In the West, we are most familiar with knockoffs of physical products, but in China retail knockoffs are proliferating such as the large national franchise Dairy Fairy which is a knockoff of the American Dairy Queen chain [6]. In some cases however, the differences between a knockoff and the original brand name and logo may be so subtle that the knock-off effectively functions as an illegal counterfeit. For example, through legal action Dairy Queen was able to shut down a knockoff called Dairy & Queen [6]. Legally, knockoffs are considered in violation of copyright law if they are deemed likely to cause confusion. However, overburdened prosecutors may well not pursue producers and sellers of knockoffs, choosing instead to focus on unambiguous cases of counterfeiting. Thus, knockoffs add further confusion in the marketplace.

16.4 The Non-Cannibalization Hypothesis

Because a variety of terms are used in the counterfeits literature to describe different groups of consumers, it is now necessary to introduce some definitions (see **Table 16.1**), which will be used in the following discussion.

Direct cannibalization refers to a situation where a consumer would have purchased a genuine BLG, but opts for a counterfeit of that brand instead. What we are calling the non-cannibalization hypothesis holds "that millions of fake luxury handbags are sold each year. But very few of them, if any, cannibalize the sales of the real thing" [45].

Table 16.1 Definitions of BLG Buyer Groups

	Genuine buyers Someone who has purchased an authentic BLG within the timeframe of the study being cited.	**Genuine non-buyers** Someone who has not purchased an authentic BLG within the timeframe of the study being cited.
Fake buyers Someone who has purchased a counterfeit BLG within the timeframe of the study being cited.	Omnivorous buyers	Fake-only buyers
Fake non-buyers Someone who has not purchased a counterfeit BLG within the timeframe of the study being cited.	Genuine-only buyers	BLG non-buyers

Proponents of the non-cannibalization hypothesis tend to argue that fake buyers and genuine buyers are different groups of people, so that whatever fake buyers do with their money has little direct effect on sales of BLGs (indirect effects are discussed in a later section of this chapter). For example, in a particularly interesting study, Han, Nunes and Drèze (2010) [28] divided consumers into several non-overlapping segments. Genuine-only buyers were either the high cultural capital patricians who favored BLGs with subtle logos or no visible logos at all, or the lower cultural capital parvenus who favor loud highly visible designer logos. Fake-only buyers fell into a group called Poseurs.

"Like the parvenus, (Poseurs) are highly motivated to consume for status' sake. Poseurs, however, do not possess the financial means to readily afford authentic luxury goods. Yet they want to associate themselves with those they observe and recognize who have the financial means, the parvenus, and dissociate themselves from other less affluent individuals. Hence, they are especially prone to buying counterfeit luxury goods. If brand status is important to a person, as it is with poseurs, but is unattainable, it has been shown that he or she is likely to turn to counterfeit products as cheap substitutes for the originals" (Han, Nunes and Drèze 2010, p. 17 [28]).

Note that in this model, any given consumer can only belong to one of these three groups: Patricians, Parvenus, Poseurs, or Proletarians. Poseurs are the only group that buys counterfeits, and since they could not afford authentic BLGs in the first place, this model assumes little or no direct cannibalization. The Han, Nunes and Dreze (2010) [28] model, is a more sophisticated take on what is probably the most frequently addressed research question in the counterfeits literature: how do fake buyers differ from fake non-buyers (e.g. Eisend and Schuchert-Güler 2006 [19])? Several studies, especially those conducted a decade or more ago, confirmed general expectations that intentional buyers of counterfeit

BLGs tended to be younger and lower income than non-buyers of these goods [5], [29], [53], [52]. Perhaps surprisingly, other studies found a positive correlation between income and the propensity to buy counterfeit BLGs [13], [42], [11], a finding that may reflect the fact that these studies included low income populations for whom even the purchase of a counterfeit BLG was a major expenditure, so higher income consumers were more likely to make these purchases. Studies have also found some psychographic differences, for example as compared to fake non-buyers, fake buyers tend to be less ethically oriented and more materialistic [21], [31], [43], [41], less concerned about possible embarrassment or other risks associated with the purchase of counterfeits [17], and not surprisingly, have more positive attitudes about the purchase of counterfeits [41], [51]. Social context also matters, consumers who believe that "everybody does it" with regard to purchasing counterfeits, are themselves more likely to purchase counterfeits [39].

Perhaps the single most replicable finding from these many studies, is their inability to explain much of the variance in purchases counterfeit BLGs [19]. As Penz and Stöttinger (2008) [39] note, profiling approaches based on psychographic features do not have strong explanatory power.

16.4.1 Evidence for the prevalence of the omnivorous buyer

As noted above, the conventional wisdom holds that there is "no cannibalization as the people who generally buy these fake products are not in a position to buy the original" [57]. This belief may have had more validity in the past [36], [56], and it may still hold true for the very highest and very lowest ends of the income distribution [54], but it is roundly contradicted by the data for most of today's consumers in the developed world.

As luxury has become democratized, it has moved quite far down into the middle class [46]. Davenport Lyons (2007) [15] found that in England, 41% of genuine buyers earned less than 30.000 pounds per year, which was almost identical to the 43% of fake buyers who are in that income bracket. The study concluded that "there is very little to distinguish demographically between those that have bought a fake and those that have not" (p. 6), a conclusion which was also supported in Penz, Schlegelmilch and Stöttinger (2009) [40]. We do not believe that these more recent findings showing little or no effect for income on intention to purchase counterfeits, so much contradict earlier studies which showed stronger relationship, as they do update these studies and show how market for counterfeits is changing over time.

One obvious thing that fake buyers and genuine buyers have in common is that both groups want to own BLGs. Therefore, it should not be surprising to learn that fake buyers are more likely to have also purchased authentic luxury brands than are average consumers: 64% of UK fake buyers have also purchased originals, as compared to only 42% of fake non-buyers (Davenport Lyons 2007, p. 6). Similarly, a Canadian study found that 59% of genuine buyers had also purchased a fake within the past 3 years [36]. It is even quite common for consumers to own fakes and originals from the same brand [39].

Not only are these omnivorous buyers prevalent, they are also big spenders, and their spending seems to be growing over time. In 2006, English combo buyers spent 42% more on luxury leather goods than did genuine-only buyers, and in 2007 that figure increased to 51% [15]. The change between 2006 and 2007 is even more pronounced in watches, clothes and shoes, where in 2006 genuine-only buyers spent more on luxury brands in these categories than did combo buyers, but by 2007 the position had reversed with combo buyers outspending genuine-only buyers by between 3% and 17% depending on the category [15].

These data suggest a fundamental shift in the way we think about a typical fake buyer. No doubt the former image of a typical fake buyer as someone who would like to buy the original but cannot afford to purchase even one authentic BLG, still holds for a segment of the market. But in the developed world, a more accurate image of a typical fake-buyer might consist of a middle to upper-middle class consumer with such a strong desire to own a collection of various BLGs. Although they can afford to purchase some authentic BLGs, they cannot afford to satiate their demand by purchasing only authentic BLGs. Therefore, they purchase a combination of counterfeit and authentic goods in order to create their desired wardrobe. This suggests a shift in focus for future research, from trying to figure out the differences between fake-buyers and genuine-only buyers (who, in the revised model, are very often the same person), to figuring out under what conditions a luxury consumer will buy a fake and under what conditions will that same consumer buy an authentic brand [7]. Penz and Stöttinger (2008) [39] and Wilcox, Kim, and Sen (2009) [58], move research in the right direction by looking at the different meanings that counterfeit and original BLGs have for the consumers to use them. This focus on the meanings and uses of these products, allows us to understand why some consumers buy types of goods, and use them in different social contexts and for different reasons.

16.4.2 Collateral damage

While most of the focus has been on debating the prevalence of direct cannibalization (i.e. the sale of a fake brand X product substitutes for the sale of an original brand X product), the possibility of what Pace, Gistri, Romani, Ahuvia and Masserini (2011) [37] call collateral damage has been largely overlooked. Collateral damage occurs when the sale of a fake brand X product substitutes for the sale of an original brand Y product. Why might this happen?

There are many gradations of exclusiveness and price within the luxury goods category. For example in the US, Coach handbags which are commonly viewed as the bottom of the luxury segment, range from $150 – $400, Gucci handbags at Sac's Fifth Avenue range in price from $600 – $4700, and that the extreme high end a matte crocodile Hermes bag sells for $120.000 and the Diamond Forever bag from Chanel sells for $261.000. While the entry-level BLGs are much cheaper than high-end BLGs, they are still more expensive than typical counterfeits which can cost around $100 for a handbag. When a consumer chooses a $150 Coach bag over a $600 Gucci bag, she reveals a desire to own a BLG coupled with a willingness to accept lower status benefits from the bag in exchange for saving money. Similarly, when people buy fake Gucci bags instead of authentic ones, they too are revealing a desire to own a BLG coupled with a willingness to accept lower status benefits from the bag in exchange for saving money. Hence, had the fake not been available, it is likely that many consumers would have increased their spending enough to purchase an entry level luxury brand like Coach in its place, rather than dramatically increased their spending to purchase the genuine version of the counterfeit bag. It may be that some of counterfeiter's biggest victims are not the brands being counterfeited, but rather entry level luxury brands that suffer as collateral damage.

Pace et al. (2011) [37] studied collateral damage effects by comparing female consumers' purchase intentions for handbags when faced with (a) a choice set which included both legitimate and counterfeit goods at a range of price points, versus (b) the same choice set, only without the counterfeit option. Their findings were largely supportive of the importance of collateral damage effects, in that many consumers who selected counterfeits when available, opted for authentic entry-level luxury brands when counterfeits were not available. However, the products losing the most sales to counterfeits, were non-luxury brands selling at or below the same price point as the counterfeits. Interestingly, 8% of counterfeit consumers indicated that had counterfeits not been available, they would have purchased the authentic high-end brand. It is likely that this 8% figure exaggerates the number of sales lost to high-end BLGs, since it is much easier to indicate a purchase intention for these goods than it is to actually spend the money to acquire them. Nonetheless, given the very high profit margin on high-end BLGs, this 8% figure suggests that direct cannibalization may be a financially significant problem for legitimate BLG firms.

16.5 Counterfeits Harm the Original brand by Reducing their brand Equity

It is commonly argued that although most fake buyers "would not have purchased the goods if they had had to pay full price. Nevertheless, counterfeit goods cheapen the image of legitimate brands" (Green and Bruce 1997, p. 100 [26]; see also Wilke and Zaichkowsky 1999 [59]), and reduce the original "brand's distinctiveness" (Berman 2008, p. 192). More specifically, people who hold this view often argue that the perception of rarity is a key element of the brand equity of luxury goods [8], [18], [21], [33], [55]. And therefore the "proliferation of unauthorized copies undermines the scarcity factor so critical to enhancing the desirability of these goods" (Giacalone 2006, p. 39 [23]), especially among status conscious consumers [58].

Commuri (2009) [14] studied how consumers of authentic BLGs respond to the proliferation of counterfeits. The author found that consumers of genuine items adopt one of three strategies when faced with the prospect of their favorite brands being counterfeited: flight, reclamation or "abranding". The flight strategy consisted of abandoning the brand in response to extensive counterfeiting. The flight strategy was most common among young and the newly affluent consumers. Consumers who had used the brand for a long time before counterfeiting became popular, often chose a "reclamation" strategy in which they emphasized the pioneering qualities of the brand and recommitted to their relationship with it. Finally, some of the most affluent consumers engaged in "abranding" in which they continued to use the brand but removed all visible logos so that they would not be associated with counterfeits. Abranding is consistent with findings from Wong, Tambyah and Ahuvia (2010) [60] found that some consumers of authentic BLGs refrain from buying highly counterfeited brands, for fear that someone might mistake their authentic item for a counterfeit.

We presented evidence above that a significant number of consumers purchase counterfeits unknowingly. When this happens, if the counterfeit is low quality, the authentic brand may take the blame [14]. It should also be noted that the quality of counterfeits has in some instances become "so good that even manufacturers themselves cannot tell the difference without the help of laboratory analysis" (Chow 2003, p. 10 [12]). If consumers knowingly purchase these counterfeits and discover that their quality is comparable to the original, this too is likely to decrease the brand equity of the original [14].

16.6 Could Counterfeits Increase the Original's Brand Equity?

The conventional wisdom that a fake of any given brand harms the brand equity of that same brand is the core assumption motivating much of the concern around non-deceptive counterfeits. But this view is not without its critics, who put forward four main arguments to suggest that fakes may actually help the brand equity of the brands they are imitating.

Counterfeits increase awareness and credibility of the original brand

As noted above, the proliferation of counterfeits may decrease the perception that the brand is scarce. But this is a double-edged sword, as the prevalence of counterfeits can benefit the brand by increasing awareness [57]. Being counterfeited also increases the credibility of the legitimate brand as truly being a prestigious luxury good [2]. As noted by Bian and Veloutsou (2007) [4], counterfeiting "is a sign of success for the counterfeited brand. Only well-established brands that are well recognized, liked and respected will be counterfeited" (p. 219). Vertu – Nokia's luxury brand – capitalizes on this by providing potential customers with information about Vertu counterfeits, which function as "a certificate of true luxury" (El Harbi and Grolleau 2008, p. 387 [20]) for the authentic products.

As an example of how this might work, consider the Mona Lisa, perhaps the most highly copied image in the world. All those copies tend to raise, rather than lower, the value of the original painting. The key here though, is that in the case of the Mona Lisa it is easy to distinguish between the fake and the original. When that is the case, the fakes pay homage to the originals without reducing their perceived scarcity among likely buyers [2]. However, the harder it is to distinguish between fakes and originals, the more the presence of fakes may reduce the perceived scarcity of originals. So as the quality of fakes continues to improve, the negative impact of fakes on the brands they imitate may grow worse. Based on extant research, it is reasonable to conclude that counterfeits have some effects which increase brand equity, and simultaneously have other effects which decrease brand equity, for the original BLGs. Unfortunately, the current state of research cannot tell us which of these competing affect are stronger.

Data on how counterfeits influenced the brand equity of authentic BLGs is in short supply. Some surveys have looked at the public perception that fakes decrease the desirability of the real products. Davenport Lyons (2007) [15] found that 55% of UK consumers think that fakes harm the brand equity of originals. In contrast, a Canadian study [36] found that 69% of luxury goods owners rejected the idea that fakes made the real products less desirable, and only 23% agreed with the idea that fakes harm the brand equity of the authentic products. This led Nia and Zaichkowsky (2000) [36] to conclude that "counterfeits do not seem to affect demand for originals" (p. 6). Since these two studies were conducted in different times and places it is difficult to compare the results, but it is plausible to speculate that average consumers are somewhat more likely to accept the idea that fakes harm the brand equity of the authentic product, whereas actual consumers of the luxury brands tend not to agree with this view.

While the results of Davenport Lyons (2007) [15] and Nia and Zaichkowsky (2000) [36] give us useful and interesting data about public opinion, they provide little information about the actual effects of counterfeits on the brand equity of BLGs. This is because survey data of this kind reveals respondents' lay economic theories about how brand equity works (i.e. if counterfeits were eliminated, how would consumers respond?). Unfortunately, there is little reason to conclude that because a theory is widely believed by consumers, it is necessarily true. By way of analogy, asking consumers if they think the existence of fakes causes a reduction in the desirability of the genuine products, is like asking them if they think reducing government spending will increase or decrease economic growth. Understanding popular opinion on these issues is interesting and important in its own right, but it is not the same as learning if these lay theories are correct. Getting to the truth about the influence of counterfeits on brand equity for BLGs requires data about consumer behavior, not survey data about how popular various lay economic theories are.

Counterfeits serve as "gateway" products

Fakes may help the sales of originals if they serve as what we will term "gateway" purchases, which introduce new consumers to BLGs. In this sense, counterfeits can lead to increased interest in BLGs and eventually the purchase of originals [22], [57]. This idea was supported by the Davenport Lyons (2007) [15] study which found that 29% of fake buyers said that "the fake item had made them more likely to buy the genuine one" and this survey finding was also "borne out in the focus groups" (p. 14). But to what extent this gateway effect is real, and to what extent it is a rationalization put forward by fake buyers to justify their purchase, remains to be seen empirically.

Counterfeits increase perceived scarcity and selectivity of originals

Ritson (2007) [45] suggests that counterfeits may help luxury goods firms maintain an elite image even as they sell more products into the mass market.

"Let's say you are walking down Bond Street and a young man of apparently meager income and untidy countenance barges past you with a Gucci bag (the same one you own) slung over his shoulder. You curse him under your breath, but just before you reappraise the great house of Gucci and its fine clientele, you pause and sneer "Must be fake". It is probably not, but sometimes it helps to have genuine brands mistaken for forgeries when brand equity is at stake" [16].

While this argument may sound odd at first, people frequently assimilate ambiguous information to fit pre-existing stereotypes. Hence, so long as a consumer holds the stereotype that genuine luxury brand consumers are affluent, they can easily maintain that stereotype even in the face of growing evidence to the contrary, by assuming that most middle and lower income people they see with designer logo goods are really carrying fakes.

Counterfeits spur creativity in original brands

Finally, Wilke and Zaichkowsky (1999) [59] argue that fakes and knockoffs may actual spur creativity among legitimate producers, as they strive to stay ahead of the counterfeiters. El Harbi and Grolleau (2008) [20] go further, pointing out that counterfeiters "are not only copiers, they can also be innovators. They possess skills and a creative capital that allow them to explore new directions. Their innovations can be freely and profitably re-appropriated by the high-end designers of the pirated firms" (p. 382).

Are sales of originals growing?

Looking at actual consumer behavior we find that the percent of people who bought a real luxury item in 2007 increased to 24%, a 4% increase over 2006. But over the same period, the percent of people buying fakes that year fell very slightly from 5.9% to 5.3% [15]. This suggests that the presence of fakes has not dramatically harmed the sales of BLGs. However, there are so many other variables involved in determining sales figures, that this data is far from conclusive. In addition, more reliable data can be drawn after the global economic crisis.

16.7 Conclusion

Looking over the four elements of the conventional view, we find support for element A, which holds that counterfeiting is a prevalent global phenomenon. Although making a more detailed assessment of exactly how prevalent it is, remains beyond the scope of the available data.

For element B the data suggests that while many fake luxury goods are purchased as non-deceptive counterfeits, it is also true that many consumers unknowingly purchase fake luxury goods. Therefore, consumer fraud may be a larger problem with regard to counterfeit luxury goods than is often assumed.

For element C, the assumption that fake buyers and original buyers are largely two separate groups of people is widely contradicted by available data. This suggests that purchases of counterfeits may cannibalize purchases of genuine BLGs to a greater extent than was previously thought. Furthermore, the possibility of collateral damage from fakes (i.e. sales of fake brand X reduces sales of original brand Y) is understudied in the literature.

For element D, that counterfeits reduce the brand equity of the original BLGs, we are rich with theory but poor in data. What data we have suggests that one impact of counterfeits is indeed to reduce brand equity of the originals among some consumer segments. However, we have many good arguments to suggest that the impact of counterfeits is quite complex, and while some aspects of their impact may decrease brand equity, other aspects of their impact are likely to increase brand equity. Which of these impacts is more powerful, when, and why, has yet to be demonstrated. Therefore, the issue of how the existence of fakes influences the brand equity of originals remains the subject of a fascinating debate, but a scarcity of relevant data.

References

[1] International Anti-Counterfeiting Coalition (2011): Media Kit, Available at: http://www.iacc.org/news-media-resources/media-kit.php (accessed November 14th, 2011).

[2] Barnett, J. M. (2005): Shopping for Gucci on Canal Street: Reflections on Status Consumption, Intellectual Property, and the Incentive Thesis, in: Virginia Law Review, Vol. 91, pp. 1381-1423.

[3] Berman, B. (2008): Strategies to Detect and Reduce Counterfeiting Activity, in: Business Horizons, Vol. 51, pp. 191-199.

[4] Bian, X./Veloutsou, C. (2007): Consumers' attitudes regarding non-deceptive counterfeit brands in the UK and China, in: Journal of Brand Management, Vol. 14,No. 3, pp. 211-222.

[5] Bloch, P.H.,/Bush, R.F./Campbell, L. (1993): Consumer 'accomplices' in product counterfeiting, in: Journal of Consumer Marketing, Vol. 10, No. 4, pp. 27-33.

[6] Burkitt, L./Chao, L. (2011): Made in China: Fake Stores, in: The Wall Street Journal, August 3.

[7] Casola, L./Kemp, S./Mackenzie, A. (2009): Consumer decisions in the black market for stolen or counterfeit goods, in: Journal of Economic Psychology, Vol. 30, pp. 162-171.

[8] Catry, B. (2003): The Great Pretenders: The Magic of Luxury Goods, in: Business Strategy Review, Vol. 14, 3, pp. 10-17.

[9] Chao, L. (2006): What Happens When an eBay Steal Is a Fake; We Sought Redress From Site And Sellers After Our Fendi, Gucci, Tiffany Proved Faux, in: Wall Street Journal, June 29, p. D.1.

[10] Chaudhry, P. E. (2006): Changing levels of intellectual property rights protection for global firms: A synopsis of recent U.S. and EU trade enforcement strategies, in: Business Horizons, Vol. 49, pp. 463-472.

[11] Cheung, W.-L./Prendergast, G. (2006): Buyers' perceptions of pirated products in China, in: Marketing Intelligence & Planning, Vol. 24, No. 5, pp. 446-62.

[12] Chow, D. (2003): Investigating the counterfeit nightmare: Procter & Gamble and Oki Data count the costs, in: Strategic Direction, Vol. 19, No. 3, 9-12.

[13] Chuchinprakarn, S. (2003): Consumption of counterfeit goods in Thailand: who are the patrons?, in: European Advances in Consumer Research, Vol. 6, pp. 48-53.

[14] Commuri, S. (2009): The Impact of Counterfeiting on Genuine-Item Consumers' Brand Relationships, in: Journal of Marketing, Vol. 73, No. 3, pp. 86-98.

[15] Davenport Lyons (2007): Counterfeiting Luxury: Exposing the Myths, Research carried out by Ledbury Research.

[16] Daye, D./VanAuken, B. (2007): Counterfeits: Good for Luxury Brands?, in: Branding Strategy Insider, available at: http://www.brandingstrategyinsider.com/2007/10/fakes-can-genui.html (accessed Novemebr 14th, 2011).
[17] De Matos, C. A./ Iltuassu, C. T./ Rossi, C. A. V.(2007): Consumer Attitudes Toward Counterfeits: A Review and Extension, in: Journal of Consumer Marketing, Vol. 24, No. 1, pp. 36-47.
[18] Dubois, B./Paternault, C. (1995): Observations: Understanding the World of International Luxury Brands: The Dream Formula, in: Journal of Advertising Research, Vol. 35, No.4, pp. 69-76.
[19] Eisend, M./Schuchert-Güler, P. (2006): Explaining Counterfeit Purchases: A Review and Preview, Academy of Marketing Science Review, Vol. 12, pp. 1-22.
[20] El Harbi, S./Grolleau, G. (2008): Profiting from Being Pirated by Pirating the Pirates, in: Kyklos, Vol. 61, No. 3, pp. 385-390.
[21] Furnham, A./Valgeirsson, H. (2007): The Effect of Life Values and Materialism on Buying Counterfeit Products, in: The Journal of Socio-Economics, Vol. 36, pp. 677-685.
[22] Gentry, J.W./ Putrevu, S./Shultz II, C. J. (2006): The Effects of Counterfeiting on Consumer Search, in: Journal of Consumer Behaviour, Vol. 5, No. 3, 245-256.
[23] Giacalone, J. (2006): The Market for Luxury Goods: The Case of the Comité Colbert, in: Southern Business Review, Fall, pp. 33-40.
[24] Gistri, G./Romani,S./Pace, S./Gabrielli, V./Grappi, S. (2009): Consumption Practices of Counterfeit Luxury Goods in the Italian Context, in: Journal of Brand Management, Vol. 16, No.5/6, pp. 364–374.
[25] Graham, S./Upton, M. (2009): LVMH successfully sues eBay for unauthorised sales of genuine products, in: Blake Dawson – IT, Communications & Media Update, Feb. 9.
[26] Green, W./Bruce,K. (1997): Riskless Crime?, in: Forbes, 160 (August 11), pp. 100-102.
[27] Grossman, G./Shapiro, C. (1988): Foreign Counterfeiting of Status Goods, in: Quarterly Journal of Economics, February, pp. 79-100.
[28] Han, Y. J./Nunes, J. C./Drèze, X. (2010): Signaling Status with Luxury Goods: The Role of Brand Prominence, Journal of Marketing, Vol. 74, July, pp. 15-30.
[29] Hustak, A. (1990): Ritzy Ripoffs, in: The Gazette, August, p. 31.
[30] Kearney, Í./Mitchell, V-W. (2001): Measuring consumer brand confusion to comply with legal guidelines, in: International Journal of Market Research, Vol. 43, No. 1, pp. 85-91.
[31] Kozar, J. M./Marcketti, S. B. (2011): Examining ethics and materialism with purchase of counterfeits, in: Social Responsibility Journal, Vol. 7, No. 3, pp. 393-404.
[32] Lambkin, M./ Tyndall, Y.(2009): Brand Counterfeiting: a Marketing Problem That Won't Go Away, in: Irish Marketing Review, Vol. 20, No. 1, pp. 35-46.
[33] Mason, R. S. (1981): Conspicuous Consumption: A Study of Exceptional Consumer Behavior, New York: St. Martin's Press.
[34] Matlack, C./Mullaney, T.(2006): Fed up with fakes, in: Business Week, Oct 9, p. 56.
[35] Mitchell, V.-W./Kearney, Í. (2001): A critique of legal measures of brand confusion, in: Journal of Product and Brand Management, Vol. 11, No. 6, pp. 357-379.
[36] Nia, A./Zaichkowsky, J. L.(2000): Do counterfeits devalue the ownership of luxury brands?, in: The Journal of Product and Brand Management, Vol. 9, No. 7, pp. 485-497.
[37] Pace, S., Gistri, G., Romani, S., Ahuvia, A., and Masserini, L. (2011): Collateral Damage and the Impact of Non-deceptive Counterfeiting on Legitimate Brands, Working paper.
[38] Penz, E./Stöttinger, B. (2005): Forget the 'real' thing-take the copy! An explanatory model for the volitional purchase of counterfeit products, in: Advances in Consumer Research, Vol. 32, pp. 568-575.
[39] Penz, E./Stöttinger, B. (2008): Corporate image and product similarity – Assessing major demand drivers for counterfeits in a multi-country study, in: Psychology & Marketing, Vol. 25, No. 4, pp. 352-381.
[40] Penz, E./Schlegelmilch,B. B./Stöttinger, B. (2009): Voluntary Purchase of Counterfeit Products: Empirical Evidence From Four Countries, in: Journal of International Consumer Marketing, Vol. 21, No. 1, pp. 67-84.

[41] Phau, I./Teah, M. (2009): Devil wears (counterfeit) Prada: a study of antecedents and outcomes of attitudes towards counterfeits of luxury brands, in: Journal of Consumer Marketing Vol. 26, No. 1,, pp. 15-27.
[42] Phau, I./Prendergast, G./Chuen, L. H. (2001): Profiling brand-piracy-prone consumers: An exploratory study in Hong Kong's clothing industry, Journal of Fashion Marketing and Management, Vol. 5, No. 1, 45-55.
[43] Phau, I./Sequeira, M./Dix, S. (2009): To buy or not to buy a 'counterfeit' Ralph Lauren polo shirt: The role of lawfulness and legality toward purchasing counterfeits, in: Asia-Pacific Journal of Business Administration Vol. 1, No.1, pp. 68-80. Available at: [Accessed August 18, 2011].
[44] Prendergast, G./Chuen, L. H./Phau, I. (2002): Understanding consumer demand for non-deceptive pirated brands, in: Marketing Intelligence & Planning, Vol. 20,No. 7, pp. 405-416.
[45] Ritson, M. (2007): Fakes can genuinely aid luxury brands, in Marketing, July, p. 21.
[46] Silverstein, M.J./Fiske, N. (2003): Luxury for the Masses, in: Harvard Business Review, Vol. 81, No. 4, pp. 48-57.
[47] Staake, T./Thiesse, F./Fleisch, E. (2009): The emergence of counterfeit trade: a literature review, in: European Journal of Marketing, Vol. 43, No. 3/4, pp. 320-349.
[48] Stewart, C. (2005a): Counterfeit Goods a Genuine Problem in the U.S.?, in: gpns, Feb. 22.
[49] Stewart, C. (2005b): Brand Piracy: A Victimless Crime?, in: gpns, March 1.
[50] Strategic Direction (2003): Investigating the counterfeit nightmare, Vol. 19, 3, pp. 9-12.
[51] Swami, V./Chamorro-Premuzic, T./Furnham, A. (2009): Faking it: Personality and individual difference predictors of willingness to buy counterfeit goods, in: Journal of Socio-Economics, Vol. 38, No. 5, pp. 820-825.
[52] Swee, H. A./Cheng, S. P./Lim, E. A. C./Tambyah, S. K. (2001): Spot the difference: consumer responses towards counterfeits, in: Journal of Consumer Marketing, Vol. 18, 3, pp. 219-35.
[53] Gail, T./Garibaldi, B./Zeng, Y./Pilcher, J. (1998): Consumer demand for counterfeit goods, in: Psychology & Marketing, Vol. 15, No. 5, pp. 405-421.
[54] Van Kempen, L. (2003): Fooling the eye of the beholder: deceptive status signalling among the poor in developing countries, in: Journal of International Development, Vol. 15, No. 2, p. 157.
[55] Veblen, T. (1899/1979): The Theory of the Leisure Class: An Economic Study of Institutions. The 100 greatest masterpieces of American literature. Franklin Center, Pa: Franklin Library.
[56] Wee, C.-H./Tan, S-J./Cheok, K.-H. (1995): Non-price determinants of intention to purchase counterfeit goods, in: International Marketing Review, Vol. 12, 6, pp. 19-46.
[57] Whitwell, S. (2006): Piracy: Faking it Can Be Good, in: Brand Strategy, May 8th, p. 30.
[58] Wilcox, K./Kim, H. M./Sen, S. (2009): Why Do Consumers Buy Counterfeit Luxury Brands?, in: Journal of Consumer Research, Vol. 46, No. 2, pp. 247–259.
[59] Wilke, R./Zaichkowsky, J. L. (1999): Brand imitation and its effects on innovation, competition, and brand equity, in: Business Horizons, Vol. 42, No. 6, pp. 9-18.
[60] Wong, N./Tambyah, S./Ahuvia, A. (2010): Luxury Consumption and the Pursuit of Authenticity, in Consumer Culture Theory Conference, Madison Wisconsin, June 10-13.

17 Counterfeiting of Luxury Brands: Opportunity beyond the Threat

Silvia Grappi, Ilaria Baghi, Bernardo Balboni, Veronica Gabrielli

17.1 Counterfeiting: a growing threat to luxury products

...counterfeiting is a tremendous problem we are facing,
but in spite of this, I'm a little bit worried
when I don't see Gucci fakes displayed by street vendors
(Frida Giannini, Gucci Creative Director,
D Repubblica, 2011-02-19)

Luxury has traditionally been associated with exclusivity, status and quality. During the seventeenth century, luxury was found in extraordinary commodities, such as rare pearls, perfumes, and spices from the Caribbean. During the nineteenth, with the rise of world trade, the concept of luxury became attached to the products of great craftsmen: Christian Dior, Louis Vuitton, Gucci. More recently, in the industrialized world, luxury has increasingly been perceived in terms of the brand: carefully crafted symbols which go beyond the material to invoke a world of dreams, images and signs [5]. In order to evoke exclusivity, luxury goods have to show a well-known brand identity and enjoy high brand awareness [32]. For this reason, brand management has become an important aspect of luxury goods marketing.

Brand strategy has a key role to play in the luxury market since these are products that must be associable with an exclusive image and superior brand equity. By relying on a well-known luxury brand, consumers experience a mix of sensations, feelings and cognitions. With this in mind, Kapferer and Bastien [22] suggest that a luxury brand should have a very strong personal and hedonistic component. The symbolic dimension is related to the social meanings [14][10]. The social function has two aspects: the value a luxury brand signals to others [28], and the value of that signal to the person exhibiting it [27]. Thus a Ferrari may signal wealth, prestige, and performance to others, while at the same time reinforcing the owner's self image.

The key dimensions that characterize a brand as a luxury brand, implicitly reveal people's motivation to buy, possess and show off such a brand. Buying a luxury brand gives the confidence of having chosen excellent quality [31] and is a way of improving self-esteem [33]. In addition, showing possession of a luxury brand is a way to gain social approval through a product that embodies exclusive value, meanings and images [9].

Figure 17.1 Constituent value dimension of Luxury Brand

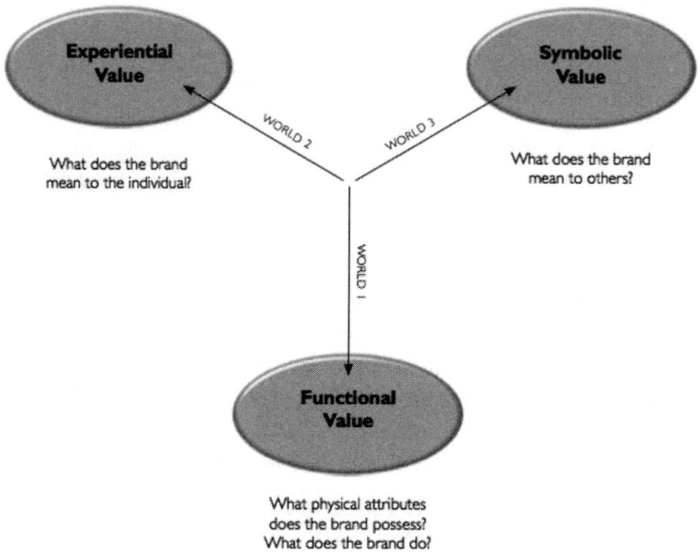

Source: Berthon, et al. 2009, p 49

The increase of people's aspiration for social approval by means of luxury brands has led to the phenomenon of counterfeiting. A counterfeit is a product bearing a copy of a trademarked brand, but not made by the legitimate manufacturer. The counterfeiter thereby infringes the rights of the holder of the trademark [6]. Counterfeiting is far from being a new phenomenon [12] but its growth over the last three decades has increased its economic impact, engendering serious concern on the part of legitimate manufacturers and governments. Estimates of the value of counterfeited and pirated goods internationally traded stand at about $250 billion for 2008 [30]. A more recent study developed by Business Action to Stop Counterfeiting and Piracy (BASCAP), an initiative of the International Chamber of Commerce, upgrades previous estimates of the global impact of counterfeiting. The new value, which includes domestically produced and digitally pirated product, ranges from $450 to $650 billion, and it is expected to double by the end of 2015 [4].

Among products infringing copyright, fraudulent luxury items represent 5% of the overall annual sales of counterfeit products [35]. In relation to the UK, BASCAP's estimation model assumes that between 2% and 4% of luxury products are counterfeits [3]. In Italy – the cradle of luxury fashion – the estimated extent of counterfeiting within the fashion industry, of which luxury brands represent an important share, stands at about €2.6 billion [11], which represents more than 30% of the Italian black market in fakes (see **Table 17.1**).

Table 17.1 Estimates of counterfeit values in Italy divided by category (2008)

Sectors	Estimates 2008 (million €)
Fashion	2,608.2
Audiovisual and software	1,646.7
Food and drink	1,153.7
Electrical components	688.7
Watches and treasures	508.5
IT product	224.1
Perfumes and make-up	114.2
Automotive	112.3
Toys	31.2
Pharmaceuticals	19.4
Total	**7,107.0**

Source: CENSIS, 2009

The increasing level of counterfeiting, all over the world, has led marketing research to devote much effort towards understanding the origin of this phenomenon. Most research has focused on investigation of the intention to buy counterfeit luxury brand products [35]. Marketing experts are essentially in agreement in identifying the reasons that induce consumers to buy counterfeit goods: the chance to wear a luxury brand at an affordable price [1], [16]; the desire to impress others [7], [20]; the desire to communicate membership of a higher social status [7], [31]and the desire to be fashionable [26], [31].

It has been often argued that counterfeit goods are a hidden competitor for original and luxury brands. Counterfeiting has been considered as a parasite that eats the core element of a luxury brand: its exclusivity [18].The risk assumed is that the image of the luxury brand is devalued and trivialized by the existence of fake and accessible copies. However marketing literature does not exhaustively clarify in what ways consumers change their evaluation of brands as a consequence of the existence of a counterfeit version [19]. To better assess the phenomenon of counterfeiting, companies should take care to understand the opinions not only of those who knowingly choose to buy counterfeit products, but also of those that, without buying counterfeit products, are aware of the existence of the fake copies of luxury goods in the market. In mainstream marketing literature the opinion has been consolidated that copied products not only ruin the special status of the original brand, but also contribute to a loss of exclusivity and uniqueness because of the increased availability of cheap imitations [13], [15], [18]. However, there is no unanimous agreement on this point: while Fournier [15] suggests that counterfeits have the potential to unsettle the inaccessibility of luxury brands, damaging the prestige of their image, Jugessur and Cohen [21] argue that fake products do not succeed in stealing customers from the most famous luxury brands (e.g. Chanel and Cartier), rather, fake products simply allow a different target of consumers to realize their dreams without spending too much. From this perspective, counterfeits do not damage luxury brands but help to nourish the desires of less wealthy people, increasing rather than satisfying their aspirations.

A similar conclusion is reached in some recent studies [23] [19] which argue that the supposed damage done by the fake product to the original brand perception is not definitively proved. When Nia and Zaichkowsky [28] explored the attitude of the owners of genuine luxury brands towards counterfeit luxury goods, most respondents said they believed that counterfeits are inferior products and that ownership of authentic products is more prestigious than ownership of counterfeits. Further, most respondents rejected the idea that the availability of counterfeits negatively affects their purchase intentions in relation to original luxury brands. Overall, Nia and Zaichkowsky's results showed that "the value, satisfaction, and status of original luxury brands were not decreased by the wide availability of counterfeits" ([28], p. 485).

Nia and Zaichkowsky's study revealed important insights into the effect of counterfeits on consumers' perception of the genuine brand, but provided only a partial perception of the phenomenon: the authors investigate only the intention to buy, and overall attitudes toward the original brand. Moreover, the focus is strictly on consumers of genuine luxury brands. They did not consider the attitudes of potential consumers in the light of the counterfeiting phenomenon. Moreover, Nia and Zaichkowsky [28] measured only conscious attitude shift arising from the existence of counterfeits. Hieke [19], however, did set out to measure the unconscious effects of counterfeiting. Unfortunately, that study's focus narrowed to the effects of mere exposure to a counterfeit product.

The present study follows the same path explored by Nia and Zaichkowsky [28] and Hieke [19], but aims to offer a better-structured investigation into the effects of the counterfeiting phenomenon on luxury brands. That is: we propose to conduct a deeper analysis by investigating, from a consumer perspective, all aspects of brand equity that might be affected by counterfeiting. Further, we propose a wider analysis based on observing not only actual consumers of genuine luxury brands but also potential ones. The assumption is that the latter type of consumer is an attractive target for luxury companies, and one that could be affected by counterfeiting practices.

To sum up, at the end of our study we hope to be able to answer the following questions:

a. As well as the undeniably negative effects on sales, might counterfeits induce positive effects on the perception of luxury brand equity in consumers' minds?

b. Are genuine prospective consumers of a luxury brand willing to play an effective role to in any crackdown on counterfeiting?

17.2 An in-depth view of luxury brand equity

In order to verify what kind of effects counterfeits induce in potential consumers we decided to adopt a model designed to depict separately all aspects of consumers' perceptions and attitudes towards a specific brand: the Customer-Based Brand Equity (CBBE) Pyramid conceptualized by Keller [23], [24], [25].

Keller states that CBBE represents "the differential effect of brand knowledge on consumer response to the marketing of the brand" [23]. In this model, the maximum degree of brand equity occurs when a brand successfully achieves six different steps, corresponding to the six blocks of the pyramid: salience, performance, imagery, judgement, feelings, and resonance (**Figure 17.2**).

Figure 17.2 Customer-based Brand Equity Pyramid

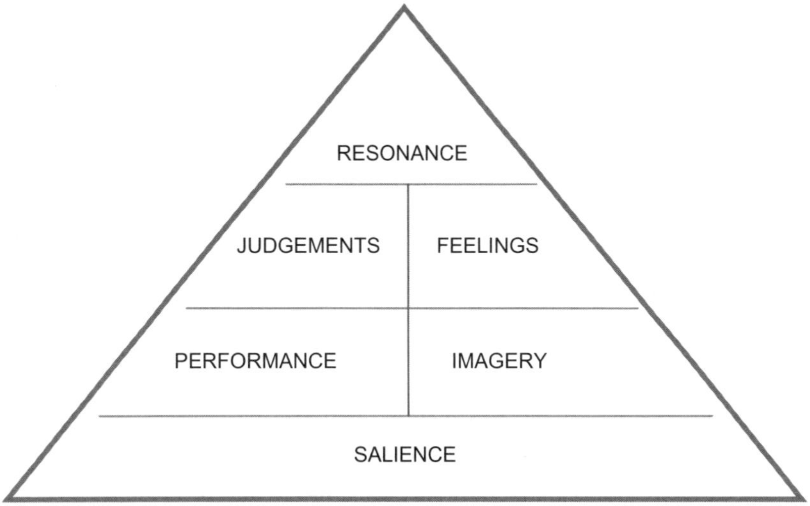

Source: Keller K.L., 2001, p.16

From the bottom to the top of the pyramid, consumers' involvement increases. That is why the base is larger than the top: for a very well-known luxury brand, many people are aware that the brand exists (salience), while far fewer develop the intention to act in favour of the brand; that is, to adopt a sort of defensive engagement on behalf of the brand (resonance). In Keller's perspective it is important to build all the pyramid blocks strongly because each of them significantly supports the upper level. In the middle of the pyramid we observe the two sides of the brand equity building process: on the right there is the perceptive and emotional; on the left there is the cognitive. These sides both feed into the behavioural results gained by the brand. Our purpose is to comprehend whether the existence of counterfeiting could induce a positive effect on some of these blocks and whether this effect could in turn lead consumers to behave more in favour of the brand.

At the base of the pyramid is *salience,* consumers' level of awareness of the brand's existence. This block could be affected by counterfeiting because the large number of copies available on the market and the high prominence of the logo on fake products can significantly amplify the visibility of the brand.

The second level of the pyramid reflects people's ability to ascribe some meanings to the brand. It is composed of two blocks: imagery and performance.

The *imagery* block brings together all the specific associations consumers spontaneously linked to the brand. They should be positive, unique, and favourable [23]. In the case of luxury brands, the imagery block is particularly focused on user profiles in terms of their characteristics, personality and values. We expected this block to be affected by counterfeiting because it encourages attempts by fake-users to lay claim to some of these personality traits, and thereby highlights the social desirability of the brand.

Performance represents the brand's functional characteristics as recognized by consumers; e.g. reliability, durability, serviceability, efficiency, effectiveness, and empathy in service delivery. For luxury products some of these evaluations could be considered as a given, reinforced by effective and empathetic service delivery, which could play an important role in after-sale contacts with consumers. We expected that the existence of counterfeits would underline the importance of functional performance, since this is not easy to replicate in a fake; it also highlights the importance of an effective customer service, which is totally absent for fakes.

At the third level of the pyramid consumers react to the brand, expressing judgements and experiencing feelings. The *judgements* block summarizes consumers' opinions about the brand in terms of quality, credibility, consideration and superiority. For luxury brands, the judgement of superiority plays a central role because of the uniqueness of the brand. Strong exposure to counterfeiting could even increase the perception of the superiority of this brand, in particular in comparison with other luxury brands.

In the same level of the pyramid, there is also the affective sector, representing *feelings*. It is composed of emotions (e.g. warmth, fun, excitement, security) which originated on a personal level; and other emotions (e.g. self-respect and appreciation) based on social approval. Thus, it is reasonable to hypothesize that counterfeits have a positive emotional impact on consumers who can afford to buy a genuine product.

The fourth block of the pyramid is *resonance.* It corresponds to the highest level of commitment by consumers, due to a personal identification with the brand. This block is determinant for companies because it is based on an intense attachment to the brand and an active effort from consumers in its favour. The attachment is expressed by a sense of community among adopters or lovers of a brand. Active effort is seen through positive behaviours such as purchase, word of mouth recommendation, searching for product news, and disparagement of counterfeiting.

To investigate the impact of counterfeiting on resonance is interesting because current consumers, and even prospective ones, could play a positive role in helping companies combat this phenomenon. In particular, a genuine consumer, strongly aware of counterfeiting could develop a feeling of protectiveness toward a brand being hurt by fakes, and increase their resonance by acting in favour of it. A positive shift on resonance occurs only if most of the previous blocks of the pyramid are altered positively.

17.3 Two empirical studies of counterfeiting's impact on CBBE

Even if current and potential consumers of a luxury brand do not give in to the temptation of counterfeit luxury products, they are still exposed to this phenomenon: they read news about fakes, they probably know someone who consumes fakes, and they have probably read information about how to detect a fake. This individual perception, shaped by different degrees of awareness of counterfeit practices, interacts with each consumer's brand perception and could alter some of the blocks of the CBBE pyramid.

Most companies try to prevent fakes from catching the eyes of consumers by, for example, publicising the large number of successful legal actions taken against counterfeiting. Their assumption is that a rise in the number of fakes in circulation could have a negative influence on consumer's brand equity. However, this approach shows that they lack an understanding of how counterfeiting impacts on actual or prospective consumers; as a result, they fail to benefit from these consumers' potential defensive engagement.

17.3.1 Sample and procedure

In order to test this different point of view we conducted two empirical studies among actual and potential consumers of two different luxury brands: Louis Vuitton and Gucci. We have used two distinct brands in order to test the robustness of our results. We selected these two brands because they both occupy high positions in consumers' mental league tables of luxury product; and they are at the top of the list of brands hit by counterfeiting: Louis Vuitton was the most counterfeited brand in 2010 in global terms, with 604 seizures worth € 74 million [36],while Gucci came third (behind Chanel), with seizures worth approximately € 26 million. Two years earlier, Gucci topped the list with seizures valued at € 592 million [37].

Our studies were conducted online. A message was posted to several online forums and blogs directly related to luxury brands and fashion. The message explained the purpose of the research and provided a link to a questionnaire consisting of several statements which referred to the different dimensions of the CBBE pyramid. Respondents were asked to state their level of agreement with the statements on a 7-point Likert scale. When available, items were taken from previous studies, such as that conducted by Washburn and Plank [34], especially those items referring to salience and resonance. In addition, three extra items were introduced for the purpose of measuring respondents' awareness of the level of counterfeiting affecting the brand: "I can recognize a fake X product"; "Brand X is one of the most affected by counterfeiting"; "I often read news about counterfeiting of product X".

We conducted our studies among Italian consumers, Italy being selected because it plays a central role in luxury fashion and therefore has a privileged point of observation of this phenomenon. However we included non-Italian brands in the studies so as not to induce an ethnocentric effect.

The questionnaire took about 15 minutes to complete, and 585 Italian consumers complied with our request. Each respondent was allowed to participate in one study only, on the basis of random assignation. Of the 585 consumers, 316 replied to questions about Gucci, and 269 to questions about Louis Vuitton. The overall participant sample has the following characteristics: 180 men (25.9%) and 515 women (74.1%); 76.9% are between 18 and 35 years old, 15.1% between 36 and 45, 6% between 46 and 55, and 2% over. Both sub-samples reflect this socio-demographic composition.

17.3.2 Results

In our analysis, we will first point out the adequacy of the multi-dimensional structure of the CBBE construct. Then, we will analyse the difference between the two brands, in relation to the CBBE structure, among different kinds of consumer. Getting down to detail, we will categorize respondents into different groups based on: (1) the nature of their brand consumption (i.e. non-brand users vs. genuine brand users); and (2) their awareness of the level of counterfeiting affecting the brand (i.e. low counterfeit awareness vs. high counterfeit awareness). Then we will reveal the eventual shifting of each of the six blocks, induced by consumers' counterfeiting awareness. This analysis will allow us to reply to the questions posed at the end of the first paragraph of this chapter. We will in parallel make the same analysis for both luxury brands (Gucci and Louis Vuitton) in order to stress differences and similarities in the registered effects.

17.3.2.1 The empirical evidence of the structure of the CBBE pyramid

Prior to analysis of the effects of counterfeiting on CBBE, an exploratory factor analysis (EFA) was performed on the data. This analysis was conducted on all respondents, without discrimination between the two brands. The objective of this step of the research is to verify the pertinence of the CBBE structure for luxury fashion products over and above the specific characteristics of each brand.

The analyses revealed six factors, accounting for 71.1% of the total variance (see **Table 17.2**). All the reliability coefficients exceeded the minimum standard for reliability [29].

Table 17.2 Exploratory factor analysis on customer-based brand equity construct

	Factor loading	Eigenvalue	Variance explained %	Reliability coefficient
FEELINGS		5.84	20.13	.93
Peaceful	.68			
Sentimental	.78			
Amused	.66			
Excited	.75			
Enthusiastic	.70			
Safe	.82			
Self-assured	.75			
Admired	.74			
Accomplished	.71			
RESONANCE		4.58	15.79	.93
BRAND X would be my first choice	.71			
I consider myself to be loyal to BRAND X	.82			
I think I belong to BRAND X lovers	.83			
I like to be seen as a consumer linked to BRAND X	.82			
I keep myself informed about BRAND X news	.79			
I am willing to talk positively about BRAND X	.63			
PROMINENCE		3.39	11.67	.85
I know what BRAND X looks like	.68			
I can recognize BRAND X among other competing brands	.79			
I am aware of BRAND X	.77			
I can quickly recall the symbol or logo of BRAND X	.79			
I have difficulty in imagining BRAND X in my mind (R)	.69			
PERFORMANCE		2.40	8.26	.85
Services offered to customers by BRAND X are of high quality	.78			
BRAND X takes care of its customers	.82			
BRAND X looks after consumers' interests	.74			

	Factor loading	Eigenvalue	Variance explained %	Reliability coefficient
IMAGERY		2.27	7.82	.73
Those consumers who possess BRAND X have a certain personality	.73			
You can always wear a BRAND X product with confidence	.78			
BRAND X products give you a certain personality	.65			
JUDGEMENTS		2.15	7.43	.88
BRAND X is unique	.75			
BRAND X products have some characteristics absent from competing offerings	.76			
BRAND X products are better than competing ones	.74			
Total variance explained			71.10	

The principal component analysis rotated by an oblique procedure. Eigenvalues greater than one. Item with a factor loading greater than .50 on their focal factor and not higher than .25 on another retained.

We used structural equation modelling (SEM) (Lisrel 8.7). In order to assess the psychometric characteristics of our measures, we ran a confirmatory factor analysis (CFA) [2] [8] (see **Table 17.3**). We did not discriminate between the two brands, considering all the respondents together. The fit of the model was acceptable: (χ^2 (df) = 1713.42 (362); RMSEA = .08; SRMR = .05; NFI = .96; NNFI = .96; CFI = .97), all factor loadings were significant and correlations between factors were below .70 [10] (see **Table 17.4**).

Table 17.3 Confirmatory factor analysis on customer-based brand equity construct

	Factor loading (*t*-value)	SMC (R^2)
FEELINGS		
Peaceful	.86 (--)	.74
Sentimental	.80 (22.14)	.63
Amused	.78 (21.34)	.61
Excited	.87 (25.90)	.76
Enthusiastic	.89 (26.88)	.79
Safe	.86 (25.48)	.74
Self-assured	.78 (21.52)	.61
Admired	.75 (19.97)	.56
Accomplished	.79 (21.86)	.63

	Factor loading (t-value)	SMC (R^2)
RESONANCE		
BRAND X would be my first choice	.89 (--)	.79
I consider myself to be loyal to BRAND X	.92 (32.13)	.85
I think I belong to BRAND X lovers	.95 (35.42)	.91
I like to be seen as a consumer linked to BRAND X	.94 (33.70)	.88
I keep myself informed about BRAND X news	.86 (27.57)	.74
I am willing to talk positively about BRAND X	.75 (21.22)	.57
PROMINENCE		
I know what BRAND X looks like	.65 (--)	.42
I can recognize BRAND X among other competing brands	.85 (15.44)	.72
I am aware of BRAND X	.86 (15.59)	.74
I can quickly recall the symbol or logo of BRAND X	.85 (15.43)	.72
I have difficulty in imagining BRAND X in my mind (r)	.66 (12.62)	.43
PERFORMANCE		
Services offered to customers by BRAND X are of high quality	.91 (--)	.83
BRAND X takes care of its customers	.92 (28.82)	.85
BRAND X looks after consumers' interests	.68 (17.93)	.47
IMAGERY		
Those consumers who possess BRAND X have a certain personality	.62 (--)	.38
You can always wear a BRAND X product with confidence	.74 (11.72)	.55
BRAND X products give you a certain personality	.75 (11.76)	.56
JUDGMENTS		
BRAND X is unique	.84 (--)	.70
BRAND X products have some characteristics absent from competing offerings	.92 (25.39)	.84
BRAND X products are better than competing ones	.85 (22.71)	.72
χ^2 (df) = 1713.42 (362); RMSEA = .08; SRMR = .05; NFI = .96; NNFI = .96; CFI = .97		

Table 17.4 Correlation value between the six blocks of the CBBE pyramid (*t*-value)

	Feelings	Resonance	Prominence	Performance	Imagery	Judgments
Feelings	1					
Resonance	.63 (11.90)	1				
Prominence	.38 (7.88)	.44 (8.99)	1			
Performance	.39 (8.47)	.44 (10.09)	.44 (8.83)	1		
Imagery	.49 (8.34)	.34 (7.40)	.34 (6.42)	.43 (7.81)	1	
Judgments	.57 (10.53)	.58 (11.50)	.40 (7.83)	.52 (9.94)	44 (7.88)	1

A second-order CFA was then conducted to assess possible hierarchical relations among the first-order factors; that is, the possibility of a second-order factor was investigated. SEM was used to assess the factors' relationships (Lisrel 8.7). The fit statistics of the model were subsequently examined. These checks revealed that, in terms of model design, it is possible to assume six first-order latent factors (prominence, performance, judgments, imagery, feelings and resonance), reflecting a second-order factor (CBBE) (see **Table 17.5**). The model's goodness-of-fit is satisfactory: χ^2 (df) = 1836.87 [10]; RMSEA = .08; SRMR =.06; NFI = .96; NNFI = .96; CFI = .97. Therefore, the second order CFA confirmed that the six factors were valid and reflect a second-order factor (CBBE).

Table 17.5 Second-order CFA: Detailed results for the measurement model and the structural model

	Factor loading (*t*-value)	SMC (R^2)
Measurement model		
FEELINGS		
Peaceful	.86 (--)	.73
Sentimental	.80 (22.13)	.64
Amused	.78 (21.39)	.61
Excited	.87 (25.77)	.76
Enthusiastic	.89 (26.79)	.79
Safe	.86 (25.41)	.74
Self-assured	.79 (21.65)	.62
Admired	.74 (19.83)	.55
Accomplished	.79 (21.73)	.62

		Factor loading (t-value)	SMC (R^2)
RESONANCE			
BRAND X would be my first choice		.89 (--)	.79
I consider myself to be loyal to BRAND X		.92 (31.85)	.84
I think I belong to BRAND X lovers		.95 (35.29)	.91
I like to be seen as a consumer linked to BRAND X		.94 (33.73)	.88
I keep myself informed about BRAND X news		.86 (27.50)	.74
I am willing to talk positively about BRAND X		.75 (21.20)	.57
PROMINENCE			
I know what BRAND X looks like		.65 (--)	.42
I can recognize BRAND X among other competing brands		.85 (15.48)	.72
I am aware of BRAND X		.86 (15.59)	.73
I can quickly recall the symbol or logo of BRAND X		.85 (15.46)	.72
I have difficulty in imagining BRAND X in my mind (r)		.66 (12.65)	.43
PERFORMANCE			
Services offered to customers by BRAND X are of high quality		.91 (--)	.83
BRAND X takes care of its customers		.92 (28.08)	.85
BRAND X looks after consumers' interests		.68 (17.85)	.47
IMAGERY			
Those consumers who possess BRAND X have		.61 (--)	.37
You can always wear a BRAND X product with confidence		.73 (11.26)	.53
BRAND X products give to you a certain personality		.77 (11.42)	.59
JUDGMENTS			
BRAND X is unique		.84 (--)	.71
BRAND X products have some characteristics absent		.92 (25.50)	.85
BRAND X products are better than competing ones		.84 (22.67)	.71
Structural model			
CBBE →	Feelings	.82 (18.02)	
CBBE →	Resonance	.90 (20.81)	
CBBE →	Prominence	.65 (11.30)	
CBBE →	Performance	.67 (14.64)	
CBBE →	Imagery	.65 (9.94)	
CBBE →	Judgments	.83 (17.48)	
	Feelings		.67
	Resonance		.80
	Prominence		.42
	Performance		.45
	Imagery		.42
	Judgments		.69

χ^2 (df) = 1836.87 (371); RMSEA = .08; SRMR = .06; NFI = .96; NNFI = .96; CFI = .97

17.3.2.2 The effects of counterfeiting on CBBE

After having confirmed the structure of the CBBE pyramid, we investigate the effects of counterfeiting awareness on each of the six blocks. **Table 17.6** and **Table 17.7** separately summarize the results among actual and potential users across the two brands.

In the Louis Vuitton group of respondents, the median value of the counterfeit awareness related to this brand is 5.00 (M=5.40 SD=1.20), while in the Gucci group it was 4.33 (M=4.22 SD=1.34). These values were adopted in order to distinguish, within the respondent consumers (genuine users and non-users) those who are highly aware of counterfeiting and those not so aware of it. We performed a t-test to compare the value registered for each pyramid block between these two groups of consumers. An eventual statistically significant difference between these two values will identify the impact of counterfeiting awareness on the reference block, as shown in **Table 17.6** for Louis Vuitton and in **Table 17.7** for Gucci.

Table 17.6 The effects of counterfeiting awareness on Louis Vuitton genuine users and non-users

	Louis Vuitton non-users			Genuine Louis Vuitton users		
	Low counterfeit awareness (N = 77)	High counterfeit awareness (N = 35)		Low counterfeit awareness (N = 21)	High counterfeit awareness (N = 54)	
	M (SD)	M (SD)	t-test (sig.)	M (SD)	M (SD)	t-test (sig.)
Prominence	4.43 (1.17)	5.36 (.67)	-5.21 (.00)*	5.38 (.63)	5.58 (.40)	-1.54 (.13)
Imagery	3.71 (1.35)	4.02 (1.65)	-1.06 (.29)	3.68 (1.37)	4.54 (1.33)	-2.46 (.02)*
Performance	4.36 (.99)	5.08 (1.06)	-3.44 (.00)*	4.64 (1.47)	5.43 (1.04)	-2.57 (.01)*
Judgments	3.18 (1.40)	3.56 (1.37)	-1.33 (.19)	3.83 (1.38)	4.83 (1.40)	-2.73 (.01)*
Feelings	2.16 (1.29)	2.62 (1.60)	-1.49 (.14)	2.78 (1.10)	3.62 (1.41)	-2.46 (.02)*
Resonance	1.45 (.59)	1.91 (.84)	-2.95 (.00)*	2.31 (1.20)	3.94 (1.75)	-4.58 (.00)*

* Significant positive difference corresponding to high counterfeiting awareness level

Table 17.7 The effects of counterfeiting awareness on Gucci genuine users and non-users

	Gucci non-users			Genuine Gucci users		
	Low counterfeit awareness (N = 134)	High counterfeit awareness (N = 65)		Low counterfeit awareness (N = 34)	High counterfeit awareness (N = 81)	
	M (SD)	M (SD)	t-test (sig.)	M (SD)	M (SD)	t-test (sig.)
Prominence	4.06 (1.04)	4.88 (. 82)	-6.06 (.00)*	4.82 (.81)	5.33 (.60)	-3.73 (.00)*
Imagery	3.39 (1.41)	3.73 (1.32)	-1.64 (.10)	3.50 (1.29)	4.02 (1.45)	-1.82 (.07)*
Performance	4.01 (1.20)	4.34 (1.13)	-1.88 (.05)*	4.60 (1.16)	4.84 (1.13)	-1.00 (.32)
Judgments	2.88 (1.34)	3.16 (1.36)	-1.41 (.16)	3.54 (1.47)	4.13 (1.50)	-1.93 (.05)*
Feelings	2.49 (1.32)	3.07 (1.28)	-2.98 (.03)*	3.19 (1.33)	3.81 (1.57)	-2.02 (.04)*
Resonance	1.42 (.67)	1.77 (.95)	-2.65 (.01)*	2.48 (1.20)	3.20 (1.73)	-2.56 (.01)*

* Significant positive difference corresponding to high counterfeiting awareness level

For the group responding in relation to Louis Vuitton, the results show that, within the sub-group of consumers that do not purchase the brand (Louis Vuitton non-users), respondents with a high perception of the level of counterfeiting affecting the brand rate the brand higher in *prominence, performance,* and *resonance* compared to the respondents with a low perception of the level of counterfeiting. Of the other Louis Vuitton sub-group (genuine users), respondents with a high perception of the level of counterfeiting affecting the brand rate the brand higher in all of the dimensions except *prominence* ($t = -1.54$; $p = .13$). A point worth noting here is that the largest gap between these two groups concerns the *resonance* of the brand, particularly emphasizing that, within the genuine users, the more they consider the brand liable to counterfeit practices, the more they are disposed to behave in favour of it (e.g. to talk positively about the brand; to keep themselves informed about it; etc.).

For the group responding in relation to Gucci, the results show several differences among consumers. In the sub-group of respondents that do not purchase the brand (Gucci non-users), those with a higher perception of the level of counterfeiting affecting Gucci rate the brand higher in *prominence, performance, feelings* and *resonance.* Considering the other sub-group (genuine Gucci users), respondents with a high perception of the level of counterfeiting affecting Gucci rate the brand higher in all CBBE dimensions except *performance.* The gap between these two sub-groups concerning the *resonance* of the brand underlines that, among the genuine users, the more people consider the brand under pressure from counterfeiters, the more they are disposed to behave positively in its favour.

To summarize, in both groups (Louis Vuitton and Gucci), potential consumers highly aware of counterfeiting, compared to potential users less well-informed about counterfeiting, (1) show a higher awareness of the existence of the brand and claim a higher ability to recall and recognize it (*prominence*); (2) ascribe a more favourable evaluation of the ability of the brand to meet their needs (*performance*); and (3) show a stronger identification with the brand (*resonance*). Further, within the Gucci analysis, results show an additional difference between the two groups of respondents: brand non-users with a high perception of the level of counterfeiting affecting Gucci exhibit stronger emotional reactions (*feelings*) towards the brand as well.

Considering the genuine users, in both cases (Louis Vuitton and Gucci) consumers with a high perception of the level of counterfeiting affecting the brand declare a stronger relationship and a higher personal identification with the brand (*resonance*) compared to genuine users less well-informed about counterfeiting. This difference is explained not only by the cognitive elements associated with the brand (*judgments* and *performance* – with an exception in the case of the Gucci experimental group, where performance is not different between the two groups of respondents), but also by the affective elements connected to the brand (*imagery* and *feelings*).

17.4 Conclusions

In the light of the results collected by these two studies conducted within one of the most "fashion victim" countries (Italy), we might attempt a preliminary answer to the following crucial question:

> *"Apart from undeniably negative effects on sales, might counterfeiting also induce positive effects on luxury brand equity in the minds of consumers?"*

Respondents to this double study seem to say *"yes"*. Genuine consumers are shown to be particularly sensitive to the counterfeiting phenomenon. The more they perceive the existence of fakes of a brand they possess, the stronger their cognitive evaluations and affective relationship with this brand, and the stronger their personal identification with it. Surprisingly, several positive effects have also been registered within potential, as well as actual, consumers.

These results do not in any way imply that counterfeiting is a positive phenomenon. It is an illegal practice that has many negative effects on economics and on labour. Nonetheless it has to be said that when facing this phenomenon, companies usually adopt mistaken policies.

The first mistake is to focus attention on the negative effects on the luxury brands hit by fakes, (e.g. loss of sales). Our research demonstrates that, contrary to expectation genuine consumers are not sensitive to the temptation to completely substitute original with fakes. Previous research [17] demonstrated that consumers who buy fakes (so called "hybrid consumers") usually do it in addition to their purchase of genuine products. That is, for a luxury brand consumer a fake does not usually replace a genuine purchase. Probably, regular buyers of fakes are not potential buyers of genuine products. The alternative might be an unbranded product or a middle-segment brand. This is why the fight against counterfeits need not be perceived as solely on behalf of luxury brands. In turn, upper level brands hurt by fakes could register positive side effects. In potential consumers' mind the existence of fakes enhances the visibility of the brand and thereby increases their ability to recall it and include it in the consumer consideration set; furthermore it highlights the uniqueness of product characteristics and the ability of the brand to look after consumers, totally absent in fake alternatives. These effects have been tested in two of the most-faked luxury fashion brands (Louis Vuitton and Gucci). Among genuine users the extent of the positive effects is even stronger. Counterfeiting awareness also highlights the distinctive characteristics of genuine users (in terms of consumers' lifestyle and brand personality); thus genuine customers feel more enthusiastic, excited and accomplished at being able to possess and show off a product so desired as to be highly copied.

The second mistake made by companies facing counterfeiting is to shut out genuine and potential consumers from the fight against the practice. The following question arises:

> *Are actual and prospective consumers of a luxury brand willing to play an effective role in the battle to defeat counterfeiting?*

The significant impact of counterfeit awareness on consumers' resonance, registered in these two studies across both actual and potential consumers implies a positive answer to this question. In fact, the more consumers consider the brand as being under attack by counterfeit practices, the more they are disposed to behave in favour of it (to talk positively about it, to keep themselves informed about it, etc.). In particular, with actual consumers, this tendency is grounded in complex, interrelated, and different dimensions (i.e. both affective and cognitive), while with potential consumers it is mainly based on cognitive evaluations related to the brand (*performance* – with an exception in the case of the Gucci experimental group, where *feelings* come alongside *performance* in supporting the difference in *resonance,* but no other elements play a role in giving details of the differences between the two groups).

To conclude we can say that, when aware of counterfeiting, both actual and potential users develop stronger and more lasting relationships with the brand, and these relationships are the ground on which sound and steady fortresses can be built against the counterfeit phenomenon. Consumers of this type can reveal great potential in helping brands to protect themselves against counterfeit practices. Companies that embrace this kind of consumer engagement considering counterfeiting not only as a problem, would have an opportunity to enrich their relationship with consumers.

References

[1] Albers-Miller, N. D. (1999): Consumer misbehavior: why people buy illicit goods, in: Journal of Consumer Marketing, Vol 16, No. 3, pp. 45-57.
[2] Bagozzi, R.P./Foxal, G. R. (1996): Construct validation of a measure of adaptive-innovative cognitive styles in consumption, in: International Journal of Research in Marketing, Vol. 13, No. 3, pp. 201-213.
[3] BASCAP (2009): The impact of counterfeiting on governments and consumers, Frontier Economics Ltd, London
[4] BASCAP (2011): Estimating the global economic and social impacts of counterfeiting and piracy, Frontier Economics Ltd, London.
[5] Berthon, P./Pitt, L./Parent, M./Berthon, J.P. (2009): Aesthetics and ephemerality: Observing and preserving the luxury brand , in: Californian Management Review, Vol. 25, No. 1, pp. 45-67
[6] Bian, X./Veloutsou, C. (2007): Consumers' attitudes regarding non-deceptive counterfeit brands in the UK and China, in: Journal of Brand Management, Vol. 14, No. 4, pp. 211-222.
[7] Bloch, P.H./Bush, R.F./ Campbell, L.(1993): Consumer `accomplices' in product counterfeiting: a demand-side investigation, in: Journal of Consumer Marketing, Vol. 10, No. 4, pp. 27-36.
[8] Bollen, K. A. (1989): Structural equations with latent variables, Wiley, New York.
[9] Bushman, B.J. (1993): What's in a name? The moderating role of public self-consciousness on the relation between brand label and brand preference, in: Journal of Applied Psychology, Vol. 78, No. 5, pp. 57-61.
[10] Campbell, D. T./Fiske, D. W. (1959): Convergent and discriminant validation by the multitrait-multimethod matrix, in: Psychological. Bulletin 56, pp. 81-105.
[11] CENSIS (2009): Il fenomeno della contraffazione nel mondo e le ricadute sul mercato italiano: gli scenari e le strategie di contrasto, Censis, Roma.
[12] Chaudhry, P.E./Walsh, M.G. (1996): An assessment of the impact of counterfeiting in international markets: the piracy paradox persists, in: Columbia Journal of World Business, Vol. 31, No. 3, pp. 34–49.
[13] Commuri, S. (2009): The impact of counterfeiting on genuine-item consumers' brand relationship, in: Journal of Marketing, Vol. 73, No. 2, pp. 86-98.
[14] Dubois, B./Duquesne, P. (1993): The market for luxury goods: Income versus culture, in: European Journal of Marketing, Vol. 27, No. 1, pp. 35–44.
[15] Fournier, S. (1998): Consumers and their brands: Developing relationship theory in consumer research, in: Journal of Consumer Research, Vol. 24, No. 2, pp. 343–73.
[16] Gentry, J.W./Putrevu, S./Schulz, C. (2006): How now Ralph Lauren? The separation of brand and product in a counterfeit culture, in: Advances in Consumer Research, Vol. 28, No. 3, pp. 258-267
[17] Gistri, G./Romani, S./Pace, S./Gabrielli, V./Grappi, S. (2009): Consumption practices of counterfeit luxury goods in the Italian context, in: Journal of Brand Management, Vol. 16, No. 5/6, pp. 364-374.
[18] Hellofs, L. L./Jacobson R. (1999): Market share and customers' perceptions of quality: When can firms grow their way to higher versus lower quality?, in: Journal of Marketing, Vol. 63, No. 3, pp. 16–25.
[19] Hieke, S. (2010): Effects of counterfeit on the image of luxury brands: An empirical study from the customer perspective, in: Journal of Brand Management, Vol. 18, No. 2, pp. 159-173.
[20] Hoe, L./Hogg, G./Hart, S. (2003): Fakin' it: Counterfeiting and consumer contradictions, in: European Advances in Consumer Research, Vol. 6, Darach Turley and Stephen Brown, eds. Provo, UT: Association for Consumer Research, pp. 60–67.
[21] Juggessur, J./Cohen, G. (2009): Is fashion promoting counterfeit brand?, in: Journal of Brand management, Vol. 16, No. 2, pp. 383-394.
[22] Kapferer J./Bastien, V. (2009): The specificity of luxury management: Turning marketing upside down, in: Journal of brand management, Vol. 16, pp. 311-322.
[23] Keller, L.K. (1993): Conceptualizing, measuring, and managing, Customer-Based Brand Equity, in: Journal of Marketing, Vol. 57, No. 2, pp. 1-22.

[24] Keller, L.K. (2001): Building Customer-Based Brand Equity, in: Marketing Management, Vol. 10, No. 2, pp. 14-19.
[25] Keller, K.L. (2009): Buildings strong brands in modern marketing communication environment, in: Journal of Marketing Communications, Vol. 15, No. 2-3, pp. 139-155.
[26] Leisen, B./Nill, A. (2001): Combating product counterfeiting: An investigation into the likely effectiveness of a demand-oriented approach, in AMA Winter Educators' Conference Proceedings, Vol. 12, R. Krishnan and M. Viswanathan, eds. Chicago: American Marketing Association, pp. 271–77.
[27] Mendel, N./ Petrova, P.K./Cialdini, R.B. (2006): Images of success and the preference for luxury brands, in: Journal of Consumer Psychology, Vol. 16, pp. 59-67.
[28] Nia, A./Zaichkowsky, J. L. (2000): Do counterfeits devalue the ownership of luxury brands?, in: Journal of Product & Brand Management, Vol. 9, No. 7, pp. 485-497.
[29] Nunnally, J. C./Bernstein, I. H. (1994): Psychometric theory, 3rd edition, McGraw-Hill, New York.
[30] OECD (2009):The economic impact of counterfeiting and piracy: an update, OECD Publishing, Paris.
[31] Penz, E./Stöttinger, B. (2008b): Original brands and counterfeit brands – do they have anything in common?, in: Journal of Consumer Behaviour, Vol. 7, No. 3, pp. 146 – 163 .
[32] Phau, I./Teah, M., (2007): Devil wears (counterfeit) Prada: A study of antecedents and outcomes of attitude towards counterfeit of luxury brands, in: Journal of Consumer Marketing, Vol. 26, No. 1, pp. 15-27
[33] Silverstein, M. J./Fiske, N. (2000): Trading up: The new American luxury, New York.
[34] Washburn, J.H./Plank, R. (2002): Measuring brand equity: An evaluation of a Consumer-Based Brand Equity Scale, in: Journal of Marketing Theory and Practice, Vol. 10, No. 1, pp. 46-61.
[35] Wilcox, K./Kim, H.M./Sen, S. (2009): Why do consumers buy counterfeit luxury brands?, in: Journal of Marketing Research, Vol. 46, No. 2, pp. 247–259.
[36] World Customs Organization (2010): Customs and IPR Report 2010, WCO, Bruxelles
[37] World Customs Organization (2008): Customs and IPR Report 2008, WCO, Bruxelles

Part 7

The Management of Luxury and Luxury Brands

18 Luxury SMEs Networks

Gaetano Aiello, Raffaele Donvito, Silvia Ranfagni, Laura Grazzini[1]

1 The authors share the final responsibility for the contents of this chapter. However Gaetano Aiello wrote paragraph 1.1; Raffaele Donvito wrote paragraph 1.2; Silvia Ranfagni wrote paragraph 1.3; Laura Grazzini wrote paragraphs 1.4.2 and 1.4.3.2; Raffaele Donvito and Silvia Ranfagni wrote together paragraphs 1.4.3.1; the authors wrote together paragraphs 1.4.1, 1.4.3 and 1.5.

18.1 Luxury Italian SMEs

As is known, the concept of luxury is characterised by a multidimensional structure of meanings particularly difficult to define (see Brioschi 2000 [14], Aiello & Donvito 2006 [3]; Wiedmann et al 2009 [100]). Nevertheless, common elements in the various theoretical positions can be found in the concepts of heritage also linked to a specific territory, exclusivity, uniqueness, high price and aesthetic value. In fact, luxury and luxury products are often the result of production processes that require artistic-craft skills and instruments that are difficult to replicate and strongly rooted in a specific area of origin.

Elegance, culture of design, tradition, refined style are recognised as distinctive features of Italian manufacturers and are at the basis of the Made in Italy identity. This identity has deep-running and far-off roots, founded in the history, culture and geographical fascination of this country. In the collective imagination [38] it is a synonym of Italian quality which combines tradition, traditional vocations and the local territory along with innovation, technology, design and craftsmanship. Made in Italy distinguishes also Italian luxury products that have constructed a brand image based on the same items characterizing the Italian country of origin; it implies the innate skills of Italian creative artists, whether they are stylists, architects, engineers or jewellers. The Made in Italy system is composed not only by the top brands, which have made a clear contribution to making the country and its products famous abroad, but also by small and medium enterprises from a range of different market sectors (clothing-fashion, home furnishings, agri-food, automation-mechanics-plastics-rubber) who find themselves facing the possibility of creating exclusive luxury products in addition to accessible luxury goods. Scholars emphasises that the strategic management of luxury brands is characterised by certain aspects, which derive primarily from the way customers relate to it; luxury brands are described as brands to which consumers assign a premium price more for their psychological value (symbolic or hedonistic) than for their economic or functional value. Particularly Dubois and Czellar (2002) [28] distinguish between luxury brands and prestige brands, defining prestige as the subjective evaluation relating to the high social status of a person or object as a brand that can incorporate it. Luxury brands, on the other hand, are defined on the subjective perceptions of aesthetics or a sumptuous lifestyle. As a result, luxury brands are associated with a smaller nucleus of goods and services with respect to prestige brands. Made in Italy enterprises appear to be able to position themselves credibly as both luxury and prestige brands, thanks to their real values, tangible and objective (production facilities, distribution channels, etc.), their virtual values, intangible and subjective (awareness, the image of leading products, enterprise history, etc.), and their symbolic values, cultural, emotional and affective [4]. In particular, Made in Italy luxury firms qualify for a specific "system of excellency" [22] that encompasses product quality, design originality and creativity that is inspired by the features also emphasised by Kapferer (1997) [63], according to whom "Luxury defines beauty; it is art applied to functional items. Like light, luxury is enlightening. [...] They offer more than mere objects: they provide reference of good taste".

As mentioned above, Italian luxury products are made by large and very well-known global design brands that base their brand personality on calibrated communication and distribution processes. However, it is also qualified by a significant number of small and medium (and even micro) enterprises (SMEs), highly rooted within their territorial frame and characterized by a strong craftsmanship and product culture often accompanied by a low level of marketing skills. These enterprises, which will be the focus of our attention from here on, are defined by the centrality of the figure of the entrepreneur [88] as capable of creating, defining and orientating business activities. These figures are individual capitalist entrepreneurs who invest their own financial resources, labour and knowledge at their own risk, regardless of the enterprise's legal form or structure. They put their own capital of distinctive skills and relations on the line and promote enterprises in which the economic network is closely linked with the social network [89], [9]. In addition to running the enterprise, entrepreneurs are directly involved with the operations, so the ownership of the enterprise is not separated from the entrepreneur, and the entrepreneur involves their family in the business (or other figures with whom they conduct social trust relationships). Lastly, the enterprise is rooted in a specific territorial environment. In industrial or managerial capitalism, the entrepreneur or manager is separated from the ownership, direct work and even management tasks through delegation, while a large part of the enterprise's activities and value chain are often separated from its territory of origin. In financial capitalism, the exercise of shareholder rights is separate from the management of the company, and operates according to the logic of short-medium term yield.

In the type of firms where personal capitalism has a strong hold, (see Welsh e White 1981 [99]; Mc Cartan-Quinn and Carson 2003 [72], Ferrero 2001 [32]) the field for marketing and management models often appears to be restricted. These models [69] tend to differ considerably from the management style usual in this type of firm, where the entrepreneurs themselves are directly responsible for the development of the organisation [24]. Nevertheless, Grandinetti (1989) [46] suggests that, while the separation between marketing and small enterprises can theoretically be justified because marketing is mainly seen as a functional tool, it cannot be justified where it holds strategic value, unless we accept that SMEs do not use strategy and do not actively promote market relations. If we consider SMEs not only on an individual basis but also as part of a connective network of relations with other enterprises and institutions, it becomes clearer that the most suitable marketing approach for SMEs is relationship marketing one, which appears more coherent and closer to the real "modus operandi" of these enterprises [31], [47]. Moreover, the relational activities of small enterprises imply deep and intense interpersonal relations, and are therefore rooted in the culture of the small entrepreneur more than the components of the marketing mix [21], these interpersonal relations contribute also to increase the firm level of tacit knowledge and to define the entrepreneur's strategic representation of the market. The marketing implemented by SMEs is mainly a reactive marketing (rather than proactive), empirical and unstructured [69], [71] in which the prevailing personal dimension of relations is backed up by social and territorial mechanisms. Ultimately, this is an example of "entrepreneurial" marketing [17], [18], [48], [49], [25]. Some researchers (Varaldo, Dalli and Resciniti 2006, p.1 [97]), related to the Italian economic-industrial system, talk about an "Italian-style marketing", that is "virtual and submerged, with no explicit definition or identification and often without even any formalisation in terms of structures and organisational resources". For SMEs, marketing is an ongoing learning process, capable of constructing value with capital resources, especially intellectual, and through the starting up and maintenance of flexible and customized customers relationships. Essentially, marketing can facilitate the valorisation of SMEs cognitive and expertise capital (especially those with a craft vocation). More specifically, marketing supports SMEs to understand the market (regarding its relational and reticular nature) and to direct investment of intellectual capital and personalisation processes towards customer demand.

We can therefore assert that Italian Luxury SMEs have the following main features: i) the strong component of empirical-manual tacit knowledge; ii) the use of highly specialized manual labour as a distinctive resource; iii) the high level of customization capability and production flexibility; iv) the overall creativity endowment [47]; v) the presence of an "entrepreneurial marketing" perspective [17], [18], [69], [48], [25] often not formalized in terms of organizational structures and resources [97]. Lastly, as already stated, these SMEs are qualified by the close interdependence between the entrepreneur, the enterprise itself and environmental factors (the territory). For this reason, it is necessary to analyse not only the SMEs individually but also considering the interrelationships that arise between them and the different actors of the territorial area in which they are rooted.

18.2 Country of origin and territory: features of Made in Italy

Italian SMEs, particularly those operating in luxury sector, gain part of their competitive positioning directly from their country of origin (COO). As known, the image of a product's country of origin is defined as the representation, reputation and stereotype that consumers associate with products from a specific country [75], [76]. According to Roth and Romeo (1992) [87], country image is determined by a series of factors including innovation, design, prestige and workmanship. Country image is connected to a product's country of origin which, in turn, influences consumer perceptions and behaviours [95], [59]. Aaker (1991) [1] and Keller (1993) [64] demonstrate how country of origin can influence brand equity by creating associations between the brand and a particular country.

Country of origin. Country of origin is considered a real source of competitive advantage, especially in markets characterized by high levels of competition [6]. In fact, marketers may use country of origin cues in order to add value to their products and brands differentiating them by specific positioning. Consumers show a growing tendency to use the country with which they associate the origin of a product as a means to synthesise its characteristics and to guide their purchase choices [15], [10]: many scientific contributions [95], [44] emphasise the impact that COO has on consumer perceptions and behaviour through the image of the product's country of origin. If correctly managed, COO can therefore become an important competitive lever on both the business and the country level [59]. Schooler's (1965) [91] studies highlight that products made in less-developed countries are not seen as quality products; Reierson's research found that country of origin exists both in case of general and of specific products. Other studies underline that country stereotypes have an impact not only on product evaluation, but also on purchase intention [41].

From a conceptual point of view, it must be emphasised that the notion of "country of origin" is neither univocal or elementary. While the concept of COO was originally linked to that of the generic "Made in Country" (see Nebenzahl et al., 1997 [78]) the spectrum of meanings has gradually been extended; in fact, we are aware of how distinctions can be made between "country of manufacture" (COM), i.e. the country in which the production phases are performed, "country of assembly" (COA) i.e. the country where the parts of the product are assembled, and "country of design" (COD), the country where the product was conceived, planned and designed (cfr. [90], [78], [60]). This separation of meanings is still inspiring researchers today to understand the effects of the interaction between these "declinations" on consumer purchase and decision-making processes, together with the effects of the interaction between these types of country of origin and brand; we believe that exactly for this reason it becomes even more interesting to understand the role of the antecedent of COO – i.e. country image. In this context, particular attention must be paid to the relation between brand of hybrid product and country of origin in consumer behaviour. Some authors introduce the concept of brand country image (BCI) to summarize the interaction between the brand image and the identity of a country [82] in a new form. The brand country image (BCI) is the image of a brand through the image of a country. The interaction between country image and brand image can be bilateral: on one hand national symbols and attributes characterizing the origin country surround the brand contributing to define its image, on the other hand, strong brand may affect the country image and produce spill-over effects on other national brands [66].

Research has also turned its attention towards the phenomena of interaction between COO and brand, with particular attention for individuals' perception and purchase intentions. In this context, Häubl (1996) [56] holds that purchase intentions are influenced by both the brand and the COO: in fact, customers take into account the brand connotations and the COM/A while formulating their purchase decision (see also Tse, Gorn, 1993 [94]; Ahmed, d'Astous, 1996 [2]). Aaker (1991) [1] and Keller (1993) [64] show how COO can influence brand equity by generating associations between the brand and a certain country; even a "foreign-sounding" brand name is capable of affecting the brand equity of a brand [68]. In the same way, the brand, by playing its role of synthesis, can sometimes surrogate the COO, thanks to the association that that brand name inspires with a specific country [11]. Consumers therefore infer the COO directly from the brand name [55]; this is an example of the so-called "Country of brand origin" (o country-of-brand – COB) effect, i.e. the influence of the country associated with the brand placed on the product [58]. The interactions between the brand image and country image can then be articulated in a series of scenarios; the favourable/unfavourable image of a brand can be reduced if the product is perceived as designed or physically created in countries with a negative/positive image [61]. At the same time, a decidedly strong and affirmed brand can reduce the importance of the COO [79]. Some empirical findings [67] show how COO information (especially regarding the COM) do not have an important effect on the evaluation of branded products if the information are coherent with the brand origin; however, if the product is physically created in a country with a weaker image with respect to the country associated with the brand, the COM information create a negative effect on the product's value, and effect which tends to be even more unfavourable for brands with low brand equity. Häubl and Elrod (1999) [57] have also noted that perceptions of a product are more favourable if the brand and the country of origin are coherent. In the same way, research done by Busacca et al. (2006) [15] show how the effect of the interaction between brand image and country of origin image varies in intensity and direction according to the level of perceptive consonance between the two elements. The perception that consumers have of quality is influenced not only by the brand name, but also by where the products are manufactured and assembled [57]. Moreover consumer brand associations change as the brands are made in different countries [55], [77]. Consumer's brand loyalty is linked to the image of the brand's country-of-origin [65], [80]: favourable country image can lead to brand popularity and thus to consumer brand loyalty. Thus the creation of brand association and of brand loyalty can be affected by the levels of COO that Chao (2001) [19] identifies in the country of design (including manufacture), the country of assembly and the country-of-origin for parts. The higher the congruence between the components of the COO, the greater the capacity of COO to increase brand equity and influence consumer attitude towards a brand. In fact, the country of origin becomes a demographic variable of the brand that contributes to outlining the features of the brand personality [92]. The analysis of the interaction between these two elements is particularly important for global brands, which often offer the market products with different COD and COM; once again, it becomes important to understand if and how much consumers consider the brand name as an indication of the same COO. Pecotich and Ward (2007) [81] believe that a brand gradually takes on the role of synthesis of the attributions of offer as the customer increases familiarity with the brand itself; therefore, the more familiarity grows (often higher with global brands) the less consumers are likely to consider other information that is more or less extrinsic, such as the price or country of origin. Still according to Pecotich and Ward [81], a brand that has succeeded in constructing a perception of familiarity with consumers is also capable of improving the perception of the country of origin with which it is associated, to the point of neutralising any negative effect often associated with emerging economies.

Italy country image and features of Made in Italy. On an international level, Italy country image appears to be based on factors such as "history and culture", "creativity", "design", "tourism", "well-being" and "lifestyle" [29]. Such an image is inevitably compared to "Made in Italy" production characteristics, which, in turn, are based on four principal macro sectors: clothing/fashion, furniture/household goods, agri-food production, automotive/machinery/rubber and plastic products [8], [73], [20], [39], [40]. "Made in Italy" thus expresses the culture and characteristics of the Italian spirit well, and its products represent significant symbols of Italy image [96], contributing to the creation of a positive country image abroad.

It is necessary to emphasise that alongside these four macro-sectors, there are also other sectors with a high technology content (cruise ships, helicopter and aerospace, military equipment, chemical and pharmaceutical speciality products, diagnostic analysis, biomedical). Lastly it is relevant to consider the Italian tourism system which is based in turn on another four sectors: environment, art, architecture, accommodation [39]. Considering the excellence in these sectors, "Made in Italy is a good expression, in its various components and manifestations, of the Italian culture and character and its products represent important symbols of the image that this country boasts in the world" (Varaldo 2001, p.26 [96]); it is precisely these products created in these sectors that contribute to create a positive image of Italy abroad [8].

Recent studies on Italian demand, relating to the effects of the fragmentation of design and production of Made in Italy products [7]; suggest that in "hybrid" products, i.e. with different COD and COM, the role of production origin (COM) is particularly important in creating a favourable opinion in consumers' minds. In particular, research conducted on a product with an Italian-Chinese COO, showed that the positive opinion of a hybrid product depends more on the origin of the Italian production (COM) than on the origin of Italian design (COD). At the same time, according to others [35], Made in Italy appears to overcome the meaning of a simple indication of country of origin, becoming a synthesis tool capable of linking aesthetic and quality components with technological contents; Made in Italy takes on the role of a container of intangible and evocative values linked to the "Italian style and way of life".

However, Pratesi (2001) [83] also found the presence of cultural, social and political factors in the country image associated with Made in Italy; these factors are essentially traceable to six main areas: the lack of a true national identity, or the presence of a prevalently individual pride that is rarely collective, the excessive self-referencing of the Italy system, the limited tendency towards innovation, the reduced capacity for continuity and perseverance in the choices made (political and economic) in addition to the inefficient valorisation of the Italian cultural offer. In addition to the weight of these factors, Made in Italy is also facing the risk of a fall in investment from the Italian entrepreneurs who animate it [30]. In fact, according to the district view, while products relating to Made in Italy have consolidated their position on global markets thanks to competitive district advantages and the fact that the products are of medium-high quality targeted at higher income classes of consumers (see Azzolini 2003 [5]), it appears to be ever more important for Made in Italy enterprises to introduce adequate brand strategies, involving customers in their innovation processes [22], [27].

18.3 Territory as network and creativity in network interactions

Territory as network. To better understand the COO effect related to SMEs it appears relevant to analyse the territory in which those SMEs are embedded. A territory may be considered not only as a geophysical space, but also as a multidimensional construct characterized by the interconnections which exist among the actors ascribable to that territory [42]; that is, the territory is characterized by a marked presence of networks. In order to represent the territory as a network, we adopted the perspective of the IMP Group, for which the network is a set of nodes connected to each other by their relationships [50]. Each node can represent an actor in the network, within which it utilizes resources to carry out certain activities, developing links to other actors. The model elaborated by Håkansson (1987) [51] allows us to analyse the multipolar nature of the interactions among nodes: it presents the reticulate structure as the result of interrelations between networks of actors, of resources, and of activities. This concept of network is useful for studying the specific spaces of interaction, including territorial ones. Indeed, the territory may be represented as a set of networks of actors that interact with each other and share resources, developing spatially rooted activities [70]. The territory is important insofar as it can give substance to this interaction; actors, resources, and activities are, in fact, situated in a relational space in which the interactions that occur influence the actors' positions in the network [62], [37]. Håkansson, Tunisini and Waluszewski (2002) [54] define the territory as a set of resources, noting the presence of relationships linking the resources of a particular territory with those of another territory. It follows that a territory is not only made up of material components, but also of intangible components which have become established over time and which are the fruit of the actors' relationships with each other and of the territory's own socio-cultural heritage [26]. These intangibles promote a shared territorial image upon which the territory's identity is based [93].

Creativity in network interactions. Lastly, the chapter considers the concept of creativity as the outcome of interactions among multiple actors (especially Luxury SMEs) integrated into the network "milieu". Creativity on an individual level is the ability of knowing how to create, producing something new and original [43]. However, creativity can be the result not only of these individual capabilities or an organization's internal relationships, but also of the interactions between pluralities of actors within a network. Network creativity is always, therefore, the generation of something new and original [13], created by interaction-driven cognitive relationships in which reciprocal helpfulness, trust, and collaborative ability exists between the partners [51], [52]. This chapter therefore considers creativity outputs in terms of luxury goods but also in terms of new marketing strategies and policies made by Luxury SMEs and Luxury SMEs networks. Creativity in network interactions. The paper presents the concept of creativity as the outcome of interactions among multiple actors integrated into the network milieu. Creativity at the individual level is the capability of knowing how to create, producing something new and original [43]. Regarding this, Bohm (1998) [13] sustains that "creativity is the outcome from original thinking based on the will to develop new ideas rather than to conform what is taken for granted or commonly shared knowledge." At the company level, creativity is a strategic resource [36], [101], the existence of which is not strictly correlated to the creativity of individuals. In other words, individuals' creativity is important to the firm, but a group of creative people does not necessarily make for a creative organization [102]. This means that what determines an organization's creativity is not the quantity of creative people within it, but the way in which the interactions between the individuals and between the other elements generate the conditions to render an organization capable of creating something highly innovative [98]. The authors agree with Binning (1991) [12], when he defines creativity as "the aptitude of a system to evolve". From this definition, widely adopted in the literature [98], [45], it follows that all social systems, comprised of interacting parts, are capable of creativity.

Creativity can be the result of not only the organization's internal relationships, but also of the interactions amongst a plurality of actors within a network. Adopting a conceptual framework similar to the one used to move from the level of individual creativity to that of organizational creativity, it is possible to shift from the firm's perspective to a network perspective, in which creativity is no longer a resource belonging only to individual interacting companies, but now belonging to the entire network (network creativity). The network produces creativity if it is made up of relationships among actors based on exchanges of knowledge (knowledge as a resource and as the result of the creative process) and a strong relational dynamism (openness to change), and guided by a co-evolutionary reticulate path that produces changes in the interacting components. Network creativity, the outcome of which is, in any event, the generation of something new and original (Bohn, 1998), presupposes interaction-driven cognitive relationships in which there exists reciprocal helpfulness, trust, and collaborative ability between the partners [51], [52].

18.4 Luxury SME networks in the Florence area: the results of an empirical analysis

18.4.1 Research aim, phases and methodology

This part of the chapter shows the empirical research run on a set of small and medium enterprises producing luxury goods, located in the Florence area. The aim of this research conducted by the Business Science Department for Confindustria Firenze (industrial association) is to understand how networks can form a valid tool for creative SMEs with strong roots in the territory but a reduced brand visibility to develop in new markets and increase their competitiveness in an international context. Florentine luxury goods manufacturers currently hold great importance, not only for the contribution to the economic growth of the territory, but also and above all for the role that they perform in promoting the image of Florence in foreign markets. Often, their small size and unstructured marketing skills represent limits that do not allow these enterprises to develop and consolidate their position in international markets. A possible alternative to overcome these barriers is to aggregate SMEs around network initiatives. This typology of network is constructed on solid trust relations between partners, and depends on shared goals to be achieved through the network project, and lastly is based on the identification of specific economic advantages that would otherwise be impossible to achieve.

As already stated, there is a strong relation between companies and territory; this means that luxury companies may use the territory as a platform for initiating and fostering their creative relationships [45]. More specifically a territory can be articulated in "project networks" [85] that include different actors which interact realizing specific initiatives, generally within a certain timeframe [23], [84]. These actors include not only companies, but also institutions, artistic and cultural organizations; each of them makes its own resources and competences available, conserving its core activities. The creation of territorially based creative project networks is encouraged by the presence of an active marketing place, which permits the construction, maintenance, and strengthening of advantageous exchange relationships between internal, but also external actors [16]. In this chapter, we assume the "project networks" perspective to identify the activities and projects around which Florentine Luxury SMEs can aggregate their resources and skills in order to mobilize business networks. These do not have hubs or centres, even if within them it is possible to identify network mobilizers [53], [74] and so drivers of project network, characterized by network "closeness centrality [that] reflects the distance between an actor and all other actors in the network. An actor with high closeness centrality can access other members of the network via the lowest possible number of links. Thus, the actor is not too far removed from the majority of the other actors in the network, as is consistent with the intuitive notion of being in the centre".

The identification of Florentine luxury business networks required a structured field research, performed through focus groups and in-depth interviews. The focus groups examined the possible trajectories both for internationalisation and aggregation followed by a set of particularly active Luxury SMEs; these firms have been selected because of their tendency to mobilize formal business networks. The in-depth interviews were performed on a sample (defined by analysing the results of the focus groups and with the support of Florentine industrial and trade associations) of twenty-five luxury enterprises (eleven from the fashion sector and nine from the home furnishings sector); particularly the enterprise representatives were interviewed using a semi-structured questionnaire which, in addition to collecting profile information (size, turnover, strong and weak points, etc.), allowed for the identification and analysis of the development strategies undertaken, the relation between these strategies and the potential deriving from being part of business networks. As we will see, the combination of the focus groups and the in-depth interviews allowed us to outline some business network hypotheses divided by business, outlet markets, organisation and economic-business features. Before presenting the analysis of the main results, the following paragraph provides some details on the enterprises typology that characterise the Florence area, in order to provide a background of the empirical findings.

18.4.2 The Florence area and the SMEs analyzed

The Florence area. Much of the excellence of "Made in Italy" production exported all over the world is concentrated in the Florence area. Luxury leather goods and clothing, manufacturing, mechanics industry form the central axis of Florentine economy, the third most important industrial centre in Italy and the eighth province on a national level by number of enterprises. Despite the economic crisis, the Florentine economy has been slightly less affected, above all thanks to a stronger hold in international markets (FirenzeBusiness 2010). In fact, with a per capita GDP of approximately 30,400 Euro in current value, the province of Florence is at eleventh position in Italy, with a value of over 21.8% with respect to the national average, and in first place among the provinces in central Italy, while it also forms one of the most important production nodes in Europe. The Florentine economy is far more open to international exchanges with respect to the rest of the region: turnover is approximately 7% higher than the regional average, and 8% higher than the national average. This is formed of a tight network of dynamic and flexible small and medium enterprises (more than 105,000), integrated with larger enterprises and forming the cornerstone of the local economy. Among the various types of business, those associated with fashion (in particular shoes, bags, belts and leather clothing) permeate the Florentine area, leading exports in the whole region and ensuring world leadership in the sector for know-how and style. Florence is the second Province in Italy for leather goods exports, with a volume of 522 million Euro for shoes and 976 million Euro for travel goods and bags. There is a network of small-medium sized enterprises, linked through production with the larger Tuscan based groups such as Salvatore Ferragamo, Gucci and Prada, and to international names such as Fendi, Louis Vuitton, Chanel, Dior and Céline, which, although maintaining management centres abroad, have chosen the Florentine district to locate their design and production facilities. Around Florence, there are also leather-finishing businesses, ceramics and home furnishing, glass and wooden product manufacturers. As mentioned above, our study will concentrate on small-medium sized Florentine enterprises operating in the luxury goods sector.

The profile of the SMEs analysed. The set of enterprises analysed in this study is formed by small and medium enterprises operating in the fashion and home furnishings sectors.

From an organisational point of view, these enterprises are distinguished by the presence of family groups as both owners and entrepreneurs, and by their reduced sized in terms of number of employees and turnover. In particular, approximately 60% of the enterprises interviewed had less than 10 employees; 27% had 50 units or less, while only 13% had between 50 and 100 units. In terms of turnover, 25% of the sample had a yearly turnover of 2 million Euro or less; 56% had a turnover of 10 million Euro or less; while lastly, only 19% recorded an annual volume of sales of 50 million Euro or less.

All the enterprises interviewed count on product quality craftsmanship and Made in Italy. These factors, together with the search for brand awareness, fashion contents, innovation and a competitive quality-price ration form the strengths of the analyzed firms. These advantages come from specific critical factors such as organisational skills, flexibility and production reliability, uniqueness and creativity of the product together with the high level of care in managing customer relations. However, these firms have also several weakness, such as the not structured marketing function, the low negotiating power and the difficulties in accessing credit (banks and investors); these factors combined with the high production costs in terms of labour and the presence of a small number of customers, which in some cases was reduced to a single client (customer dependency).

18.4.3 Main results

This part of the chapter finally presents, with reference to the enterprises analysed, a) the development strategies under planning; b) the tendency and the desire to actively participate in business networks; c) the concept of luxury assumed by the management. The study, with respect to the fashion and home-furnishings sector, has also identified several hypotheses for business networks (project networks) which the enterprises will attempt to activate soon.

a. Regarding the development strategies, nearly all the enterprises interviewed (90% of the sample) is orientated towards increasing their presence on international markets, concentrating their commercial efforts in emerging markets, especially China and Russia, followed by India and Brazil, while the Middle and Far East were seen as no less important. There is a widespread desire to develop a direct presence in these markets by increasing and starting relationships with retail operators, opening directly operated point of sales and giving greater strategic importance to the web channel. The need to guarantee a wider offer in foreign markets induced some of the enterprises interviewed to extend their business into similar sectors, developing new production processes and strategically rethinking their logistics activities.

b. Regarding the tendency to actively participate within a network, the overall attitude found in the sample of enterprises was positive. The network is considered as a useful tool to extend business activities, to share resources and risks. This openness was however mitigated by direct and indirect experience of the scarce level of efficiency of other network systems; these experiences refer to business consortia that were attributed with a merely passive role in promoting the enterprises in the context of international trade fairs. Overcoming this inefficiency can make the network a valid tool for increasing the competitiveness of small and medium enterprises in foreign markets; the activities that the interviewees would like to see performed by a potential network were of both a strategic and operative nature. These included the creation of a brand representing the group, the development of contacts with final and distributors, the development of commercial relations and the management of marketing activities, and in particular communication activities to valorise the image of the groups in new markets. From the study, it is clear how commercial type activities that can be performed through the network are dominant; in particular the sample selected emphasise the importance of being part of business networks in order to create purchase groups, have shared commercial premises abroad and to start up market knowledge exchange processes. In order to create this system, the enterprises are willing to put their resources and development capacity together for both the productive and commercial development of the network itself.

c. With reference to the perception of the concept of luxury, a certain convergence was noted among interviewees in attributing the significance of desirability, exclusivity and prestige. Luxury was also taken to mean high price, emotionality, rarity and elitism. A luxury product does not evoke extravagance, sophistication, snobbism or showiness; instead, these adjectives were perceived as far-off from the linguistic code of luxury.

Luxury SMEs Network hypotheses (project networks). As mentioned above, the study also allowed us to formulate some hypotheses for project networks with reference to the fashion and home-furnishing sectors. Before presenting the various hypotheses, we would like to repeat that the enterprises interviewed agreed in considering the network as an indispensable tool for increasing their market power. They see the network as a way to come closer to the final consumer (thereby favouring greater customer intimacy), developing forms of direct presence and valorising their brand identity. The network is considered as a shared environment where resources and skills can be pooled in order to overcome the typical weak points of small-medium sized enterprises and allow them to achieve strategic goals of growth in new markets that would otherwise be impossible. Examples of an integrated network include the creation of a group brand, the opening of shared showrooms and the production of a range of integrated and complementary products.

18.4.3.1 Network hypotheses (project networks) in the home-furnishings sector

The following network hypotheses were proposed for the home-furnishings sector:

- creative network;
- incoming network;
- distribution network.

We would now like to pause to analyse each of these three alternative networks in order to evaluate their aims, reference market, organisational characteristics and comparative economic aspects.

Creative network. This network aims to develop an integrated range of restoration services and artistic furnishing products. The achievement of these aims would involve a network of enterprises to be constructed with capacities and skills that can be integrated ("contracts"), the identification of large orders and the integrated design of the product. The products in this creative network appear to be mainly associated to countries that show considerable development prospectives in the next few years, including China, Russia (ex-Soviet republics), Saudi Arabia and North America. The reference segments within these markets are formed of big players from the artistic and cultural worlds. These include political, religious and cultural institutions in addition to private customers and owners of luxury properties. From an organisational point of view, it is apparent that this network can also be composed of enterprises from different business areas, linked by a central governance, that allows them to set up and develop commercial interactions aiming to complete the order. Moving on to a synthetic analysis of the economic-business elements, it is clear how this project produces advantages, thanks to the acquisition of large orders and an increase in the prestige of individual enterprises. However, this is countered by considerable costs, due to the coordination and negotiation between actors in the network together with the costs of managing the order.

Incoming network. This second type of network aims to develop the attraction of intermediate and affluent clients through income activities in the Florentine territory. In this case the excellence of a territory such as Florence or Tuscany, rich in cultural, artistic, eno-gastronomic and entrepreneurial assets with the excellence of the luxury enterprises that are part of it. The incoming network is responsible for the development of mixed activities to select new potential market segments, identifying entrepreneurial and territorial resources, designing a package of business-leisure offers and managing relations with intermediate/affluent clients in the territory. The reference markets for this network are formed of countries such as Russia, the Middle East, Far East and Chine; in terms of segments it is possible to distinguish between intermediate clients such as buyers and opinion leaders (designers and technical firms) which constitute contact activators for product commercialisation and final clients. In organisational terms, the actors dedicated to design and create the incoming offer, i.e. tour-event organisers and actors in the territorial offer system such as accommodation structures, cultural institutions and service providers (transport) take on an important role as potential coordinators. Regarding an analysis of the economic factors, it may be noted how advantages include a growth in turnover through the identification of new clients, the loyalisation of existing customers, greater brand notoriety for individual enterprises and the possibility of accessing public funding. About the economic aspect, however, there are considerable costs for designing and implementing the package, along with personalisation and hosting costs.

The distribution network aims to set up a direct presence in foreign markets, through the development of the interaction with distributors and of direct stores. In this case the activities required of the network are particularly complex. They include the analysis of offer and demand to define the retail distribution formula (planned shopping centres, malls, etc.), the identification of the format (shop in shop, store) and the sign, the definition of the product assortment and the management of the distribution format. In terms of the geographic areas chosen, the markets of China, Russia and the Middle East take on a central role, as important markets for Made in Italy luxury products. In terms of target, the aim is to act in two strategic segments: on one hand, final customers looking for an emotional and exclusive luxury, and on the other the big buyers or specialised retailers (intermediate clients), which could improve their commercial offer with prestige products. From an organisational point of view, the participating subjects will be luxury enterprises, held together by a central governance, aiming to promote and sell their products. Regarding the sources of revenue and costs that the initiative entails, it is possible to short the distribution channel producing a growth in marginality, and expanding markets; at the same time, however, the costs for market analysis and the initiative start-up costs will have to be sustained.

18.4.3.2 Network hypotheses (project networks) in the fashion sector

There are two network hypotheses for the fashion sector:

- services network;
- "total look" network.

The services network would be created with two specific aims: facilitating access to foreign markets for Florentine Made in Italy enterprises and maintaining the identity of individual enterprises. In other terms, the project network would seek to offer the possibility, through the network's integrative character, of promoting products to markets that would not otherwise be accessible to individual enterprises. For this reason multiple activities would be required of the network. They include the analysis of foreign markets, the organisation of promotional events and trade fairs; the activation of relations with buyers and specialised large-scale distribution. Other activities include the management of foreign sales networks and the financial services, necessary for accessing foreign markets. The markets comprise the BRIC countries (Brazil, Russia, India and China), the Middle East and North America. Regarding the market segments that the network intends to target, the top position is held by luxury fashion brands. The Florentine enterprises involved in this project network create not only finished products, but also semi-finished products. The contact with luxury brand can produce new orders and additional development possibilities. In addition to top labels, another two important segments are industrial and traditional retailers. From an organisational point of view, the network will take on a structure in which the various enterprises along the production chain are governed by a central organization, which will provide them with the knowledge and resources necessary to access new markets and acquire new clients. In terms of sources of revenue, the project would entail the possibility of increasing the turnover for each enterprise, thanks to both the acquisition of new clients and the greater penetration for current clients. Furthermore, it would allow for the development of knowledge of new markets, necessary to consolidate their competitive position. In this case, however, a series of costs would also be necessary and essential for the success of the network itself, such as staff costs, marketing research costs to analyse markets and to make scouting activities. Other costs include market relations management and lastly resources to be used for creating projects.

The "total look" network would be created with the aim of developing a new Florentine luxury Made in Italy brand. The intention would therefore be to create a range of luxury products characterised by distinctive features such as craftsmanship, design and quality, which would be recognisable through the development of a brand. To do this it would therefore be necessary to define the brand personality, in terms of characteristics, values and marks, and its competitive positioning, through market analysis. This in addition to the importance of defining the outlines of the total look, using targeted range choices in addition to develop and manage adequate marketing mix policies. As in the cases above, the areas of greatest interest are the so-called emerging economies (BRIC), followed by the Middle East, North America and Italy. The market segments that we intend to reach belong to luxury markets, with particular reference to the operators in specialised distribution, along with specialised multi-brand shops and to the direct channel. Ideal participants would include both enterprises that create finished and semi-finished products, which work closely with designers to create the total look line to address to markets exploiting the integrated action of the network in terms of strategic planning and promotion. The advantages of this proposal are linked to the possibility of generating turnover for each enterprise by acquiring new clients. However, developing this enterprise project network would require enterprises to create a shared vision of a new brand as an added value and as a tool to acquire power along the production chain. There would also be a series of start-up costs for the initiative in addition to funding and operating costs.

18.5 Final remarks

In conclusion, we have tried to demonstrate how the different project networks have the main characteristic of being a form capable to support the growth of the firms involved. These enterprises, as emphasised several times above, offer products that could be successful in specific market even if they have critical factors that reduce their chances to access new business opportunities. It is therefore required to implement actions for increasing Florentine luxury enterprises awareness in international markets. In particular it is important to valorise luxury Italian SMEs together with the territory they belong; innovation, design, prestige and workmanship are strictly connected to their country of origin. So far, the luxury Italian SMEs could develop their differentiation capabilities only if the territory maintains intangible resource and specific creative knowledge. The openness to other countries and territory through collaborative interaction could be the way to reinforce creativity network and develop new businesses.

References

[1] Aaker, D.A. (1991): Managing Brand Equity New York, The Free Press.

[2] Ahmed, S. A./d'Astous, A. (1996): Country-of-Origin and Brand Effects: A Multi-Dimensional and Multi-Attribute Study, in: Journal of International Consumer Marketing, Vol. 9, No. 2, pp. 93-115.

[3] Aiello, G./Donvito, R. (2006): L'evoluzione del concetto di lusso e la gestione strategica della marca. Un'analisi qualitativa delle percezioni sul concetto, sulla marca e su un prodotto di lusso, in: the proceedings of 5th International Conference Marketing Trends, 2006, Paris-Venice, ed. Jean-Claude Andreani and Umberto Collesei, Paris-Venice: Marketing Trends Association.

[4] Allérés, D. (1997): Luxe...strategies-marketing, Economica, Parigi.

[5] Azzolini, R. (2003): Alcune considerazioni sui distretti industriali del Made In Italy negli anni Novanta, in: Convegno Nazionale Una politica industriale contro il declino, Roma.

[6] Baker, M.B./Ballington, L. (2002): Country-of-origin as a source of competitive advantage, in Journal of Strategic Marketing, Vol. 10, pp. 157-168.

[7] Balboni, B./Gabrielli, V./Grappi S. (2009): L'effetto del Country-of-Design e del Country-of-Manufacture sulla valutazione di prodotto: un'analisi sperimentale sul consumatore italiano, in: the proceedings of 8th International Conference Marketing Trends, 2008, Paris-Venice, ed. Jean-Claude Andreani and Umberto Collesei, Paris-Venice: Marketing Trends Association.

[8] Becattini, G. (2000): Dal distretto industriale allo sviluppo locale. Svolgimento e difesa di una idea, Bollati Boringhieri, Torino.

[9] Becattini, G. (2004): Per un capitalismo dal volto umano. Critica dell'economia apolitica, Bollati Boringhieri, Torino.

[10] Bertoli, G./Busacca, B./Molteni, L. (2005): L'interazione fra "made in" e marca nel processo di scelta del consumatore: un'indagine empirica nel contesto italiano e statunitense, in: the proceedings of 4th International Conference Marketing Trends, 2005, Paris-Venice, ed. Jean-Claude Andreani and Umberto Collesei, Paris-Venice: Marketing Trends Association.

[11] Bhaskaran, S./Sukumaran, N. (2007): National culture, business culture and management practices: consequential relationships?, in: Cross-Cultural Management, Vol. 14, No. 1, pp. 54-67.

[12] Binning, G. (1991): Dal nulla. Sulla creatività dell'uomo e della natura, Garzanti, Milano.

[13] Bohm, D. (1998): On Creativity, Lee Nichels, Routledge, London.

[14] Brioschi, A. (2000): Comunicare il lusso, in: the proceedings of 1st International Conference Marketing Trends, 2000, Paris-Venice, ed. Jean-Claude Andreani and Umberto Collesei, Paris-Venice: Marketing Trends Association.

[15] Busacca, B./Bertoli, G./Molteni, L. (2006): Consumatore, marca ed "effetto made in": evidenze dall'Italia e dagli Stati Uniti, in: Finanza marketing e produzione, Vol. 24, No. 2.
[16] Caroli, M.G. (1999): Il marketing territoriale, Franco Angeli, Milano.
[17] Carson, D.J. (1985): The evolution of marketing in small business, in: European Journal of Marketing, Vol. 19, 5.
[18] Carson, D.J./Coviello, N. (1996): Qualitative issues at the marketing/entrepreneurship in SMEs. An Innovative Approach, Prentice-Hall, London.
[19] Chao, P. (2001): The moderating effects of country-of-assembly, country-of-parts and country-of-design on hybrid product evaluation, in: Journal of Advertising, Vol. 30, 4, pp. 67-81.
[20] Chiarvesio, M./Di Maria, E. (2009): Internationalization of supply networks inside and outside clusters, in: International Journal of Operations & Production Management, Vol. 29, 11, pp. 1186-120.
[21] Conti, E. (2005): Approcci di marketing per la piccola impresa: il caso Eva Garden s.r.l., in: Piccola Impresa, Vol. 18, 1, pp.155-182.
[22] Corbellini, E./Saviolo, S. (2004): La scommessa del Made in Italy e il futuro della moda italiana, Etas, Milano.
[23] Cova, B./Hoskins, S. (1997): A twin-track networking approach to project marketing, in: European Management Journal, Vol. 15, 5, pp. 546-556.
[24] Cozzi, G. (1985): Soggettività e organizzazione nel marketing delle piccole e medie imprese, in: Economia e politica industriale, Vol. 47.
[25] Cozzi, G. (2006): Note in margine alla lezione di Isa Marchini sul marketing imprenditoriale nelle imprese minori, in: AA.VV., Scritti in onore di Isa Marchini, Franco Angeli, Milano.
[26] Cresswell, T./Hoskins, G. (2008): Place, persistence, and practice: evaluating historical significance at Angel Island, San Francisco, and Maxwell Street, Chicago, in: Annals of the Association of American Geographers, Vol. 98, 2, pp. 392-413.
[27] Di Maria, E./Finotto, V. (2008): Communities of Consumption and Made in Italy, in: Industry and Innovation, Vol. 15, 2, pp. 179–197.
[28] Dubois, B./Czellar, S. (2002): Prestige Brands or Luxury Brands? An Exploratory Inquiry on Consumer Perceptions, in: Brand Management Track, HEC School of Management and University of Geneva, pp. 4-5.
[29] Esposito, G.E. (2006): Elogio della diversità: Made in Italy, Hoepli, Milano.
[30] Eurispes, (2005): Il made in Italy: quel che distingue ancora l'Italia nel mondo, in: Rapporto Italia 2005, pp. 613-626.
[31] Ferrero, G. (1992): Struttura, strategia e processi innovativi nelle piccole imprese, Edizioni LINT, Trieste.
[32] Ferrero, G. (2001): Nuove tendenze negli studi sulla piccola impresa, in: Piccola impresa, Vol. 14, 1, pp. 3-7.
[33] Firenzebusiness, (2010): Area Metropolitana, Promofirenze, in: www.firenzebusiness.it.
[34] Firenzebusiness, (2010): Settori Chiave, Promofirenze, in: www.firenzebusiness.it.
[35] Fondazione Manlio Masi (2007): La sfida della qualità. Il futuro delle aziende italiane sui mercati internazionali, in: VI Forum del Comitato Leonardo, Roma, p. 109.
[36] Ford, C.M./Gioia D.A. (2000): Factors influencing creativity in the domain of managerial, Harper Collins, New York.
[37] Ford, D./Gadde L.E./Hakansson, H./Snehota, I./Waluszewski, A. (2008): Analysing business interaction, in: 24th IMP-conference in Uppsala, Sweden.
[38] Fortis, M. (1996): Crescita economica e specializzazioni produttive, sistemi locali e imprese del Made in Italy, in: Vita e Pensiero, Milano.
[39] Fortis, M. (2005): Il Made in Italy tra commercio leale e innovazione industriale, in: II Conferenza Nazionale sul Commercio con l'Estero, Roma.
[40] Fortis, M./Realacci E. (2009): ITALIA: geografie del nuovo made in Italy, in: Symbola e Fondazione Edison, p. 9.

[41] Gaedeke, R. (1973): Consumer attitudes towards products made in developing countries, in: Journal of Retailing, Vol. 49, pp. 13-24.
[42] Gavinelli, L. (2007): Il territorio come sistema di relazioni: il ruolo dell'Ecomuseo, in: IV convegno annuale della Società Italiana Marketing, Roma.
[43] Gil, R./Spiller P. (2007): The organizational dimensions of creativity: motion picture production, in: California Management Review, Vol. 50, n.1.
[44] Giraldi, A. (2010): Country-of-Origin – una rassegna analitica dei principali lavori dal 1962 al 2008, in: the proceedings of 9th International Conference Marketing Trends, 2010, Paris-Venice, ed. Jean-Claude Andreani and Umberto Collesei, Paris-Venice: Marketing Trends Association.
[45] Goldenberg, J./Mazursky, D. (2003): Creatività ed innovazione di prodotto, Etas Libri, Milano.
[46] Grandinetti, R. (1989): Il marketing delle grandi e delle piccole imprese: dalla separazione alla convergenza, in: Economia e politica industriale, Vol. 63, pp. 83-94.
[47] Grandinetti, R./Rullani, E. (1996): Impresa transnazionale ed economia globale, La Nuova Italia Scientifica, Roma.
[48] Guercini, S. (2003): La Conoscenza del Vertice di Impresa, Milano, Franco Angeli.
[49] Guercini, S. (2005): Marketing imprenditoriale, marketing manageriale e conoscenza di mercato del vertice d'impresa, in: Mercati e Competitività, Vol. 1, pp. 143-164.
[50] Håkansson, H. (1982): International marketing and purchasing of industrial goods, John Wiley & Sons, Chichester, U.K.
[51] Håkansson, H. (1987): Industrial technological development, Croom Helm.
[52] Håkansson, H./Huysman, M./Meijer, A. (2001): Inter-Organizational interaction and organizational Teaching, in: Hakansson H./Johanson J., Business network learning, Pergamon.
[53] Håkansson, H./Snehota, I. (1990): No business is an island: the network concept of network strategy, in: Scadinavian Journal of Management, Vol. 4, No. 3, pp. 187-200.
[54] Håkansson, H./Tunisini, A./Waluszewski, A. (2002): Place as a resource in business networks, in: 18th IMP Conference, Dijon, France.
[55] Han, C.M./Terpstra V. (1988): Country-of-origin effects for Uni-National and Bi-National products, in: Journal of International Business Studies, Vol. 19, No. 2, pp. 235-255.
[56] Häubl, G. (1996): A cross-national investigation of the effects of country of origin and brand name on the evaluation of a new car, in: International Marketing Review, Vol. 13, No. 5, pp. 76-97.
[57] Häubl, G./ Elrod T. (1999): The impact of congruity between brand name and country of production on consumer's product quality judgments', in: International Journal of Research in Marketing, Vol. 16, No. 3, pp. 199-215.
[58] Insh, G.S./McBride J.B. (2004): The impact of Country of Origin Cues on consumer perceptions of product quality: a binational test of the decomposed country of origin construct, in: Journal of Business Research, Vol. 57.
[59] Jaffe, E./Nebenzahl I.D./Pastore A. (a cura di) (2008): Made In..., Baldini Castoldi Dalai, Milano.
[60] Jaffe, E./Nebenzahl, I.D. (2001): National image and competitive advantage: The theory and practice of country-of-origin effects, Copenhagen Business School Press, Copenhagen.
[61] Jahanson, J.K./Nebenzahl I.D., (1986): Multinational production: effect on brand value, in: Journal of International Business Studies, Vol. 17, pp. 101-126.
[62] Johanson, J./Mattsson, L.G. (1992): Network positions and strategic action – an analytical framework, in Axelsson, B./Easton, G. (Eds), Industrial Networks. A New View of Reality, Routledge, London, pp. 205-17.
[63] Kapferer, J.N. (1997): Strategic Brand Management, Kogan Page, 2d edition, UK.
[64] Keller, K.L. (1993): Conceptualizing, measuring and managing customer-based brand equity, in: Journal of Marketing, Vol. 57, pp. 1-22.
[65] Kim, C. (1995): Brand popularity and country image in global competition: managerial implications, in: Journal of Product and Brand Management, Vol. 4, No. 5, pp. 21-32.

[66] Kleppe, I.A./Iversen N.M./Stensaker I.G. (2002): Country image in marketing strategies: conceptual issues and an empirical Asian illustration, in: Journal of Brand Management, Vol. 10, pp. 61-74.
[67] Koubaa, Y. (2008): Country of origin, brand image perception, and brand image structure, in: Asia Pacific Journal of Marketing and Logistics, Vol. 20, No. 2, pp. 139-55.
[68] Leclerc, F./Schmitt, B.H./Dube, L. (1994): Foreign branding and its effects on product perceptions and attitudes, in: Journal of Marketing Research, Vol. 31, No. 2, pp. 263-270.
[69] Marchini, I. (1995): Il governo della piccola impresa. Le basi delle conoscenze, Vol. I, Aspi/Ins-Edit, Urbino.
[70] Martin, R. (2003): Putting the economy in its place: on economics and geography, in: Proceedings of the Cambridge Journal of Economics Conference.
[71] Mattiacci, A./Ceccotti F. (2005): Lo sviluppo del marketing nella PMI: riflessioni a margine di un'esperienza d'impresa, in: Micro & Macro Marketing, Vol. 1, pp. 65-106.
[72] McCartan-Quinn, D./Carson, D. (2003): Issues which Impact upon Marketing in the Small Firm, in: Small Business Economics, Vol.21.
[73] Micelli, S./Di Maria, E. (2000): Distretti industriali e tecnologie di rete: progettare la convergenza, FrancoAngeli, Milano.
[74] Mouzas, S./Naudé, P. (2007): Network mobilize, in: Journal of Business & Industrial Marketing, Vol. 22, No. 1, pp. 62–71.
[75] Nagashima, A. (1970): A comparison of Japanese and US attitudes toward foreign products, in: Journal of Marketing, Vol. 34, No. 1, pp. 68-74.
[76] Nagashima, A. (1977): A comparative 'made-in' product image survey among Japanese businessmen, in: Journal of Marketing, Vol. 41, No. 3, pp. 95-100.
[77] Nebenzahl, I.D./Jaffe E.D. (1996): Measuring the joint effect of brand and country image in consumer evaluation of global products, in: International Marketing Review, Vol. 13, No. 4, pp. 5-12.
[78] Nebenzahl, I.D./Jaffe, E.D./Lampert, S.I. (1997): Towards a theory of country image effect on product evaluation, in: Management International Review, Vol. 37, pp. 27–49.
[79] Papadopoulos, N./Heslop, L.A. (1993): Product-Country Images: Impact and Role in International Marketing, International Business Press, New York, NY.
[80] Paswan, A.K./Kulkarni S./Ganesh G. (2003): Loyalty towards the country, the state and the service brands, in: Journal of Brand Management, Vol. 10, No. 3, pp. 233-251.
[81] Pecotich, A./Ward, S. (2007): Global branding, country of origin and expertise. An experimental evaluation, in: International Marketing Review, Vol. 24, No. 3, pp. 271-296.
[82] Phau, I./Prendergast, G. (2000): Conceptualizing the country-of-origin of brand, in: Journal of Marketing Communication, Vol. 6, pp. 159-170.
[83] Pratesi, C.A. (2001): Il marketing del Made in Italy. Nuovi scenari e competitività, Franco Angeli, Milano.
[84] Purchase, S./Olaru D. (2004): Substance in Business-t-Business Relationships, in: Journal of Business-to- Business Marketing, Vol. 11, No. 3, pp. 23-52.
[85] Ramos, C./Ford D. (2009): Contextual factors of network picture formation: an empirical examination, in: 25th IMP Conference, Euromed Management, Marseilles, France.
[86] Reierson, C. (1966): Are foreign products seen as national stereotypes?, in: Journal of Retailing, Vol. 42, pp. 33-40.
[87] Roth, M.S./Romeo, G.B. (1992): Matching product category and Country Image Perceptions: A framework for managing Country of Origin Effects, in: Journal of International Business Studies, Winter, Vol. 23, No. 3, pp. 477-497.
[88] Rullani E./Bonomi A. (2005): Il capitalismo personale. Vite al lavoro, Einaudi, Milano.
[89] Rullani, E. (1999): Crescita e successione la metamorfosi del capitalismo personale, in: CUOA Rivista, n. 2.
[90] Samiee, S. (1994): Customer evaluation of products in a global market, in: Journal of International Business Studies, Vol. 25, No. 3, pp. 579–604.

[91] Schooler, R.D. (1965): Product bias in the central American common market, in: Journal of Marketing Research, Vol. 3, pp. 394-401.
[92] Thakor, M.V./Kohli, C.S. (1996): Brand origin: conceptualisation and review, in: Journal of Consumer Marketing, Vol. 13, No. 3, pp. 27-43.
[93] Trueman, M.M./Cornelius, N./Killingbeck-Widdup, A. (2007): Urban corridors and the lost city: overcoming negative perceptions to reposition city brands, in: Journal of Brand Management, Vol. 15, No. 1, pp. 20-31.
[94] Tse, D.K./Gorn, G.J. (1993): An Experiment on the Salience of Country-or-Origin in the Era of Global Brands, in: Journal of International Marketing, Vol. 1, No. 1, pp. 57-76.
[95] Usunier, J-C. (2006): Relevance in business research: The case of country-of-origin research in marketing, in: European Management Review, Vol. 3, pp. 60-73.
[96] Varaldo, R. (2001): Il marketing del made in Italy: quadro d'insieme, in Carlo Alberto Pratesi, (a cura di) Il marketing del Made in Italy. Nuovi scenari e competitività Franco Angeli, Milano 2001.
[97] Varaldo, R./Dalli D./Resciniti R. (2006): Marketing-non-marketing all'italiana: virtù, limiti e prospettive, in: the proceedings of 5th International Conference Marketing Trends, 2006, Paris-Venice, ed. Jean-Claude Andreani and Umberto Collesei, Paris-Venice: Marketing Trends Association.
[98] Vicari, S. (1998): La creatività dell'impresa, Etas Libri, Milano.
[99] Welsh, J.A./White J.F. (1981): A small business is not a little big business, in: Harvard Business Review, pp.18-32.
[100] Wiedmann, K.P./Hennigs N./Siebels A. (2007): Measuring consumer's luxury value perception: a cross cultural framework, in: Academy of Marketing Science Review, Vol. 11, No. 7, pp. 1-21.
[101] Williamson, B. (2001): Creativity, the corporate curriculum and the future: a case study, in: Futures, Vol. 33, pp. 541-555.
[102] Woodman, R.W./Sawyer J.E./Griffin R.W. (1993): Toward a theory of organizational creativity, in: The Academy of Management Review, Vol. 1. Doe, J./Doe, J. (2007): Title, in: Smith, S./Smith, A. (eds.), Book Title, 3rd edition, Chicago, pp.. 3-20.

19 Managing the Value of Luxury: The Effect of Brand Luxury Perception on Brand Strength

Nadine Hennigs, Klaus-Peter Wiedmann, Steffen Schmidt, Sascha Langner, Thomas Wüstefeld

19.1 Introduction

Against the backdrop of the tremendous growth in the market for luxury goods, the concept of luxury seems to be everywhere apparent and is routinely used in our everyday life. In a combination of the rising demand for luxury products in emerging markets (e.g., the BRIC countries) and the so-called *"democratization of luxury"* [12], [40], luxury brands are available to a wider range of consumers than ever before. In consequence, the luxury industry has evolved from the exclusive domain of elite consumer groups to a new luxury sensibility marked by a change in the way consumers perceive luxury [44], [45].

However, even if various attempts to define *"luxury"* or *"brand luxury"* exist, there is no single, definite meaning of the luxury concept available [7]. Thus, luxury takes on many different forms for different people and is dependent on the mood and experience of the consumer [44]: *"Luxury is a subjective and multidimensional construct"* (p. 3). Reasoning this, luxury brand managers face the challenge to constantly improve their understanding of the complexity and dynamics of consumers' subjective perceptions of the luxury concept that affects their buying behavior. Acting in a market that is confused about the content of real brand luxury and confronted with the insatiable global appetite for counterfeits [46], insights in the perception of core brand luxury values and reasons why consumers desire and/or buy genuine luxury brands are critical success factors.

Incorporating relevant theoretical and empirical findings, the aim of the present study is to advance current understanding of brand luxury by focusing on the consumer's perspective. In particular, we examine the antecedents and outcomes of consumers' perception of a luxury brand and effects resulting on attitudinal brand strength. Our paper is structured as follows: first, we analyze existing literature on the concepts of brand luxury and brand strength; second, we develop a conceptual model focusing on the relationship between brand luxury perception and affective, cognitive and intentional brand strength; and third, we present the methodology and results of our empirical study. Based on a structural modeling approach, we identify the most important effects of the perceived brand luxury on the affective, cognitive and intentional brand strength. Finally, the results of our study are discussed with regard to future research and managerial implications.

19.2 Construct Definition and Literature Review

19.2.1 Brand Luxury Perception

By using or simply displaying them, luxury brands fulfill esteem-oriented psychological needs of consumers. The psychological satisfaction gained by particularly branded products is the basic element differentiating non-luxury from luxury goods [29]. Literature shows that luxury brands are defined by their exclusivity and rarity. Although luxury goods offer the highest price quality ratio [27], their ratio of functionality to price is mostly regarded to be low. Still, the ratio of price to intangible and situational utility offered by luxury products is usually perceived as high [30]. From a consumer's perspective, luxury brands are therefore defined by their capability to offer exclusivity and very high quality, and their ability to deeply integrate these factors in their brand identity and brand awareness [32]. As the concept of brand luxury perception is a complex and multidimensional construct, it has to be analyzed following an integrative understanding of luxury. In this paper, we fall back on the luxury brand definition of Vigneron and Johnson [41] and the brand luxury concept of Wiedmann, Hennigs, Siebels [44], [45] who developed and tested a holistic model of brand perception mainly based on ten factors (*price, usability, utility, uniqueness, quality, self-identity value, hedonism, materialism, conspicuousness and recognition value*).

19.2.2 Attitudinal Brand Strength

Following the findings of Wiedmann et al. [43], p. 4-5, we define brand strength as a brand's core ability to differentiate from its competitors. Brand strength integrates all relevant elements of a brand that create customer value through meaningful associations and/or positively affects consumers' responses to marketing efforts [25]. As brand strength strongly depends on customers' perceptions regarding a great variety of brand related factors like e.g. functional product attributes, product category, or competitive positioning [37], [1], in this article, we mainly analyze brand strength by determining the strength of consumers' attitudes towards a brand. Brand related attitudes combine all positive and negative evaluations in consumers' minds [24] including relevant inner guidelines for behavior – especially product choices. Therefore, attitudes are also described as "*tendencies to evaluate an entity [attitude object] into some degree of favour or disfavour, ordinarily expressed in cognitive, affective and behavioural responses*" [13], p. 155, and can be divided into three main characteristics: belief-based (cognitive), emotion-based (affective) and intention-based (behavioral) components [34].

19.3 The Conceptual Model and Related Hypotheses

Conceptualizing the strength of luxury brands, we refer to Keller [25], p. 59, who argues that "the power of a brand lies in what customers have learned, felt, seen, and heard about the brand as a result of their experiences over time". In this context, brand perception and strength directly relates to feelings, cognitions and behavioral responses regarding a specific brand [43]. Focusing on related outcomes of brand luxury perception and their effects on the attitudinal components of brand strength, **Figure 19.1** shows our proposed conceptual model illustrating the relationships between brand luxury perception and affective, cognitive and intentional brand strength.

Figure 19.1 The Conceptual Model

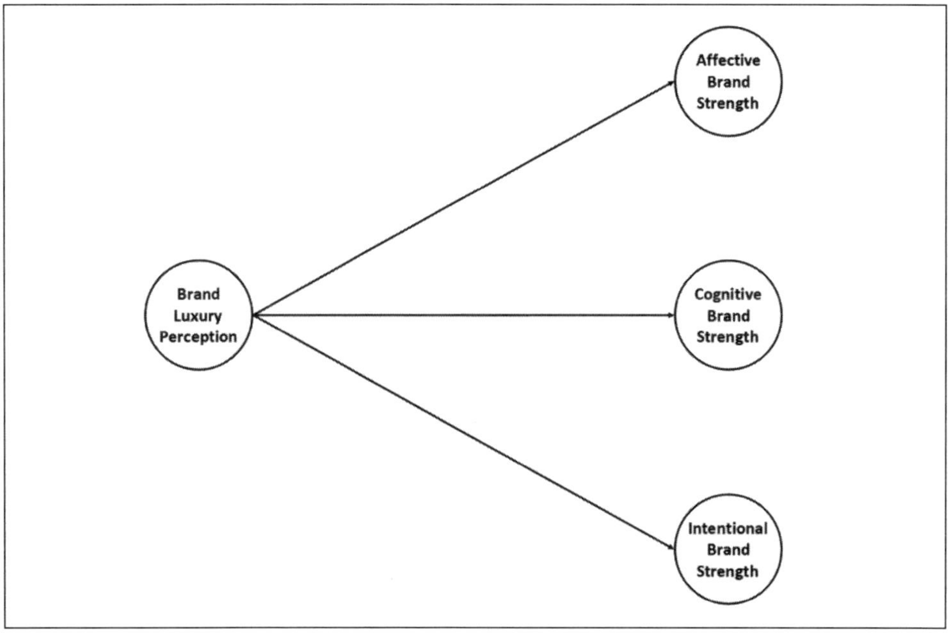

The combined cognitions, beliefs and knowledge regarding a luxury brand form the *cognitive component* of its brand strength. This involves functional elements and specific product related attributes as well as symbolic values associated with the brand [47]. The *affective component* of brand strength includes all relevant emotional characteristics consumers associate with a brand. It is best described as the level to which a customer gets personally involved with a brand, measured in devotion, affection and sympathy [28], [18]. The *intentional component* of brand strength at last integrates both cognitive and affective elements of a brand. Regarded as the behavioral dimension, it includes price sensitivity, customer loyalty, and recommendation behavior as well as purchase intentions [23], [47]. Summing up all relevant consumers' attitudes and evaluations related to brand perception and brand strength, we hypothesize the following:

Hypothesis 1: A positive brand luxury perception has a significant positive effect on cognitive brand strength in terms of distinctiveness, awareness and quality.

Hypothesis 2: A positive brand luxury perception has a significant positive effect on affective brand strength in terms of trustworthiness, affinity and affection.

Hypothesis 3: A positive brand luxury perception has a significant positive effect on intentional brand strength in terms of consumers' buying intentions, brand loyalty, and recommendation behavior.

19.4 Methodology

19.4.1 The Questionnaire

Already existing and tested reflective measures from previous studies were used to capture all three dimensions of consumers' perceived brand strength (e.g., [4], [18], [15], [43]). Moreover, to measure brand luxury perception, we used the brand luxury scale by Wiedmann et al. [45] and adapted it to a brand-oriented instead of a customer-oriented focus. In detail, the formative index development for brand luxury perception followed the four steps guideline by Diamantopoulos & Winklhofer [11]: content specification, indicator specification, indicator collinearity and external validity. All questionnaire items were rated on a five-point Likert scales (*1=strongly disagree, 5=strongly agree*). Moreover, we followed the guidelines of clarity, length, directionality, lack of ambiguity and avoidance of jargon (e.g., [9], [36]) to provide an appropriate item generation, adaption as well as adjustment. With regard to the brand context, all items were specified to the brand CHANEL, one of the world's leading luxury brands and a broad product range encompassing both haute couture fashion as well as accessible goods such as accessories, makeup, jewelry, fragrances, skincare etc. To check the reliability and validity of the used measures as well as the overall item quality, the first version of our questionnaire was face validated using exploratory and interviews marketing researchers.

19.4.2 The Sample

We employed a quantitative Internet survey with a snowball sampling method within the group of the general public interested in luxury goods to investigate our proposed research model. Trained marketing students recruited (prospective) interviewees contacting their relative social network via e-mails and links on selective web pages (e.g., Facebook profile pages) including an invitation to actively participate in the online survey. From May to June 2011, a total of 912 respondents completed the questionnaire. We excluded all interviewees who did not know the brand as well as all questionnaires with non-response items. This procedure led to final sample of 779 valid cases. **Table 19.1** describes the sample characteristics.

Interviewees were mostly aged 18 to 29 years (43.8%), female (58.3%) and single (67.8%) with higher education (A-level and university degree). The (slightly) over-representation of young female, single and highly educated respondents is indicative of the fact that many female students are particularly interested in a luxury brand like CHANEL and may also be attributed to the greater Internet usage of young to middle-aged people.

Table 19.1 Demographic Profile of the Sample

Variable		n	%
Age	18 – 24 years	341	43.8
	25 – 29 years	142	18.2
	30 – 39 years	116	14.9
	40 years +	176	22.6
	No answer	4	0.5
Gender	Male	323	41.5
	Female	454	58.3
	No answer	2	0.3
Marital status	Single	528	67.8
	Married	195	25.0
	Widowed	28	0.6
	Divorced	5	3.6
	No answer	23	3.0
Education	Not graduated from high school	8	1.0
	Lower secondary school	26	3.3
	Intermediate secondary school	130	16.7
	A-Levels	402	51.6
	University Degree	211	27.1
	No answer	2	0.3
Occupation	Full time	284	36.5
	Part-time	46	5.9
	Pensioner / retiree	24	3.1
	House wife / husband	20	2.6
	Job training	44	5.6
	Student	326	41.8
	Scholar	21	2.7
	Seeking work	9	1.2
	No answer	4	0.6

19.5 Results and Discussion

19.5.1 PLS Path Modeling and Related Evaluation Criteria

The structural equation modeling technique Partial Least Squares (PLS) was considered as the appropriate method to test our derived research hypotheses. Concretely, the analysis software SmartPLS 2.0 [33] with no replacement and a bootstrapping procedure (probing individual sign changes) was used. Moreover, the catalogue of non-parametric criteria for assessing the reliability and validity of the measures in the PLS estimation model as suggested by Chin [8] were followed. In particular, the measurement models must exhibit a satisfactory degree of reliability and validity for a worthwhile assessment of the structural model's properties [21].

19.5.2 Evaluation of the Formative Measurement Model

The manifest variables defined as formative indicators for the brand luxury perception construct are presented in **Table 19.2**. Reasoning that common evaluation criteria catalogues for reflective measures cannot be applied for formative measures [20], we followed the suggestions of Diamantopoulos et al. (2008) [10] to assess the formative measurement model of brand luxury perception.

Table 19.2 Manifest Variables of the Formative Measurement Model

Brand Luxury	
LX_Price	*"This brand is worth its price."*
LX_Usability	*"This brand makes life more attractive."*
LX_Uniqueness	*"This brand is very exclusive."*
LX_Quality	*"This brand is very valuable."*
LX_Self_Identity	*"This brand emphasizes the personality of its owner."*
LX_Utility	*"This brand stands for usefulness."*
LX_Hedonism	*"This brand stands for sensuality."*
LX_Materialism	*"This brand evokes the desire to possess it."*
LX_Conspicuousness	*"This brand gives its owner a social recognition."*
LX_Recognition	*"This brand stands for people who succeeded in their life."*

In more detail, the statistical evaluation criteria are shown in **Table 19.3** and **Table 19.4**. In this study, the maximum variance inflation factors (VIF) for brand luxury perception is 1.72 (cf. **Table 19.3**). This value lies below the common threshold of 10 [11]; thus, multi-collinearity (indicator collinearity) does not seem to be a problem in this study. With respect to the assessment of external validity of each formative indicator, the significance of the correlation with a global factor (encompassed by two seven-point semantic differentials; *1=not at all luxury* to *7=very luxury* and *1=not all premium* to *7=very premium*) that summarizes the entity of the accordant formative construct was examined. In support of external validity, all formative indicators are significantly correlated with this overall factor (cf. **Table 19.3**).

Table 19.3 Test for Multicollinearity and External Validity

Formative Indicators	VIF	Spearman's rank correlation coefficient
LX_Price	1.53	0.474 **
LX_Usability	1.54	0.310 **
LX_Uniqueness	1.56	0.592 **
LX_Quality	1.67	0.483 **
LX_Self_Identity	1.27	0.296 **
LX_Utility	1.18	0.062 *
LX_Hedonism	1.49	0.387 **
LX_Materialism	1.70	0.425 **
LX_Conspicuousness	1.69	0.434 **
LX_Recognition	1.72	0.462 **

Significance: ** $p = 0.01$; * $p = 0.05$;

Moreover, most formative indicators' weights are significant and above .1, but 4 of 10 indicators could not meet these restrictions (cf. **Table 19.4**). Nevertheless, none of them could be omitted from further analyses without losing information to cover all facets of the corresponding construct from a theory-driven point of view [6].

Table 19.4 Bootstrapping Results for the Outer Weights

Formative Indicator → LV	Original Sample	Sample Mean	Standard Deviation	Standard Error	T Statis-tics
LX_Conspicuousness → Brand Luxury	0.011	0.011	0.033	0.033	0.345
LX_Hedonism→ Brand Luxury	0.202	0.202	0.036	0.036	5.581**
LX_Materialism → Brand Luxury	0.370	0.371	0.036	0.036	10.273**
LX_Price → Brand Luxury	0.068	0.067	0.037	0.037	1.859
LX_Quality → Brand Luxury	0.220	0.220	0.035	0.035	6.368**
LX_Recognition → Brand Luxury	0.415	0.410	0.044	0.044	9.452**
LX_Self_Identity_Value → Brand Luxury	0.014	0.013	0.028	0.028	0.487
LX_Uniqueness → Brand Luxury	-0.036	-0.036	0.036	0.036	1.017
LX_Usability → Brand Luxury	0.109	0.109	0.035	0.035	3.083**
LX_Utility → Brand Luxury	0.179	0.180	0.030	0.030	5.897**

Significance: ** p = 0.01; * p = 0.05

19.5.3 Evaluation of the Reflective Measurement Models

Table 19.5 presents all manifest variables that are reflective indicators for the three measurement scales of brand strength. The factor loadings of all three factors are sufficiently high, with .74 being the smallest loading as shown in **Table 19.6**. These results are in supportive of item (indicator) reliability [5], [22].

Table 19.5 Manifest Variables of the Reflective Measurement Models

Affective Brand Strength	
BS_affective_01	*"This brand suits me completely."*
BS_affective_02	*"This brand keeps to its promise."*
BS_affective_03	*"I find this brand very pleasant."*
Cognitive Brand Strength	
BS_cognitive_01	*"This brand is very famous."*
BS_cognitive_02	*"In my opinion, the quality of this brand is very high."*
BS_cognitive_03	*"This brand is very distinctive."*
Intentional Brand Strength	
BS_intentional_01	*"I intend to buy brand XY in the future."*
BS_intentional_02	*"I am very faithful to brand XY."*
BS_intentional_03	*"The products of brand XY are worth a higher price than other products."*
BS_intentional_04	*"I would recommend brand XY to my friends."*

Moreover, average variance extracted (AVE) estimates range from 59% to 71%, the Cronbach's alphas range from .65 to .86 and the composite reliability scores range from .81 to .91 (cf. **Table 19.6**); thus, all values exhibit commonly recommended thresholds in terms of internal consistency [3], [31]. Even though cognitive brand strength has not met the Cronbach's alpha criterion at a well satisfying level, it is still in an acceptable range for exploratory research.

Table 19.6 Assessing the Reflective Measurement Models

	Factor Loadings	Average Variance Explained (AVE)	Cronbachs Alpha	Composite Reliability	Fornell-Larcker-Criterion (AVE > Corr²)
Affective Brand Strength	0.78 – 0.84	67%	0.748	0.856	0.67 > 0.65
Cognitive Brand Strength	0.74 – 0.78	59%	0.653	0.810	0.59 > 0.58
Intentional Brand Strength	0.76 – 0.89	71%	0.864	0.907	0.71 > 0.65

In addition, discriminant validity was assessed using the Fornell-Larcker criterion: the AVE of each latent variable should be higher than the latent variable's highest squared correlation with any other latent variable [17]. All three tested latent variables satisfy the criterion requirements. Hence, this result suggests discriminant validity and all reflective measures can be regarded as reliable and valid.

19.5.4 Evaluation and Discussion of the Structural Model

Next, the inner model was evaluated by examining the coefficients of determination of the endogenous latent variables (R^2) as well as the predictive relevance of the exogenous latent variable. According to Chin [8], the coefficients of determination of the three endogenous latent variables (R^2) for brand strength reveal moderate to substantial levels, with values ranging from .40 to .58, as shown **Table 19.7**. Furthermore, Stone-Geisser Q-square values were assessed [19] [38] employing a blindfolding procedure (cross-validated redundancy) [39]. All Stone-Geisser Q-square values are higher than zero, suggesting a strong predictive power of the established PLS-based model.

Table 19.7 Assessing the Structural Model

Endogenous LV	R^2	Q^2
Affective Brand Strength	0.499	0.324
Cognitive Brand Strength	0.584	0.331
Intentional Brand Strength	0.396	0.279

In order to evaluate the postulated hypotheses, a nonparametric bootstrapping procedure (individual sign changes, 779 cases and 3200 subsamples) was applied to assess the significance of the relations between the latent variables, as presented in **Table 19.8**. In hypotheses 1 to 3, we proposed a positive effect of brand luxury perception on the corresponding constructs of brand strength. The empirical results indicate significant positive effects of brand luxury perception on all three dimensions of brand strength ($p < .01$). Consequently, these findings provide full support for hypotheses 1 to 3.

Table 19.8 Bootstrapping Results for the Structural Relations

Exogenous LV → Endogenous LV	Original Sample	Sample Mean	Standard Deviation	Standard Error	T Statistics
Brand Luxury → Affective Brand Strength	0.706	0.710	0.021	0.021	33.548**
Brand Luxury → Cognitive Brand Strength	0.765	0.764	0.019	0.019	39.472**
Brand Luxury → Intentional Brand Strength	0.629	0.634	0.026	0.026	24.633**
Significance: ** p = 0.01; * p = 0.05					

With regard to the indicators' relative importance in respect to forming the summed scale that represents the latent variables brand luxury perception, the outer weights explain the latent variable with small to high impact as shown in **Figure 19.2**. In this study, the formative indicators *recognition* (.42), *materialism* (.37), *quality* (.22) and *hedonism* (.20) provide the highest impact on brand luxury perception. That implies, at least for our study sample, that our investigated luxury brand CHANEL above all has to address tangible needs (e.g., *materialism* and *quality*) as well as intangible needs (e.g., *recognition* and *hedonism*) to maximize the brand strength in terms of positive customer attitudes and behavior.

Figure 19.2 The Structural Model

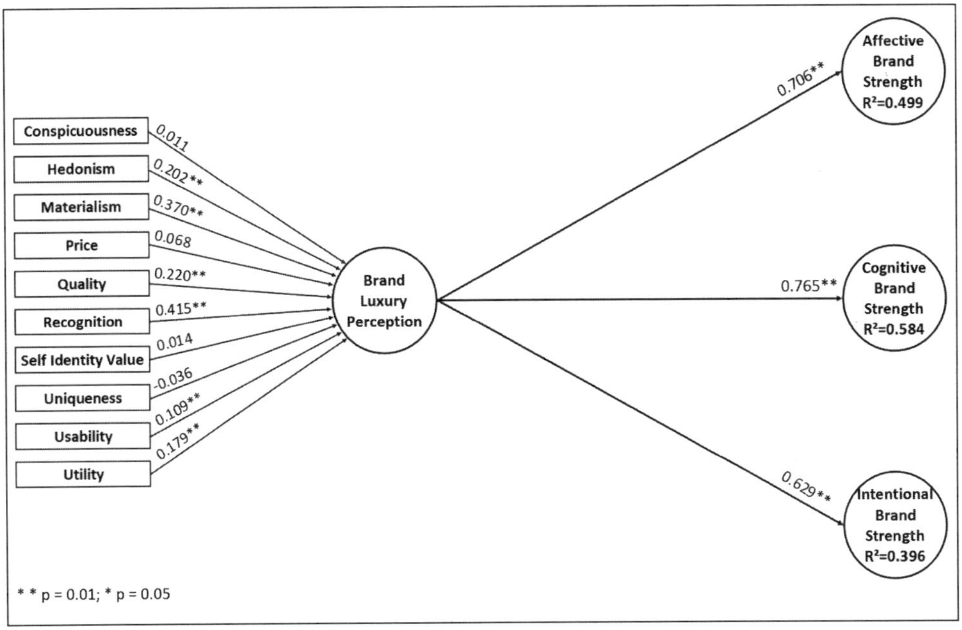

19.5.5 PLS-based Importance-Performance Analysis

Furthermore, an importance-performance analysis (IPA) on an indicator level was conducted to demonstrate the diagnostic power of our model [35]. Based on the PLS estimates for the relations in the structural model (i.e., the importance of each manifest variable) and the constructs' average value (i.e., the performance of each manifest variable), this analysis helps luxury brand managers to identify (perceived) areas of improvement, and then to develop effective marketing mix programs [14], [26]. Computed as outer weights multiplied with the adequate path coefficients, the total effects of the estimated relations determine the importance of the formative indicators [42]. To improve the interpretation of the results, the indicators were rescaled to scores ranging from 0 to 100. Then, average scores as performance values were computed [16], [2]. With regard to an optimized resource allocation, luxury brand managers should prioritize these indicators that show a relatively higher importance and a relatively lower performance [42]. **Figure 19.3** illustrates a priority map based on the results of the IPA for intentional brand strength. Our conducted IPA identifies three main areas: First, segment A with the key drivers *recognition* as well as *materialism* should be prioritized. Then, *quality* and *hedonism* as drivers of segment B should be approached in question to an optimal resource allocation. The relative high importance, but low performance level of these both segments suggests that resource inputs in these areas (e.g., enhanced brand communication with primarily emphasis on those key drivers) have the highest impact on consumer perceptions of intentional brand strength, and hence on the overall business performance.

Figure 19.3 Priority Map

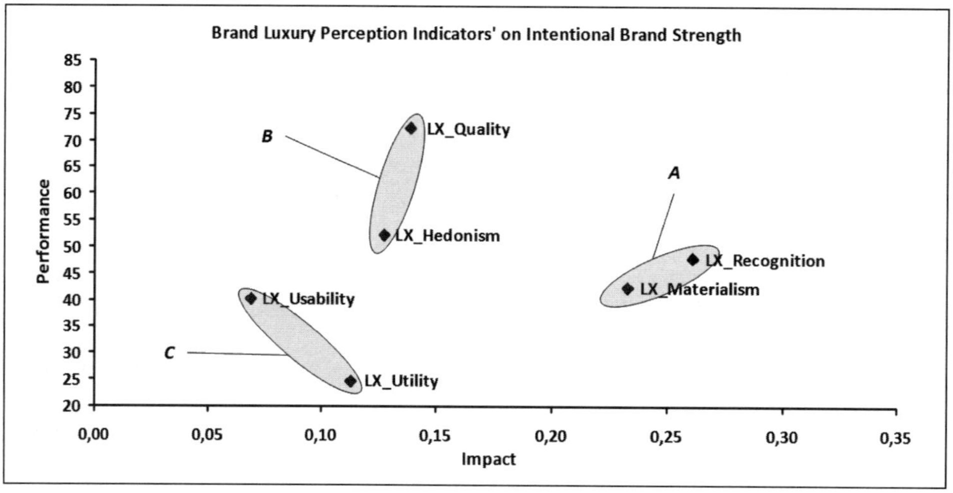

19.6 Conclusion

19.6.1 Theoretical Contribution and Research Implications

The primary goal of this paper was to empirically explore a conceptual framework of brand luxury and its effect on attitudinal components of brand strength. A better understanding of the drivers of brand luxury perception and related effects in the eyes of consumers is valuable for both researchers and marketers in the domain of luxury brands. Overall, referring to the question of what constitutes brand luxury perception and suggesting that brand luxury positively affects all components of brand strength as determining factor of consumer purchase and loyalty intentions, our empirical study reveals: First, the construct of perceived brand luxury can be measured along ten management-oriented indicators of luxury value. Second, our results show significant causal relationships between the dimensions of brand luxury and the cognitive, affective and intentional attitudes of consumers towards a luxury brand. Third, the drivers of perceived brand luxury differ in their relative importance on overall brand luxury. In sum, this paper builds upon and extends well-established foundations in the field of luxury brands as it provides deeper insights in quantifiable effects that originate from the value that luxury brands offer aiming at the consumer's mind and heart.

With reference to further research, a study investigating the causal relationship between brand luxury and brand strength focusing on different luxury brands or a comparison of the evaluation of genuine and counterfeit luxury products could enhance the measurement and management of brand luxury. Besides, it can be assumed that diverse cross-cultural consumer groups may differ in their perception and evaluation of brand luxury and brand strength. Therefore, the analysis of possible cross-cultural differences and similarities in terms of a replication of this study might advance the current understanding of the content and strength of brand luxury on a global level.

19.6.2 Implications for Luxury Brand Management

From a managerial point of view, our study may form the basis of a structured understanding of the perceived value of luxury associated with perceived brand strength. Facing a global economy, where various product categories compete based on the consumer's desire for luxury, the concentration on the specific drivers of luxury brand perception is a key component to market success. To provide luxury brand managers with explicit benchmarks for evaluating their specific brand's performance, we used the PLS-based importance performance analysis (IPA) to demonstrate the diagnostic value of our results. Based on these insights, luxury brand managers will benefit from allocating resources to the most important drivers of brand luxury and brand strength that currently show comparably low performance levels rather than distributing the resources based on the 'watering can' principle. Besides, given that low-cost counterfeits are easily available, a market communication approach emphasizing the main benefits that a given luxury brand addresses might strengthen the consumer's longing for owning *"the real thing"* instead of a fake product. Even if counterfeits are available at a lower price level and allow their buyers to be in tune with fashion without spending an exorbitant amount of money, only the authentic luxury product addresses and satisfies the diverse individual tangible and intangible consumer needs. As revealed in our study focusing on CHANEL, aspects such as *materialism*, *quality*, *recognition* and *hedonism* have been shown to maximize brand luxury and accordingly brand strength. It is assumed that referring to other luxury brands, these drivers may vary. Thus, to create and maintain sustaining brand luxury, managers have to uncover and manage the unique core and authentic essence of their brand that drives its luxury value in the perception of their consumers.

References

[1] Aaker, D. A./Biel, A. L. (1993): Brand Equity and Advertising: Advertising's Role in Building Strong Brands. Hillsdale. NJ: Lawrence Erlbaum Associates.
[2] Anderson , E. W./Fornell, C. (2000): Foundations of the American customer satisfaction index, in: Total Quality Management, Vol. 11, No. 7, pp. 869-882.
[3] Bagozzi. R. P./Yi. Y. (1988): On the evaluation of structural equation models, in: Journal of the Academy of Marketing Science, Vol. 16, No. 1, pp. 74-94.
[4] Bemmaor, A. C. (1995): Predicting behavior from intention to buy measures: The parametric case, in: Journal of Marketing Research, Vol. 32, No. 2, pp. 176-191.
[5] Carmines, E. G./Zeller, R. A. (1979): Reliability and validity assessment. Beverly Hills. CA: Sage Publications.
[6] Cenfetell,. R./Bassellier, G. (2009): Interpretation of Formative Measurement in IS Research, in: MIS Quarterly, 33(4), pp. 689-707.
[7] Chevalier, M./Mazzalovo, G. (2008): Luxury brand management. Singapore: Wiley.
[8] Chin, W. W. (1998): The partial least squares approach to structural equation modeling, in: Marcoulides, G. A. (Ed.): Modern Methods for Business Research. Mahwah. NJ: Lawrence Erlbaum Associates, pp. 295-358.
[9] DeVellis, R. F. (1991): Scale Development: Theory and Applications. Newbury Park. CA: Sage Publications.
[10] Diamantopoulos, A./Riefler, P./Roth, K.P. (2008): Advancing formative measurement Models, in: Journal of Business Research, Vol. 61, No. 12, pp. 1203-1218.
[11] Diamantopoulos, A./Winklhofer, H. M. (2001): Index construction with formative indicators: An alternative to scale development, in: Journal of Marketing Research, 38(2), pp. 269-277.
[12] Dubois, B./Laurent, G. (1995): Luxury possessions and practices: An empirical scale, in: European Advances in Consumer Research, Vol. 2, pp. 69-77.
[13] Eagly, A. H./Chaiken, S. (1993): The Psychology of Attitudes. Fort Worth. TX: Harcourt Brace Jovanovich.
[14] Eskildsen, J. K./Kristensen, K. (2006): Enhancing importance-performance analysis, in: International Journal of Productivity and Performance Management, Vol. 55, No. 1, pp. 40-60.
[15] Feldwick, P. (1996): Do we really need brand equity?, in: Journal of Brand Management, 4 (1), pp. 9-28.
[16] Fornell, C./Johnson, M. D./Anderson, E. W./Cha, J./Bryant, B. E. (1996): The American customer satisfaction index: Nature, purpose and findings, in: Journal of Marketing, Vol. 60, No. 4, pp. 7-18.
[17] Fornell, C./Larcker, D. F. (1981): Evaluating structural equation models with unobservable variables and measurement error, in: Journal of Marketing Research, Vol. 18, No. 2, pp. 39-50.
[18] Francois, P./MacLachlan, D. L. (1995): Ecological validation of alternative customer-based brand strength measures, in: International Journal of Research in Marketing, Vol. 12, No. 4, pp. 321-332.
[19] Geisser, S. (1974): A predictive approach to the random effect model, in: Biometrika, Vol. 61, No. 1, pp. 101-107.
[20] Hair, J. F./Ringle, C. M./Sarstedt, M. (2011): PLS-SEM – Indeed a silver bullet, in: Journal of Marketing Theory & Practice, 19, pp. 139-151.
[21] Henseler, J./Ringle, C. M./Sinkovics, R. R. (2009): The use of partial least squares path modeling in international marketing, in: Advances in International Marketing, Vol. 20, pp. 277-319.
[22] Hulland, J. (1999): Use of partial least squares (PLS) in strategic management research: A review of four recent studies, in: Strategic Management Journal, Vol. 20, No. 2, pp. 195-204.
[23] Jacoby, J./Chestnut, R. W. (1978): Brand Loyalty Measurement and Management. New York: John Wiley and Sons.
[24] Judd, C. M./Downing, J. W./Drake, R. A./Krosnick, J. A. (1991): Some dynamics properties of attitude structures: Context-induced response facilitation and polarization, in: Journal of Personality and Social Psychology, Vol. 60, No. 2, pp. 193-202.

[25] Keller, K. L. (2007): Strategic Brand Management: Building. Measuring and Managing Brand Equity. Upper Saddle River. NJ: Prentice-Hall.
[26] Martilla, J. A./James, J. C. (1977): Importance performance analysis, in: Journal of Marketing, 41 (1), pp. 77-79.
[27] McKinsey (1990): The luxury industry: An asset for France. Paris: McKinsey.
[28] Morgan, R. M./Hunt, S. D. (1994): The commitment-trust theory of relationship marketing, in: Journal of Marketing, Vol. 58, No. 3, pp. 20-38.
[29] Nia, A./Zaichkowsky, J. L. (2000): Do counterfeits devalue the ownership of luxury brands?, in: Journal of Product & Brand Management, Vol. 9, pp. 485-497.
[30] Nueno, J. L./Quelch, J. A. (1998): The mass marketing of luxury, in: Business Horizons, Vol. 41, pp. 61-68.
[31] Nunnally, J. C./Bernstein, I. H. (1994): Psychometric theory (3rd ed.). New York. NY: McGraw-Hill.
[32] Phau, I./Prendergast, G. (2000): Consuming luxury brands: The relevance of the "rarity principle.", in: Journal of Brand Management, Vol. 8, pp. 122-138.
[33] Ringle, C. M./Wende, S./Will, A. (2005): SmartPLS 2.0 M3. http://www.smartpls.de.
[34] Rosenberg, M. J./Hovland, C. I./McGuire, W. J./Abelson, R. P./Brehm, J. W. (1969): Attitude Organization and Change: An Analysis of Consistency among Attitude Components. New Haven. CT: Yale University Press.
[35] Slack, N. (1994): The importance-performance matrix as a determinant of improvement priority, in: International Journal of Operations and Production Management, Vol. 44, No. 5, pp. 59-75.
[36] Spector, P. E. (1992): Summated Ratings Scales Construction: An Introduction. Newbury Park. CA: Sage Publications.
[37] Srivastava, R. K./Shocker, A. D. (1991): Brand Equity: A Perspective on its Meaning and Measurement. Boston. MA: Marketing Science Institute. Report no. 91-124.
[38] Stone, M. (1974): Cross-validatory choice and assessment of statistical predictions, in: Journal of the Royal Statistical Society, Vol. 36, No. 2, pp. 111-147.
[39] Tenenhaus, M./Vinzi, V. E./Chatelin, Y.- M./Lauro, C. (2005): PLS path modeling, in: Computational Statistics & Data Analysis, Vol. 48, No. 1, pp. 159-205 .
[40] Truong, Y. (2009): New luxury brand positioning and the emergence of masstige brands, in: Journal of Brand Management, Vol. 16, No. 5/6, pp. 375-382.
[41] Vigneron, F./Johnson, L. W. (1999): A review and a conceptual framework of prestige-seeking consumer behavior. Academy of Marketing Science Review, 1, pp. 1–15.
[42] Völckner, F./Sattler, H./Hennig-Thurau, T./Ringle, C. M. (2011): The role of parent brand quality for service brand extension success . Journal of Service Research, Vol. 13, pp. 379-396.
[43] Wiedmann, K.-P./Hennigs, N./Schmidt, S./Wuestefeld, T. (2011): The importance of brand heritage as a key performance driver in marketing management, in: Journal of Brand Management, online, pp. 1-13.
[44] Wiedmann, K.-P./Hennigs, N./Siebels, A. (2007): Measuring consumers' luxury value perception: A cross-cultural framework, in: Academy of Marketing Science, Vol. 11, pp. 1-21.
[45] Wiedmann, K.-P./Hennigs, N./Siebels, A. (2009): Value-Based Segmentation of Luxury Consumption Behavior, in: Psychology & Marketing, Vol. 26, No. 7, pp. 625-651.
[46] Wilcox, K./Kim, H. M./Sen, S. (2009): Why do consumers buy counterfeit luxury brands?, in: Journal of Marketing Research, Vol. 46, pp. 247-59.
[47] Zeithaml, V. A. (1988): Consumer Perceptions of Price, Quality, and Value: A Means-End Model and Synthesis of Evidence, in: Journal of Marketing, Vol. 52, No. 3, pp. 2-22.

20 Personal Selling for Luxury Brands

The Effect of Cognitive and Affective Influence Strategies on Customer Value Perceptions

Jan Wieseke, Michael Mauer, Sascha Alavi

20.1 Introduction

The rise of purchasing power among the consumers in the industrialized countries results in an increasing demand for luxury goods and a sustainable growth of the luxury sector. This trend persists and can barely be slowed down by recessions or financial crisis [22]. These long term changes in the luxury market require luxury brands to develop radically new strategies to cope with the new situation. While luxury brands formerly have been focused on market niches to leverage growth potentials, prospectively luxury companies need to target a broad array of customer segments [4]. Thus, luxury firms in the course of a strategic shift are required to abandon the differentiation strategy in favor of an expansion strategy [4].

For the sales department as a multiplier of strategic initiatives, the strategy change of luxury firms poses a major challenge. Traditional approaches in luxury sales management might not produce results any longer and new methods and tools are required to manage the strategic shift from exclusive niches to mass markets. Researchers of luxury brand sales management have to support the practice by developing solutions that assists sales management to succeed in the new impending tasks. A major task of salespersons in general, but particularly in luxury brand sales is to communicate the brand and product value to the customer [31]. However, currently neither conceptual nor empirical works exist that investigate sales strategies for luxury brands to effectively communicate brand value. Therefore, in this work we intend to examine the effect of affective and cognitive sales strategies on customer value perceptions for luxury brands. We understand value perception as the total benefits of a product, assessed from the customer's perspective. In particular, we aim to compare the effects of sales strategies on value perceptions for luxury and manufacturer brands.

To investigate luxury sales strategies, we conducted a study in an automotive sales context. We generated a dyadic data set of customer and salespersons of several car dealerships. We tracked 351 customer-salesperson interactions which are matched to 72 salespersons. The car dealerships distribute two automotive brands, whereby one brand is classified as manufacturer brand, while the other as luxury brand. In this vein, 42% of the salespersons sell the luxury brand, while 58% sell the manufacturer brand. This setting allows us to compare the effects of affective and cognitive sales strategies for luxury and manufacturer brands.

In the following, we start by discussing the subject of personal selling, sales strategies and the importance of value perceptions for luxury brands. We then introduce our conceptual research model. Subsequently, the theoretical foundations and hypotheses are elaborated. Eventually, we empirically test the hypotheses and discuss the results in the light of recent findings in luxury sales research.

20.1.1 The Role of Personal Selling for Luxury Brands

As frontline employees and as such the major interface to the customer, it is the critical task of salespersons to represent the brand to the customer [14]. This means that the salesperson is the primary communicator of information and emotions related to the brand, as well as the brand's values. Although the importance of salespersons as boundary spanners pertains to luxury and manufacturer brands, for luxuries the salespersons bears even more responsibility. Since the purchase of luxury brands often relates to the sphere of conspicuous consumption motivated by social motives, the importance of face-to-face contact to the salesperson even increases in the luxury selling process [17].

However, more fundamental differences between selling of a luxury and manufacturer brand exist. First of all, it has to be put forward that brand strength is a critical feature that distinguishes a luxury from a manufacturer brand, as the functional utility is often similar for both brand types [19]. Thus, owing to the heightened relevance of the brand for luxury products, the communication of brand values by the salesperson for luxury brands gains highest priority.

However, to effectively communicate the brand's value the salespersons requires a profound understanding of the concepts and personality inherent to the brand. Therefore, luxury firms are bound to make considerable investments in the training of the skills of the salespersons. Knowledge about the brand has to be transferred to salespersons by extensive training and salespersons have to identify themselves with the brand personality. The latter is achieved by internal marketing, corporate visions and corporate culture that reflect the brand's personality as well as by promoting leaders as multipliers of brand personality.

Another important difference between sales for a luxury and manufacturer brand is that for luxury brands salespersons should avoid using selling pressure at all costs in order to sustain the image of exclusivity of the luxury brand. Aggressive persuasion attempts or high price discounts impede the communication of values and rather corrode the luxury brand quality and premium image. For this reason in the luxury segment revenue-based compensation is rather uncommon.

20.1.2 Research on Personal Selling for Luxury Brands

The remarks above illustrate that potentially considerable differences between personal selling for a luxury and manufacturer brand exist. Nevertheless, personal selling and sales management for luxury products has barely been investigated. While starting from the early 80ies a plethora of works examining sales management in general has been published, research investigating luxury brand sales management is scarce [34]. In the light of the significant differences in selling for luxury and manufacturer brands, this represents an important research gap in the sales management literature. In this work we aim to address this research gap by examining salesperson strategies for value communication, which is the primary task of salespersons in the interaction with the customer.

In the sales management literature the communication of the brand value to the customer is covered by the research area of influence and persuasion tactics. Extant works on influence tactics show that influence tactics can be classified in two categories: affective strategies and cognitive strategies [28]. Affective strategies communicate the brand value to the customer by engendering emotions and connecting them to the product or highlighting emotional utilities [21]. Customer's value perceptions for cognitive strategies result from the communication of information, product advantages or product attributes [21]. In practical selling guides for luxury marketing recommendations for emotional approaches are pervasive. Luxury brands are supposed to emphasize their emotional side and should primarily communicate emotionally, while refraining from highlighting product advantages [17]. However, these intuitions have not been empirically verified and findings from research on attitude change indicate that cognitive strategies might be more instrumental in increasing customer value perceptions for luxury brands [23]. Thus, considerable ambiguity with regard to benefits of affective and cognitive sales strategies for customer value perceptions prevails. In order to develop a better understanding of the consequences of affective and cognitive sales strategies for customer value perceptions of luxury brands, we subject the above notions to empirical testing.

20.2 Conceptual Model: Sales Strategies & Customer Value Perceptions

Figure 20.1 Conceptual Research Model

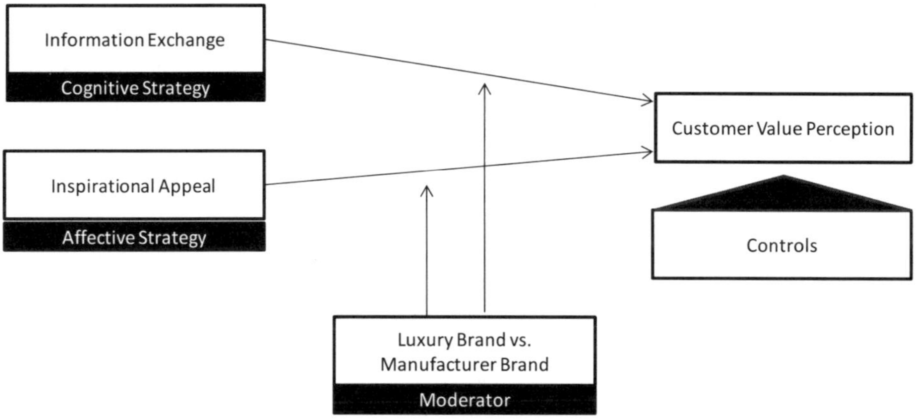

Source: Author's illustration

The research model conceptually links cognitive and affective influence strategies and brand type to customer value perceptions. As a typical cognitive strategy, information exchange is included in the model, while inspirational appeals are assessed as affective influence strategy. Information exchange refers to the extent of information and utility communication by the salesperson [11]. Inspirational appeals are defined as persuasion attempts that are based on inducing positive emotions for a product [21]. We expect that the effect of information exchange and inspirational appeals on customer value perceptions depends on whether a luxury or manufacturer brand is the focus of the selling interaction. We thus predict a moderating influence of brand type, i.e. either luxury or manufacturer brand, on the relationship between influence strategy and customer value perception.

20.2.1 Research on Selling and Persuasion Strategies

Research on selling and persuasion strategies has a long standing tradition and lead to substantial progress in the understanding of interpersonal influence processes. The foundation of this research stream are the works of French and Raven (1959) [12] who conceptually distinguish five bases of power and Kelman (1958) [18] who explores basic processes of social influence. From these seminal works as a starting point three major literature fields on persuasion developed: First, Frazier and Summer's (1984) [11] taxonomy of basic sales influence tactics, second, the Elaboration-Likelihood Model by Petty and Cacioppo (1986) [28] and third, Cialdini`s (2001) [5] six principles of influence. In the following, we confine ourselves to discussing sales influence tactics and the Elaboration-Likelihood model, since these are essential for the argumentation in the chapter at hand. Information Exchange and inspirational appeals as focal strategies of the work at hand are conceptually derived from Frazier and Summers (1984) [11] . Moreover, these strategies are based on the cognition-affect dichotomy and aim at changing attitudes, instead of directly changing behavior by enforcing compliance. Thus, for this work the Elaboration-Likelihood model (ELM) is particularly relevant, as it explicitly comprises the consequences of cognitive and affective strategies on attitude change. Next, we introduce the bases of power and the entailing sales influence strategies by Frazier and Summers (1984) [11] as the theoretical backbone of the study. Then, our study's hypotheses are derived based on the ELM and literature on attitude change.

20.2.1.1 Basic Research on Sales Strategies

French and Raven (1959) [12] differentiate five basic power categories. *Reward power* results from the perception of the influence target that the power holder might grant benefits to the target. On the contrary, *coercive power* depends on the capacity of the power holder to punish the influence target. *Legitimate power* derives from the expectation that the power holder is authorized to prescribe behavior, e.g. on strength of the hierarchical position. *Referent power* results from the personal appeal of the power holder (e.g. due to social status) which arouses the desire for association. Eventually, *expert power* derives from specialized knowledge of the power holder. In a subsequent work the initial categories have been complemented by *information power* [30] which is based on the access of the power holder to particular information.

Drawing on French and Raven's bases of power Frazier and Summers (1984) [11] conceptualize influence strategies for sales channels, which form the theoretical backbone for numerous subsequent publications. The initial model comprises the influence strategies of information exchange, recommendations, promises, threats, legitimation and requests. Consecutive works complemented these strategies with inspirational appeals and ingratiation [27], [21]. Frazier and Summers differentiated these strategies according to the extent of pressure used in coercive and non-coercive strategies.

While threats, legitimations, requests and promises are classified as coercive strategies which exert selling pressure on the customer, recommendations and ingratiation are seen as non-coervice tactics. As the focal selling strategies of the work at hand inspirational appeals and information exchange are considered as non-coercive. We selected those strategies as study object as extant works acknowledge first, that information exchange and inspirational appeals are intensely employed in sales interactions and second, have a distinct cognitive and affective character, making the cognition-affect dichotomy most pronounced. In the following, these strategies are discussed in more detail.

Information exchange refers to the communication of information about the purchase object by the salesperson to the customer, without providing explicit recommendations for actions or products. Thereby, the customer should systematically process the information and thus become deliberately convinced and change his attitude towards the product. Conversely, inspirational appeals aim at achieving attitude changes by heuristic processing, unconsciously linking emotions to a product. Inspirational appeals address ideals and social values to arouse positive affect which is transferred to the product [35]. In the following, we discuss the Elaboration-Likelihood model which comprises the effects of cognitive and affective influence strategies on attitude changes and is thus helpful in substantiating the hypotheses.

20.2.1.2 The Elaboration-Likelihood Model and Routes to Attitude Change

The Elaboration-Likelihood Model (ELM) of persuasion by Petty and Cacioppo (1986) [28] comprises mental processes that lead to attitude changes and describes factors that influence the sustainability of the attitude changes. It is regarded as one of the most prominent approaches for the investigation of persuasive communication [1]. The ELM primarily distinguishes between a central and a peripheral type of information processing which both might lead to attitude changes. The central route encompasses systematical, deliberate evaluation of information, connected with increased cognitive effort, while with the peripheral route information is evaluated heuristically, based on affective judgement. Thus, the central route is predominantly of cognitive nature and attitude changes result from a deliberate mental examination with the information, while the peripheral route is primarily of affective nature and attitude changes are achieved by emotional evaluations without extensive mental processing [28]. Whether information is processed cognitively or affectively depends on the mental capacity and motivation of the individual to exert cognitive effort.

20.2.1.3 Cognitive and Affective Sales Strategies and Attitude Change

There is a general agreement in research on attitude formation that an attitude can be based on either affect or cognition [8], [32]. Hereby, affect refers to the negative or positive emotions that an individual attaches to an attitude object. Conversely, cognitions in this context describe positive or negative attributes that are linked to the attitude object. To illustrate, an attitude of a consumer towards a new car brand might be based on affect, when he experienced advertising of the brand which emphasizes emotions like the joy of driving. On the contrary, when the consumer is confronted with advertisements which highlight the functional advantages of the car brand like higher fuel efficiency or reduced maintenance, his attitude is likely to be based on cognitions.

The base of an attitude (cognitive base or affective base) is critical for the impact of influence strategies on a potential attitude change [26], [9], [23]. Prior research shows that for attitudes based on affect, a rather cognitive argument is more instrumental in inducing attitude change, while for attitudes based on cognitions an affective argument is more effective in inducing an attitude change [23], [24]. Referred to as mismatching of attitude base and persuasive message, this phenomenon has been found to be particularly relevant for well-established attitudes [23], [24]. The mismatching effect is attributed to two reasons: First, arguments matched to the attitude base might lead to a reactance reaction and counterarguing, as the way the individual has usually thought about the object is directly questioned [3]. This means that matched arguments (e.g. cognition based attitude and cognitive argument) might challenge the adequacy of an individual's evaluation and induce counterarguing. Conversely, mismatched arguments do not threaten previously held thoughts and thus do not evoke counterarguing. Second, attitude change has been found particularly strong and sustainable when the persuasive message complements the attitude base, leading to a comprehensive, univalent pattern of affect and cognition [20].

Drawing on the mismatching literature, we propose that information exchange should be more effective in changing value perceptions for luxury brands, while inspirational appeals should be more effective in changing value perceptions for manufacturer brands. Our argumentation is based on the logic that value perceptions of luxury brands are likely to be based on affect, while value perceptions of manufacturer brands are prone to be based on cognitions. For luxury brands the hedonic value dimension often dominates the functional value dimension, while for manufacturer brands the functional value dimension dominates the hedonic dimension [19]. Extant research confirms the notion that the value and image of luxury brands is strongly linked to emotions [33], [34], [7], while manufacturer brands tend to adopt a positioning that stresses product advantages and utilities [17]. Further supporting this notion, luxury brands heavily rely on an emotional advertising strategy, while manufacturer brands generally aim at differentiating themselves in advertising by promoting product advantages [17].

Hence, in line with the literature on the mismatching effect which predicts that cognitive (affective) strategies are better suited to alter affect (cognition) based attitudes, we expect the following: The cognitive influence strategy information exchange is more effective in improving value perceptions for luxury brands, whose perception should be primarily based on affect. Conversely, inspirational appeals as affective influence strategy should be more apt to enhance value perceptions for manufacturer brands, whose perception is likely to be based on cognitions. Thus:

H1: The relationship between information exchange and customer value perception is more positive for luxury brands than for manufacturer brands.

H2: The relationship between inspirational appeals and customer value perception is more positive for manufacturer brands than for luxury brands.

20.3 Methodology of the Study

To test the proposed hypotheses a two-level regression model was calculated with the consideration of variables on the salesperson-level and on the interaction-level with the customer.

20.3.1 Description of the Data Sample

The sample consisted of 351 seller-buyer interactions in an automotive retail context. Data were collected in seven German car dealerships. This sample was appropriate to test the hypotheses because the dealerships followed a hybrid strategy and distributed a luxury brand as well as a manufacturer brand. Data was obtained from salespersons and their customers before and after sales conversations with the help of questionnaires. The questionnaires were personally administered to the salespersons and customers by the research team to achieve the best possible response and matching rates. We also interviewed the salespersons for stable traits and characteristics. Customers and salespersons were matched by code-numbers. After seven weeks of data collection 351 interactions were obtained. 72 salespersons and 351 customers were surveyed, which means that a mean of 5.75 interactions was recorded per salesperson, ranging between 1 and 17 interactions per salesperson. Hereby, 42% of the salespersons sell the luxury brand, while 58% sell the manufacturer brand. The response rate of the salespersons was 100% as participation was obligatory for them. For customers non-response was addressed by surveying customers who originally did not want to participate relying on special incentives. Based on t-test all variables in the study were compared for the customers that initially refused to participate and the main sample. No significant differences were revealed, indicating that non-response bias is unlikely to be an issue. The customers received a voucher for a free car wash for their participation and an amount of 10€ was donated to an UNICEF-project for earthquake victims in Japan for each participant.

20.3.2 Measure Description and Reliability Diagnostics

The majority of scales employed in the questionnaires were drawn from prior research and only minor modifications were conducted to suit the study's context. All variables relevant to the study were indicated by the customer with the exception of brand type, salesperson tenure and job identification. Those were indicated by the salesperson. To assess reliability and validity of the scales, Cronbach's alpha and confirmatory factor analysis are calculated and Fornell and Larcker's (1981) [10] criterion was tested. All Cronbach's alpha values of the scales exceeded the recommended threshold of 0.7 [2]. Fornell and Larcker's criterion was fulfilled as well. The average extracted variance exceeded the squared correlations between all pairs of constructs. All items, operationalizations and the results of the confirmatory factor analysis are reported in Table 1.1.

Table 20.1 Measures Used in the Study

	Operationalization / Items	IR	CR	AVE	CA
Inspirational Appeals[a] based on McFarland/ Challagalla/Shervani 2006	Made a sales pitch which tried to appeal to your emotions.	0.42	0.83	0.62	0.82
	Tried to get you excited about what he or she was selling.	0.82			
	Described the use of his or her products or services with enthusiasm and conviction.	0.65			
Information Exchange[a] based on McFarland/ Challagalla/Shervani 2006	Ensured you received all product and sales materials relevant to your purchasing decisions.	0.45	0.79	0.56	0.77
	Talked about the possible applications of his or her products or services.	0.84			
	Presented information related to your various purchase options.	0.41			
Customer Value Perception[b]	This automobile has a very [low/high] value.	-	-	-	-
Brand Type	Luxury Brand (code=1) vs Manufacturer Brand (code=0)	-	-	-	-
Controls					
Job Identification[a] based on Homburg/ Wieseke/Hoyer 2009	I strongly identify with my job as salesperson.	0.91	0.95	0.83	0.96
	I feel good to be a salesperson.	0.84			
	The job of a salesperson fits well to me.	0.77			
	I feel attached to the group of salespersons.	0.83			

	Operationalization / Items	IR	CR	AVE	CA
Selling Orientation[a] based on Thomas, Soutar, Ryan 2001	The salesperson exerted pressure on me, although the car was not right for me.	0.4	0.84	0.52	0.81
	The salesperson tried to sell me more equipment for the car than I needed.	0.71			
	The salesperson took more time to persuade me than inquiring my needs.	0.33			
	The salesperson recommended products to me that I do not need.	0.72			
	The salesperson talked more himself than listening to me.	0.45			
Relationship Length	Number of Years of Aquaintance	-	-	-	-
Salesperson Tenure	Number of Years Working in Automobile Sales	-	-	-	-

[a] Likert scale anchored by 1 = totally disagree and 7 = totally agree

[b] Semantic Differential

IR=Item reliability, CR=Composite Reliability, AVE=Average Variance Extracted, CA=Cronbach's Alpha

20.3.3 Source: Author's illustrationTwo-Level Regression as Analytical Approach

The data of the study was measured on two levels (salesperson- and customer-level) and customers' data were nested within salespersons. In order to take the nested data structure into account we employ a two-level model to test the hypotheses. The advantage of the multilevel regression is that it accounts for the dependence of the observations per cluster (see Hofmann 1997 [15]). To check whether a multilevel approach is justified the amount of between group variance is compared to the total variance for the focal constructs, i.e. the intraclass correlation is calculated (ICC). The ICC[1] indicates how much within-group variance can potentially be explained by a between-variable. ICC[1] for the baseline model was 0.125 for customer value perceptions, indicating that a multilevel approach is necessary

[29]. Prior to the estimation all continuous independent variables have been grandmean centered to reduce potential multicollinearity problems and to provide a more appropriate estimation of the moderator relationships in the research model across the two levels (see Hofman & Gavin 1998 [16]). For the multilevel regression analysis the Mplus software (Version 6; Muthen & Muthen 2010 [25]) was used, because this program permits the consideration of multilevel structures. We initially entered the study's focal predictors of customer value perceptions, i.e. information exchange, inspirational appeals and brand type into the model and as controls salesperson selling orientation, relationship length, salesperson job identification and tenure. As a second step, we add the interaction to the modell following Ganzach's (1997) [13] hierarchical procedure for the calculation of interaction effects. The final multilevel model was as follows:

Level 1 (Customer level)

$$CVP = \beta_0 + \beta_1(INF)+\beta_2(INSP)+\beta_3(SO)+\beta_4(RL)+r$$

Level 2 (Salesperson level)

$$\beta_0= \gamma_{00} + \gamma_{01}(BT)+\gamma_{02}(JI)+\gamma_{03}(TEN)+u$$

$$\beta_1 = \gamma_{10} + \gamma_{11}(BT)$$

$$\beta_2= \gamma_{20} + \gamma_{21}(BT)$$

$$\beta_3 = \gamma_{30}$$

$$\beta_4 = \gamma_{40}$$

where CVP = Customer Value Perception
INF = Information Exchange
INSP = Inspirational Appeal
SO = Selling Orientation
RL = Relationship Length
BT = Brand Type (Luxury vs Manufacturer Brand)
JI = Job Identification
TEN = Tenure

20.3.4 Results of the Two-level Regression Analysis

Prior to the analysis, correlations, means, medians and standard deviations of the employed variables were calculated (**Table 20.2**).

Table 20.2 Intercorrelation Matrix, Means and Standard Deviations

		1	2	3	4	5	6	7	8
1	Inspirational Appeals								
2	Information Exchange	0.44*							
3	Customer Value Perception	0.12	0.21*						
4	Brand Type (Luxury vs Manufacturer Brand)	0.04	0.51	-0.16*					
	Controls								
5	Job Identification	0.03	-0.05	0.03	0.04				
6	Selling Orientation	0.08	-0.12	-0.11	0.07	0.04			
7	Relationship Length	-.02	0.15*	0.05	-0.1	-0.09	-0.08		
8	Salesperson Tenure	0.07	0.11	-0.06	0.14*	-0.04	-0.05	0.06	
	Mean	3.54	4.82	5.18	-	6.31	1.39	1.42	11.24
	SD	1.62	1.65	1.41	-	1.18	0.74	4.23	9.44
	* p<.05 (two-tailed)								

Source: Author's illustration

In H1 and H2 we hypothesized interaction effects between brand type and sales strategy used, i. e. that both variables jointly influence customer value perceptions. For the interpretation of interaction effects the pattern of results in terms of sign and magnitude of simple effect of predictor and moderator as well as the interaction term has to be considered [6]. Thus, to illustrate the results we plot the interaction effects. We found support for H1 which predicted that information exchange is more effective for luxury brands to improve customer value perceptions than for manufacturer brands. The cross-level interaction effect between brand type and information exchange is significant ($\gamma = .64$, $p < .01$). As the brand type dummy is coded with 1 for the luxury brand and with 0 for the manufacturer brand the positive sign of the interaction effect indicates that value perceptions are higher for luxury brands and high levels of information exchange.

Figure 20.2 Interaction Plot of Sales Strategy and Brand Type

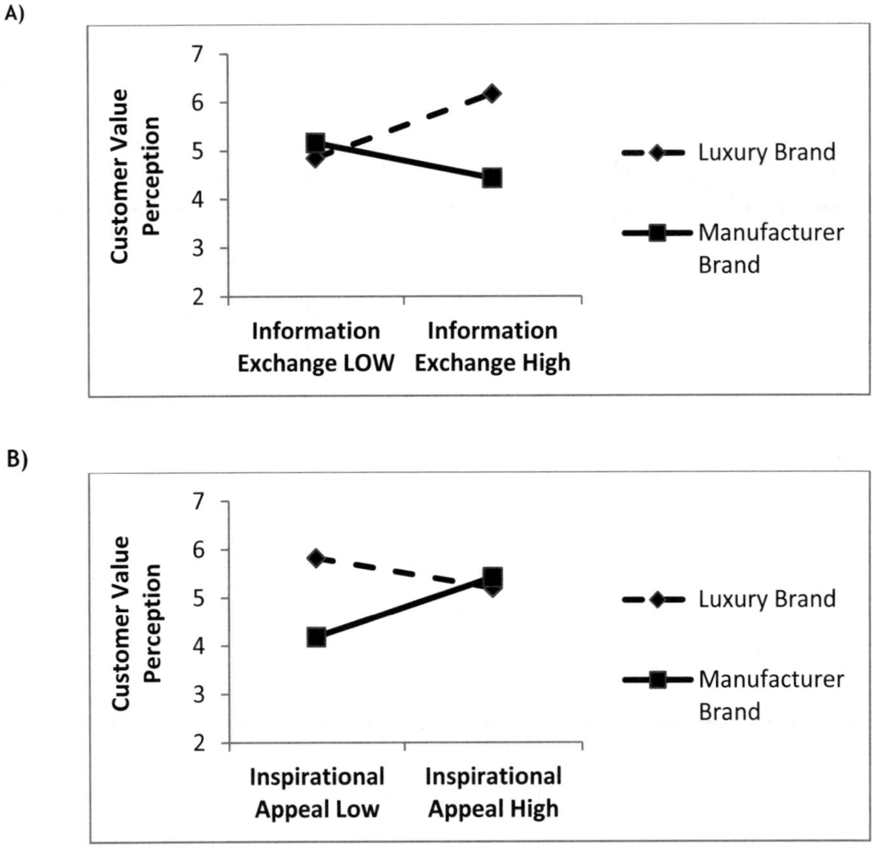

Source: Author's illustration

Figure 20.2 A shows the interaction plot for H1. We can also confirm H2 in which we postulated that inspirational appeals should produce more positive value perceptions for manufacturer brands than for luxury brands. The cross-level interaction effect of brand type and inspirational appeal is negative and significant ($\gamma = -.57$, $p < .01$), indicating that value perceptions are lower for luxury brands and high levels of inspirational appeal. Figure 20.2 B shows the interaction plot for H2. The results pertaining to the controls are displayed in Table 20.3.

Table 20.3 Results of Two-level Regression Analysis

	Dependent Variable	
	Customer Value Perception	
Independent Variables	Hypotheses	γ(SE)
Simple (Conditional) Effects		
Information Exchange		-0.22ns
Inspirational Appeal		0.39**
Brand Type (Luxury vs Manufacturer)		0.71ns
Interaction Effects		
Brand Type x Information Exchange	H1: +	0.62***
Brand Type x Inspirational Appeal	H2: -	-0.57***
Controls		
Job Identification		0.21ns
Selling Orientation		-0.65***
Relationship Length		0.13***
Salesperson Tenure		-0.05*

Source: Author's illustration

20.4 Discussion & Managerial Implications

In research, sales management for luxury brands is basically a black box. Only sparse literature exists that examines luxury sales management either empirically or conceptually. This lack of research is especially lamentable, since luxury sales is currently undergoing a challenging strategic reorientation and for this reason is in need of support and solutions provided by research. The primary task of salespersons in general, particularly for salespersons of image sensitive luxury brands, is the communication of the value of the brand to the customer. Therefore, the main aim of our study is to investigate the effect of sales strategies on customer value perceptions for luxury brands and manufacturer brands. Specifically, we compare the influence of information exchange as a cognitive strategy and inspirational appeals as an affective strategy on customer value perceptions for luxury and manufacturer brands. To achieve this aim, we conducted a study in an automotive distribution setting and interviewed salespersons and customers of several car dealerships. The car dealerships distribute a luxury as well as manufacturer brand which allows us to juxtapose the effects of the influence strategies for both brand types. In total, we tracked 351 customer-salesperson interactions which have been conducted by 72 salespersons.

The results of our multilevel regression analysis show that, contrary to common recommendations in practical selling guides, affective sales strategies are more instrumental in improving customer value perceptions for manufacturer brands than for luxury brands, while conversely, cognitive sales strategies are more effective for luxury brands. Information exchange is more suitable to improve customer's value perceptions for luxury brands, while inspirational appeals are more apt to communicate value for manufacturer brands. These findings are explained by drawing on research on attitude formation and attitude change. We argue that cognitive sales strategies are more effective in changing attitudes that are based on affect, as the mismatch of strategy and attitude base prevents counterarguing and leads to more comprehensive evaluations. Since the attitude towards a luxury brand should be based on affect, cognitive strategies are more instrumental in producing favorable value perceptions for luxury brands. On the contrary, attitudes towards manufacturer brands are prone to be based on cognitions. Following the logic of the literature on the mismatching phenomenon, cognition based attitudes are more readily altered by affective strategies.

However, our work is inflicted with two restrictions that provide starting points for future research. First, we cannot generalize our findings over different industries and settings, as our study was situated in an automotive distribution context. Although we think that our theoretical case can generally be applied to diverse industries, idiosyncrasies of car dealerships might have influenced our findings. Therefore, future research might examine whether the mismatching effect of cognitive and affective strategies and attitudes transfers to other settings. Second, we only conceptually argue that attitudes towards luxury brands are prone to be based on affect, while attitudes towards manufacturer brands are likely based on cognitions. Consequently, future research might verify this notion by investigating bases of attitudes for luxury and manufacturer brands.

Our findings hold several implications for salespersons of luxury brands. It is a major task of salespersons to communicate the product value to customers and convince customers of the product value. This especially applies to luxury brands which are strongly dependent on their image of exclusivity and superior value. Hereby, a straightforward recommendation for salespersons of luxury brands who intend to convince the customer of the value of the brand, is to rather rely on cognitive strategies, like e.g. information exchange, to improve customer's value perceptions. Unlike practical selling guides that often advocate affective strategies for luxury brands, our results show that inspirational appeals as an affective strategy do not enhance customer value perceptions for luxuries. These insights illustrate that recommendations from practical selling guides that are not empirically verified should not be adopted uncritically as they might be misleading. Eventually, salespersons need to be aware that whether cognitive or affective strategies are more effective for improving value perceptions depends on the initial attitude of the customer towards the brand. When customer's attitude toward the brand is predominantly based on affect, cognitive strategies should be used and vice versa. Thus, salespersons need to have a profound understanding of the image of their brand to select an appropriate sales strategy.

References

[1] Areni, C. S./Lutz, R. J. (1988): The role of argument quality in the elaboration likelihood model, in: Advances of Consumer Research, Vol. 15, pp. 197-203.

[2] Bagozzi, R. P./Yi, Y. (1988): On the Evaluation of Structural Equation Models, in: Journal of the Academy of Marketing Science, Vol. 16, pp.74-94.

[3] Brehm, J. W. (1966): A theory of psychological reactance, 1st edition, New York.

[4] Catry, B. (2003): The great pretenders: the magic of luxury goods, in: Business Strategy Review, Vol. 14, No. 3, pp.10-17.

[5] Cialdini, R. B. (2001): Influence: Science and Practice, Boston, MA: Allyn and Bacon.

[6] Cohen, J./Cohen, P./West, S. G./Aiken, L. S. (2003): Applied Multiple Regression/ Correlation Analysis for the Behavioral Sciences, 3rd edition.

[7] Dubois, B./Laurent, G. (1994): Attitudes towards the Concept of Luxury: An Exploratory Analysis, in: Asia Pacific Advances for Consumer Research, Siew Leng Leong and Joseph Cote (eds.), Vol. 1, pp. 273-78.

[8] Eagly, A. H./Chaiken, S. (1993): The Psychology of Attitudes, in: Fort Worth, TX: Harcourt Brace.

[9] Edwards K. (1990): The Interplay of Affect and Cognition in Attitude Formation and Change, in: Journal of Personality and Social Psychology, Vol. 59, No. 2, pp. 202-216.

[10] Fornell, C./Larcker, D. F. (1981): Evaluating Structural Equation Models with Unobservable Variables and Measurement Error, in: Journal of Marketing Research, Vol. 19, pp. 39–50.

[11] Frazier, G. L./Summers, J. O. (1984): Interfirm Influence Strategies and Their Application within Distribution Channel, in: Journal of Marketing, Vol. 48, pp.43-55.

[12] French, J. R./Raven, B. (1959): The Bases of Social Power, in: Asherman, I.G./ Asherman, S.V. (eds.), The Negotiation Sourcebook, 2nd edition.

[13] Ganzach, Y. (1997): Misleading Interaction and Curvilinear Terms, in: Psychological Methods, Vol. 2, No. 3, pp. 235-247.

[14] Hartline, M. D./Maxham III, J. G./McKee, D. O. (2000): Corridors of Influence in the Dissemination of Customer-Oriented Strategy to Customer Contact Service Employees, in: Journal of Marketing, Vol. 64, 2, pp. 35-50.

[15] Hofmann, D. A. (1997): An Overview of the Logic and Rationale of Hierarchical Linear Models, in: Journal of Management, Vol. 23, No. 6, pp. 723-744.

[16] Hofman, D. A./Gavin, M. B. (1998): Centering Decisions in Hierarchical Linear Models: Implications for Research in Organizations, in: Journal of Management, Vol. 24, No. 5, pp. 623-664.

[17] Kapferer, J. N./Bastien, V. (2009): The specificity of luxury management: Turning marketing upside down, in: Brand Management, Vol. 16, No. 5/6, pp. 311–322.

[18] Kelman, H. C. (1958): Compliance, Identification, and Internalization: Three Processes of Attitude Change, in: Journal of Conflict Resolution, Vol. 2, No. 1, pp. 51–60.

[19] Lasslop, I. (2002): Identitätsorientierte Markenführung bei Luxusmarken, in: Meffert, H./Burmann, C./Koers, M. (eds.): Markenmanagement: Grundfragen der identitätsorientierten Markenführung, Wiesbaden.

[20] Lavine, H./Thomsen, C. J./Zanna, M. P./Borgida, E. (1998): On the primacy of affect in the determination of attitudes and behavior: the moderating role of affective-cognitive ambivalence, in: J. Exp. Soc. Psychol, Vol. 34, pp. 398–421.

[21] McFarland, R. G./Challagalla, G. N./Shervani, T. A. (2006): Influence Tactics for Effective Adaptive Selling, in: Journal of Marketing, Vol. 70, pp.103-117.

[22] Merten, H. L. (2009), In Luxus investieren: Wie Anleger vom Konsumrausch der Reichen profitieren, 1st edition, Wiesbaden.
[23] Millar, M. G./Millar, K. U. (1990): Attitude change as a function of attitude type and argument type, in: J. Pers. Soc. Psychol. ,Vol. 59, No. 21, pp. 7-28.
[24] Millar, M. G./Tesser, A. (1992): The role of beliefs and feelings in guiding behavior: the mis-match model, in: Construction of Social Judgment, ed. L. Martin, A. Tesser, Hillsdale, NJ: Erlbaum, in: press.
[25] Muthen, L. K./Muthen, B. O. (2010): MPLUS user's guide, 6th edition, Los Angeles, CA: Muthen & Muthen.
[26] Olson, J. M./Zanna, M. P. (1993): Attitudes and Attitude Change, in: Annu. Rev. Psychol., Vol. 44, pp. 117-54.
[27] Payan, J. M./McFarland, R. G. (2005): Decomposing Influence Strategies: Argument Structure and Dependence as Determinants of the Effectiveness of Influence Strategies in Gaining Channel Member Compliance, in: Journal of Marketing, Vol. 69, pp. 66–79.
[28] Petty, R. E./Cacioppo, J. T. (1986): The Elaboration Likelihood Model of Persuasion, in: Advances in Experimental Social Psychology, Vol. 19, pp. 123-162.
[29] Raudenbush, S. W./Bryk, A. S. (2002): Hierarchical Linear Models: Applications and Data Analysis Methods, 2nd edition, Newbury Park, CA: SAGE.
[30] Raven, B. H./Kruglanski, A. W. (1970): Conflict and Power, in: Swingle, P.(eds.), The Structure of Conflict, New York, pp. 69-109.
[31] Reynolds, K. E./Beatty, S. E. (1999): Customer Benefits and Company Consequences of A Retail-Customer Relationship, in: Journal of Retailing, Vol. 75, No. 1, pp. 11-32.
[32] van der Pligt, J./Zeelenberg, M./van Dijk, W. W./de Vries, N. K./Richard, R. (1998): Affect, attitudes and decisions: Let's be more specific, in: European Review of Social Psychology, ed. W Stroebe, M Hewstone, 8:33-66, chichester, UK: Wiley.
[33] Vigneron, F./Johnson, L. W. (1999): A Review and a Conceptual Framework of Prestige-Seeking Consumer Behavior, in: Academy of Marketing Science Review, Vol. 1999, 1, pp. 1-15.
[34] Vigneron, F./Johnson L. W. (2004): Measuring perceptions of brand luxury, in: Journal of Brand Management, Vol. 11, No. 6, pp. 484-506.
[35] Yukl, G./Tracey J. B. (1992): Consequences of Influence Tactics Used with Subordinates, Peers, and the Boss, in: Journal of Applied Psychology, Vol. 77, No. 4, pp. 525–35.

Part 8
Luxury and Luxury Marketing in the Wine Industry

21 The Hedonistic Consumption of Luxury and Iconic Wines

Alistair Williams, Glyn Atwal

21.1 Introduction

In recent years, the complexity of the wine market has led to a number of attempts to segment wine consumers [20]; [25]. However, as criticism of segmentation models has become increasingly strident, many studies have sought to examine the relationship between involvement and perceptions of wine quality. This is particularly true within the luxury wine market, where attention has increasingly turned towards involvement as a means of defining wine drinkers' engagement with the beverage [4]; [21]. This chapter will seek to explore the ways in which luxury wine products and brands can develop effective strategies that are better conceptualised not according to functional, but experiential symbolic dimensions [53].

21.2 An experimental paradigm

Contemporary luxury consumption reflects the phenomenon of postmodernism that views consumers as emotional beings concerned with achieving pleasurable experiences [2]. Hauck and Stanforth (2007, p.177) [26] point out that American consumers are in a "heightened state of emotionalism, driving them to purchase goods and services that meet not only their physical needs but also their emotional needs". A shift in consumer values has placed a greater emphasis on consumers searching for luxury brand experiences. According to Schmitt (1998, p.11) [47] , "Experiences are private – and sometimes intimate – events that may involve the entire living being: the senses, the feelings, the intellect and the body". Indeed, the increasing significance of luxury experiences was reported to be evident in Japan. In a survey, almost 50% of respondents said they would rather spend money on experiences such as vacations, wellness, restaurants than on luxury handbags or apparel [45]. This trend towards luxury brand experiences is also evident in emerging markets. A review of the luxury market in India reported that the market value of wines and spirits increased by 25% and fine dining by 40% in 2009-2010 [12].

Wine is widely considered to be a beverage associated with the socio-cultural aspects of lifestyle [10]. As such, the consumer's relationship with wine is based on an acquired need [8]. If wine consumption is viewed in this way, its consumption can to some extent be regarded as a hedonic experience, "those facets of consumer behavior that relate to the multi-sensory, fantasy and emotive aspects of one's experience with products" (Hirschman and Holbrook, 1982 p.92) [29]. The hedonic wine consumption experience has to be considered within the contingency of the consumer's wide range of potential social experiences. A hedonic consumption experience perspective is therefore particularly apt when considering the consumption of ultra-premium and iconic wines. This suggests that consumers of luxury wines cannot be viewed from the perspective of the traditional information processing model, or that of a rational problem-solving consumer [24]; [29]. Such a perspective is supported by Addis and Podesta (2005 p.386) [1] , "The new marketing language, founded upon the four C's of change, complexity, chaos, and contradiction".

21.3 The luxury wine market

Wine is an ideal product for a consideration of how quality and its relationship to luxury is conceptualized within the marketing literature. The wine product category ranges from basic, through premium to ultra-premium or luxury products. [6] considers that wines classified as 'Icons' would be consistent with luxury status. The author estimates that this icon/luxury wine market accounts for about one per cent of the global wine market and comprises those wines selling for over US$50 per bottle, and which are generally consumed by connoisseurs and are relatively scarce.

The market for luxury wines however, is very fragmented. Whilst the ultra-luxury brands, such as Châteaux Lafite-Rothschild continue to attract premium prices for their wines, many high-end wineries are seeking new strategies to remain competitive. The impact of the economic slowdown in developed markets has created new pressures within the wine industry. It was widely reported that consumers in the U.S. reduced the amount they spent on fine wines. For instance, a key market for high-end wineries is luxury restaurants. These markets have been affected as consumers are dining out less frequently. In addition, when they do dine out they are opting to purchase wine in the $60 to $70 range, where they would have previously spent $200 or $300 on a bottle of wine. This has caused a sharp decline in sales of those wines priced between the luxury and ultra-luxury categories [9].

Messages regarding the luxury wine market are mixed. Traditional markets have to a large extent been replaced by new and emerging markets such as China and other parts of Asia. It is widely reported that the market for luxury wines in China has grown exponentially [44]. China is now the largest importer of Bordeaux by volume. Exports of Bordeaux wine rose 34 percent in value and 23 percent in volume between July 2010 and June 2011 [50]. This growth is also reflected at the high end of the market. Chinese wine buyers have been instrumental in the economic recovery at the luxury end of the wine market. For example, an anonymous Chinese bidder bought 300 bottles of Château Lafite-Rothschild bundled into a single auction lot for $539,280 [33].

These market inconsistencies have important implications for the marketing strategies of luxury wines. Quelch (2008) [42] recently suggested that as the world economy slumps, one consumer segment continues to grow faster than ever, a segment he refers to as the 'middle-aged simplifier'. Quelch (2008, p.1) [42] argues that "tomorrow's consumer will buy more ephemeral, less cluttering stuff, fleeting, but expensive experiences, not heavy goods for the home". According to Quelch, these consumers are turning away from physical goods to the consumption of experiences.

Within this context, studies of leisure and recreation involvement (e.g., Havitz and Dimanche, 1999 [27]) suggest that highly involved wine consumers can be expected to explicitly perceive wine as central to their lifestyle experiences. Such highly-involved consumers are, by extension, more likely to exhibit experience behaviors, such as joining wine clubs, educating themselves about wine and collecting or cellaring preferred wines, often of high quality [35].

21.4 Contemporary consumption patterns

Today's luxury consumers are more diverse and complex than previously assumed, and this complexity is evidenced in the ways in which they consume luxury goods and services [2]. Consumption has come to form a significant role in the creation, maintenance and reproduction of aspirations and hedonism [54].

Within contemporary western society, the puritan ethic of restrained consumption has been replaced by a commercial ethic of conspicuous consumption (Featherstone, 1991). The result, it is argued, is that we now live in an artificial world where even desire itself is manufactured. This view is supported by Bocock (1993, p.4) [7] who argues, "Contemporary consumption has been seen as epitomizing this move into post-modernity, for it implies a move away from productive work roles being central to people's lives, to their sense of identity, of who they are". Moreover, this consumption is characterised by disorder and unpredictability, with consumers doing as they please. Such contradiction includes that of the usual reference systems marketers have traditionally used to segment them into usable categories. Consumers are increasingly failing to maintain the categories which have been developed for them, as a result consumers are seen as increasingly fickle and unreliable, making it difficult to pinpoint buying behaviour. Christensen et al. (2005, p.157) [11] argue that "the classical marketing approach of assessing and adapting to the needs and wants of customers is being challenged in today's social environment in which such needs and wants are unclear or highly fluid, and in which the lifestyles of consumers exhibit disconnectedness or fragmentation".

In addition to the above, within contemporary consumption every act of consumption is also an act of production and visa versa [18]. It has become clear that consumption, rather than being a passive act of unpacking and discovering the meaning intended by the produced, is an active and creative process through which consumers continuously produce and reproduce their own identity [11]. Contemporary consumers interpret and use products and messages in ways very different from their original purpose. They reshape and adapt them to their own personal use, perverting their meanings in ways not imagined by their creators [13]; [14]. As Ogilvy (1990, p.150) [37] states, "the consumer is a semiotic field of mixed messages, conflicting meanings and inconsistent impulses". Traditional marketing theory is founded on the principles of planning, implementation and control [31]. However, today's marketer has to understand that their products are creations with lives of their own, which are interpreted vicariously by consumers, creating very individual experiences and identities.

As contemporary societies change and evolve, driven by knowledge and information technologies, they are increasingly being transformed into an economy of icons. The result is a wave of new types of consumers who are beginning to emerge. These consumers demand different types of products and use them in a variety of ways, thus presenting marketers with new opportunities, which can only be realised if marketers understand who these consumers are and how they think, feel and behave. Traditional marketing envisaged consumers who consume brands, goods and services in order to aid them in finding and reinforcing a unitary or essential self-concept. Increasingly, it is being argued that in contemporary society, consumers mix and match eclectic images in a variety of ways. Such consumers change their self-concept over very short periods of time. Much of the current marketing thought tends to hold that a unity of self or self-concept, a sense of one's identity and character, can and does exist [17]. In such a perspective, the consumer seeks satisfaction of identified needs for this self, suggesting a unity of purpose, orientation and behaviour.

There is evidence to suggest that we need to reject the notions which have shaped such marketing thought, replacing it with a view of the consumer as someone who actively communicates the social reality he or she prefers to live rather than passively inheriting one constructed without his/her participation [17]. This has increasingly resulted in contemporary economic activity aiming not for output, but for experience via consumption [32]. If we accept such a view, wine marketers have to be tolerant to the non-traditional demands communicated by contemporary consumers.

21.5 Key characteristics of luxury wines

Dubois et al. (2001) [15] identified six key characteristics of luxury that they suggest consumers perceive as inherent in luxury goods or services. If we apply this to the luxury wine market, we can make the following observations:

Firstly, luxury is synonymous with excellence of quality. This means that luxury goods and services are perceived to comprise components of an exceptional nature and the production of luxury goods and services typically involves high level of expertise, acquired over time. Several systems, which classify wine, have been developed to 'guarantee' and grade the quality of the product. These include the appellations of Europe [43], which determine 'quality' by conformance to origin, and the various classification systems (which grade wine, usually according to its price) operating in places as diverse as the USA and Australia.

Old World producers have long used classifications to denote wine quality. The French have the AOC system, supplemented with specific classifications for certain regions, such as Burgundy and Bordeaux. These classifications, which cover all elements of viniculture and viticulture, are seen to guarantee the quality of the wine. This means that a consumer buying a Château Haut-Brion, knows that he or she is acquiring a Premier Grand Cru Classe, one of the five best-of-the-best Bordeaux wines, along with all that such a classification entails. Similar schemes operate throughout the Old World, including Spain (DO), Italy (DOCG) and Germany (Padikat).

In the New World, such schemes are less common and certainly less rigorous. As a result, consumers of New World wines resort to quality indicators such as the 'scores' offered by experts such as Robert Parker's Wine Advocate system. Parker uses a 50-100 point scale and wines which score 96-100 are argued to be 'extraordinary wines of profound and complex character displaying all the attributes expected of a classic wine of its variety' [39]. Such wines can expect to be highly sought after in the luxury wine market.

A second luxury attribute identified is price itself. In the minds of consumers, the price of a luxury product is assumed to be very high, especially when compared to more utilitarian goods and services. Research suggests that the consumer perception of quality is consistently associated with the price of goods or services, a characteristic referred to as the 'halo effect' [6]. Consumers judge the quality of a product according to its price when choosing between different brands, believing that price is an indicator of quality, whilst also associating higher prices with prestige. Such examples abound within the wine industry, with the role of the auction houses prevalent in ensuring that wines achieve the desired price, their vintages demand.

The third characteristic of luxury products is that they are expected to be scarce. Items that are in limited supply are seen to command respect and prestige amongst consumers [3]. An example of such a luxury wine is Champagne. The Champagne winemaking community, under the Comite Interprofessionnel du Vin de Champagne, has developed a comprehensive set of regulations for Champagne in order to protect its economic interests. This includes aspects such as regional limits for growing, grape types, viticulture and viniculture. The effect of this is to limit the Champagne market, which maintains prices.

Next, luxury goods and services are required to have history (even if in many cases that history has been developed by effective marketing). This gives luxury goods a perceived 'heritage', an authentic aspect which gives the brand a unique identity. Creating such a perception of authenticity means building a compelling story, blending many aspects of a product's 'heritage', such as who is associated with the product (both as consumer and producer), during what era's was the product authenticated and where and under what conditions the product is made. If necessary, this 'heritage' can be subtly invented or enhanced in order to convey lineage. This is illustrated by the example of Opus One. The founders of the brand were a combination of one of the most respected European wine families, the Rothschild's, of Château Mouton Rothschild, and the iconic New World wine pioneer, Robert Mondavi. In 1981, only two years after the joint venture was introduced, a single case of Opus One wine sold at auction for $24,000, the highest price ever paid for a Californian wine.

The fifth aspect of luxury involves a strong aesthetic appeal. According to Dubois et al. (2001, p.13) [15] , "This aesthetic dimension is not only expected from the goods themselves but also from the context in which they are presented as well as from the people who consume these goods". This can be demonstrated by ensuring stylistic consistency, committing to traditional production practices or by using history and culture as referents in design and marketing. An example of such a wine would be Ornellaia wines from Tuscany. Despite the winery only beginning operations in 1981, its Bolgheri DOC Superiore is one of the regions most iconic products, much coveted by consumers of luxury wines. Hand harvesting, microvinification, controlled aging and a commitment to the environment are recognized as contributing to the overall aesthetic appeal.

Finally, luxury goods and services are perceived to some extent to be superfluous, not in an innate manner, but in terms of the product being hedonistic, extravagant or indulgent. The hedonistic aspects of the product means that these goods and services are not strictly necessary, in the form in which they are being consumed. However, because of their extended resources, luxury consumers are able to go beyond simply meeting their needs in their consumption behavior, and seek to indulge their wants.

21.6 Experiential domain of luxury wines

The evidence suggests that consumers do not purchase luxury wines merely for utilitarian reasons, such as the features and benefits that the wine confers on the consumer. As Schmitt (1999, p.53) [48] argues, "from now on, leading-edge companies, whether they sell to consumers or businesses, will find that the next competitive battlefield lies in staging experiences".

One way in which marketers can achieve this is through the use of experiential marketing. Experiential marketing is a relatively new orientation, which is gaining significant momentum within the luxury sector [2]. Experiential marketing is arguably one of marketing's most contemporary orientations, with academics, practitioners and enterprises paying increasing attention to consumers' experience and its constituents in recent years [38]. Experiential marketing views consumers as emotional beings, focused on achieving pleasurable experiences. These experiences are achieved through so-called 'experience providers' [48]. As Schmitt (1999 p.53) [48] argues, "the ultimate goal of experiential marketing is to create holistic experiences that integrate individual experiences into a holistic Gestalt". Arguably, of course, the more emotional the product, the more it lends itself to an experiential marketing campaign. Experiential marketers create an experience environment and involve consumers by asking them to take part in an activity. The aim in experiential marketing is connecting customers with products through one or more meaningful and relevant experiences, while appealing to both rational and emotional behavior, in order to elicit a visceral, positive reaction from the consumer.

A firm's competitive advantage may originate from a range of sources such as product innovation, pricing strategy or marketing channels. However, it is argued that the competitive advantage which derives from memorable experiences is more difficult to imitate and thus substitute [51]. As McLuhan (2008, p.33) [36] argues, "proponents suggest that by creating memorable and complex customer experiences comprising different but interrelated elements (e.g. cognitive as well as emotional, behavioural as well as relational), marketers can obtain a competitive advantage over rival firms". The argument is made that the degree to which a company is able to deliver a desirable experience will largely determine its success in the contemporary global marketplace [48].

The marketing of experiences is demonstrated by the expansion of LVMH into the hotel sector. LVMH, owners of luxury wine brands such as Dom Pérignon Champagne recently announced their expansion into the hotel industry. Through a joint venture with Egypt's Orascom Development Holdings, they will build two hotel resorts in the Middle East. Their first two hotels are scheduled to open in 2012, in Egypt and Oman. The hotels will be branded Cheval Blanc, after the Michelin-starred resort based in Courchevel which was opened in 2006, by Group Arnault. The hotels will have a Givenchy spa and will showcase LVMH's luxury wine brands such as France's Château d'Yquem's, Spains' Numancia and Newton wine from Napa Valley. This development by LVMH is the first time the name of a luxury wine has been put upfront as a hotel brand. In short, the focus is on experiences and lifestyles, which provide sensory and emotional values to the consumer and on the creation of synergies between meaning, perception, consumption and brand loyalty.

21.7 Strategies for experientially marketing luxury wines

Experiential marketing has been increasingly investigated in recent years, with many authors concluding that consumers' can be increasingly shown to crave innovative, engaging experiences in many aspects of their lives [19]; [41]; [49]. An experiential marketing strategy demands that marketers fully channel their resources in order to develop a set of experiences that reinforce the offer. This demands a holistic approach, one which is not exclusive to the marketing department but involves the whole organisation.

This raises the question how marketers can leverage the concept of experiential marketing for luxury wine products. It can be easily argued that the luxury wine product is almost always experiential, which puts marketers of these products in a unique position to apply the principles of experiential marketing to their activities. However, simply having an intrinsically, inherently experiential offering is very different from actively and deliberately marketing that offer in an experiential manner. To achieve this goal, frameworks through which wine marketers can strategically identify, enhance and deliver their offers have to be introduced.

21.8 Realms of experience

Pine and Gilmore (1998) [40] suggest that experiences can be plotted across two bi-polar constructs – customer participation (ranging from active to passive) and connection (ranging from absorption to immersion). From the two bi-polar constructs, experiences can be categorised into four 'Realms'; namely Entertainment, Education, Escapist and Esthetic. The four Realms are not intended to be mutually exclusive, but act in combination, ensuring that the richest experiences for consumers encompass aspects of all four Realms, producing a 'sweet spot' where the Realms meet (see Figure 1).

Figure 21.1 The four realms of experience

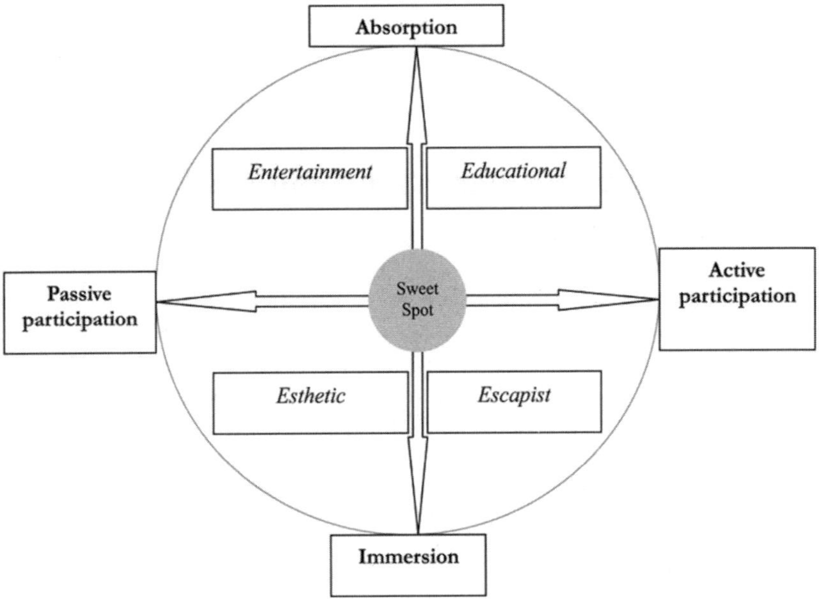

Source: Adapted from Pine and Gilmore (1998) [40]

Entertainment

Entertainment such as the staging of live events involves customers participating in a passive manner, where their connection with the activity is likely to involve absorption rather than immersion. Such events are common in the wine world, where renowned winemakers act as keynote speakers to food and wine events, corporate team building activities and wine trade conferences and seminars. Such events involve wine consumers participating in only the most passive of manners, acting as an audience to a speaker giving his or her views about the sector and luxury wines. Although entertainment may seem like an obvious application of the product or service experience, the growing challenge is to incorporate entertainment into areas outside the immediate experience. For example, Dom Pérignon enables followers of the brand to enter a virtual world of creative fantasy and artistic imagination via its so-called stories that can be viewed on its website. For example, a story includes David Lynch's personal on-site voyage around the Dom Pérignon bottle.

Education

Education involves activities in which customers are more actively involved, but are still of an absorption nature. Luxury wine companies offer not only a wealth of historical and cultural information about their products, but also tastings and presentations of these products. One of the most successful wine tasting rooms is that of Robert Mondavi wines, in Napa Valley. Guests can take part in tasting, during which a select group are accompanied by a wine educator tasting a variety of limited production, reserve and older vintage wines in an exclusive private setting. Other experiences include the opportunity to experience their luxury wines whilst listening to a wide range of live music, or viewing art exhibitions, always in the presence of accompanying experts. In a similar vein, the wine industry can enrich the overall learning experience that is tailored to the cultural needs of the luxury consumer. For example, Sopexa India launched an iPhone App called Uncork France exclusively for the Indian market that educates wine drinkers about French wine regions, grapes and food and wine pairings.

Escapist

Escapist activities involve both active participation and immersion in the activities' or activity's environment. Enthusiasts of the Spanish sparkling wines, Gran Claustre Cava and Grand Cloister Cava are offered the opportunity to visit the winery, a 13th century Romanesque Castle, which is home to an ancient wine press and current wine making facilities. What makes this experience part of the escapist realm however, is the further opportunities offered to visitors. Guests have the opportunity to work with a local Catalan celebrity chef in the estate's cooking school, preparing modern Catalan style tapas, inspired by the wines. In this experience, guests are wholly immersed in participating in the preparation and consumption of foods, which are ideally suited to pairing with these luxury Cava's.

Esthetic

Esthetic activities involve immersion but with little participation in the experience, where attention to the smallest executional details can make a tangible difference to the overall impression. For example, the role of packaging can be valued as an important esthetic attribute. For example, the American designer Denise Focis created a white leather label for Amarone Bottega (Distilleria Bottega) that sits in a case with debossed black lettering that has the style of a vintage suitcase. This is consistent with Underwood et al., (2001, p.407) "consumers' viewing a product packaging are more likely to react spontaneously to image aspects of how a product looks, feels, tastes, feels, smells, or sounds". The notion of developing an esthetic experience can be extended outside of the domain of the physical product. For example, Dom Pérignon uses celebrities such as Karl Lagerfeld and Eva Herzigova to host international launch events, such as those for their Dom Pérignon Rose Vintage. The shows are staged as an integral part of their promotional campaigns, taking place at such legendary venues as Lagerfeld's flagship hotel, the Hotel Particuliar, in Monmartre, Paris. Guests at these events are fully immersed in the glitz and glamour of the Dom Pérignon launch experience, however their participation is limited to consuming the wine and enjoying the spectacle.

21.9 Implications for luxury wine marketing strategies

What are the implications of this transition to viewing the consumption of luxury wines in terms of the experiences they engender? According to Schmitt (1999) [48] and Lindstrom and Kotler (2005) [34], a customer is attracted towards a brand based upon its sensory experience. Schmitt (1999) [48] argues that experiences are the result of living through situations, and that they are triggered stimulations to the senses, the heart, and the mind. Experiences connect products to the customer's lifestyle and place purchase occasions in a broader social context. Schmitt (1999) [48] proposed the concept of strategic experiential modules (SEMs) that managers can use to create different types of experiences for their customers. As such, experiences provide sensory, emotional, cognitive, behavioural, and relational values that replace the more functional values that marketing has traditionally espoused. Whilst each of the SEM's is seen as being independent, they are not considered to be self-contained; rather they are inter-connected and interact. Advocates of the SEM approach argue that marketers should seek to strategically create holistically integrated experiences that possess, at the same time, as many of the modules as possible. It is an emergent phenomenon, which reflects the outcome of participation in a set of activities within a social context [22]. Experiences are the feelings customers take away from their interaction with luxury wines and their atmospheric stimuli [23]. According to Berry et al. (2006 p.44) [5] when seeking to demonstrate how consumers identify experiences, they argued, "it is anything in the service the customer perceives by its presence, or absence. If the customer can see, hear, taste, or smell it, it is a clue". Luxury wine producers must manage the total customer experience is recognizing the clues it is sending to customers. This is important because the more sensory an experience, the more memorable it will be for the consumer. Most luxury wine offerings have a range of sensory elements i.e. sounds, sights, smells, touch and taste. However, the challenge is to incorporate marketing appeals that are not directly associated with the product offering. For example, the expansion of Don Pérignon branded spaces such as the Don Pérignon Bar at the Bulgari Hotel in Milan or the Don Pérignon lounge in Mumbai ensures that consumers 'feel' being part of the brand.

21.10 Conclusion

The marketing of luxury wines, too often grounded in positivistic assumptions, viewing consumption as an internalised state, has to begin to respond to alternative orientations [55]. When rival offerings become more alike than different in the concepts that dominate their design and operations they find their business commodified, and thus more likely to be purchased on the basis of price. Experiential marketing helps firms to avoid the commodification trap. If firms are to be effective in the highly competitive luxury wine industry, it is imperative that they understand contemporary markets and consumers. The luxury wine sector cannot be seen to be immune to such fundamental changes in the orientation of marketing. According to Schmitt (1998, p.11) [48], "Sophisticated customers want products and marketing campaigns that dazzle their senses, touch their hearts and stimulate their minds; that they can relate to and that they can incorporate into their lifestyles". Innovative experience design will become an increasingly important component of wine businesses core capabilities, those who go beyond product excellence and market experientially will lead the creation of value in the sector.

References

[1] Addis, M./Podesta, S. (2005): Long life to marketing research: a postmodern view, in: European Journal of Marketing, Vol. 39, No. 3/4, pp. 386-413.
[2] Atwal, G./Williams, A. (2009): Luxury Brand Marketing – The Experience is Everything!, in: Journal of Brand Brand Management, Vol. 1, No. 5-6, pp. 338-346.
[3] Atwal, G./Bryson, D./von Gersdorff, J. (2010): Deluxurufication is in fashion, in: Admap, Vol. 4, No. 10, pp. 44-45.
[4] Basset, G. (2003): The Wine Experience. Kyle Cathie: London.
[5] Berry, L.L./Wall, E.A./ Carbone, L.P. (2006): Service clues and customer assessment of the service experience: lessons from marketing, in: Academy of Management Perspectives, Vol. 20, No. 2, pp. 43-57.
[6] Beverland, M. B. (2005):. Crafting brand authenticity: The case of luxury wines, in: Journal of Management Studies, Vol. 45, No. 2, pp. 1003-1029.
[7] Bocock, R. (2003): Consumption. Routledge: London.
[8] Bruwer, J./Alant, K. (2009): The hedonic nature of wine tourism consumption: an experiential view, in: International Journal of Wine Business Research, Vol. 31, No. 3, pp. 235-257.
[9] Carlton, J./Kesmodel, D. (2009): Luxury wine market reels from downturn. WSJ, July 8 2009, http://online.wsj.com/article/SB124700844235408441.html, accessed November 11 2011.
[10] Charters, S./Pettigrew, S. (2003): I like it, but how do I know if it's any good?, in: Journal of Research for Consumers, Vol. 5, pp. 22-45.
[11] Christensen, L.T./ Torp, S./Fuat Firat, A. (2005): Integrated marketing communication and postmodernity: an odd couple, in: Corporate Communications: An International Journal, Vol. 10, No. 2, pp. 27-35.
[12] CII/A.T. Kearney (2011): India Luxury Review 2011. http://www.atkearney.com/images/global/pdf/India_Luxury_Review_2011_CII-AT_Kearney_Report-edited_v02.pdf, accessed December 10 2011.
[13] Cova, B. (1996): The postmodern explained to managers: implications for marketing, in: Business Horizons, Vol. 39, No. 6, pp. 15-23.
[14] Du Gay, P. (1996): Consumption and Identity At Work. Sage: USA.
[15] Dubois, B./Laurent G./Czellar, S. (2001): Consumer rapport to luxury: Analysing complex and ambivalent attitudes, in: Les cahiers de Recherche, Vol. 33, No. 0, pp. 1-56, http://www.hec.fr/var/fre/storage/original/application/5ecca063454eb4ef8227d08506a8673b.pdf, accessed September 12 2011.
[16] Featherstone M. (1991): Consumer Culture and Postmodernism. London: Sage
[17] Firat, A.F./Shultz, C.J. (1997): From segmentation to fragmentation, in: European Journal of Marketing, Vol. 31, No. 3/4, pp. 183-207.
[18] Firat, A.F./Venkatesh, A. (1995): Liberatory Postmodernism and the Reenchantment of Postmodernism, in: Journal of Consumer Research, Vol. 22, No. 3, pp. 239-267.
[19] Florida, R. (2002): The Rise of the Creative Class. Basic Books: New York, NY .
[20] Gil, J.M./Sanchez, M., (1997): Consumer preferences for wine attributes: a conjoint approach, in: British Food Journal, Vol. 99, No. 1, pp. 3-11.
[21] Goldsmith, R.E./D'Hauteville, F. (1998): Heavy wine consumption, in: British Food Journal, Vol. 100, No. 4, pp. 184-190.

[22] Gupta, S./Vajic, M., (2000): The contextual and dialectical nature of experiences, in: Fitzsimmons and Fitzsimmons (Eds.), New Service Development, Sage Publications: Thousand Oaks, CA.
[23] Haeckel, S.H./Carbone, L.P./ Berry, L.L. (2003): How to lead the customer experience, in: Marketing Management, Vol. 12, No. 1, pp. 18-23.
[24] Hall, C.M./Sharples, L./Cambourne, B. /Macionis, N. (2000): Wine Tourism around the World. Butterworth-Heinemann: Oxford.
[25] Hall, J./Winchester, M. (2000): Focus on Your Consumer through Segmentation, in: Australia and New Zealand Wine Industry Journal, Vol. 15, No. 2, pp. 93-96.
[26] Hauck, W.E./Stanforth, N. (2007): Cohort perception of luxury goods and services, in: Journal of Fashion Marketing and Management, Vol. 11, No. 2, pp. 175-188.
[27] Havitz, M. E./ Dimanche, F. (1999): Leisure involvement revisited: Drive properties and paradoxes, in: Journal of Leisure Research, Vol. 31, pp. 122-149.
[28] Hirschman, E.C. (2000): Consumers use of intertextuality and archetypes, in: Advances in Consumer Research, Vol. 27, eds. Hoch, S.J. and Meyer, R.J., Provo, UT Association for Consumer Research, pp. 57-63.
[29] Hirschman, E.C./Holbrook, M.B. (1982): Hedonic consumption: emerging concepts, methods and propositions, in: Journal of Marketing, Vol. 46, No. 3, pp. 92-101.
[30] Holbrook, M.B./Hirschman, E.C. (1982): The experiential aspects of consumption: Consumer fantasies, feelings, and fun, in: Journal of Consumer Research, Vol. 2 (September), pp. 132-140.
[31] Kotler, P. (2003): Marketing Management. 11th ed. Prentice Hall: NJ.
[32] Lebergott, S. (1993): Pursuing Happiness: American Consumers in the Twentieth Century. Princeton, NJ: Princeton University Press.
[33] Lee, J. (2011): Chateau Lafite Sales Tops $500,000. WSJ, September 6, http://blogs.wsj.com/scene/2011/09/06/chateau-lafite-sale-tops-500000/, accessed December 15 2011.
[34] Lindstrom, M./Kotler, P. (2005): Brand sense build powerful brands through touch, taste, smell, sight and sound. New York: Free Press.
[35] Lockshin, L./Spawton, A. (2001): Using Involvement and Brand Equity to Develop a Wine Tourism Strategy, in: International Journal of Wine Marketing, Vol. 13, No. 1, pp. 72-87.
[36] McLuhan, R. (2008): Plan now, enjoy later (experiential marketing). Marketing, February 20, pp. 33-4.
[37] Ogilvy J. (1990): This post-modern business, in: Marketing and Research Today, February, pp. 4-20.
[38] Palmer, A./Koenig-Lewis, N. (2009): An experiential, social network-based approach to direct marketing, in: Direct Marketing: An International Journal, Vol. 3, No. 3, pp. 162-176.
[39] Parker, R. (2011): The Wine Advocate Rating System. http://www.erobertparker.com/info/legend.asp, accessed October 11 2011.
[40] Pine, B.J./Gilmore, G.H. (1998): Welcome to the experience economy, in: Harvard Business Review, July/ August, pp. 68-74.
[41] Postrel, V. (2003): The Substance of Style. Harper Collins: New York, NY .
[42] Quelch, J. (2008): The Next Marketing Challenge: Selling to 'Simplifiers', in: Harvard Business School Working Knowledge, October 29, http://hbswk.hbs.edu/pdf/item/6067.pdf, accessed December 11 2011.
[43] Robinson, J. (1999): The Oxford Companion to Wine. Oxford University Press: Oxford.
[44] Robinson, J. (2011): China hosts its sales, Nov 18, http://www.ft.com/intl/cms/s/2/2901581c-100f-11e1-8211-00144feabdc0.html#axzz1gJRBaI00, accessed December 12 2011.
[45] Salsberg (2009): Japan's luxury shoppers move on. McKinsey Quarterly, August, pp. 1-8.
[46] Schmitt, B.H./Simonson, A. (1997): Marketing Aesthetics: The Strategic Management of Brands, Identity and Image. The Free Press: New York, NY.
[47] Schmitt, B. H. (1998): The power of a luxury experience. Luxury Briefing, October, http://www.exgroup.com/thought_leadership/articles/the_power_of_a_luxury_experience.pdf, accessed December 16 2011.

[48] Schmitt, B.H. (1999): Experiential Marketing, in: Journal of Marketing Management, Vol. 15, No. 1-3, pp. 53-67.
[49] Sit, J./Merrilees, B./Birch, D. (2003): Entertainment-seeking shopping center patrons: the missing segment, in: International Journal of Retail and Distribution Management, Vol. 31, No. 2, pp. 80-94.
[50] The Independent (2011): Booming China is number one Bordeaux importer, 23 September, http://www.independent.co.uk/life-style/food-and-drink/booming-china-is-number-one-bordeaux-importer-2359609.html, accessed December 18 2011.
[51] Tsaur, S./Chiu, Y./Wang, C. (2006): The visitors behavioral consequences of experiential marketing: An empirical study on Taipei Zoo, in: Journal of Travel and Tourism Marketing, Vol. 21, No. 1, pp. 25-32.
[52] Underwood, R.L./Klein, N.M./Burke, R.R. (2001): Packaging communication: attentional effects of product imagery, in: Journal of Product and Brand Management, Vol. 10, No. 7, pp. 403-22.
[53] Vickers, S.J./Renand, F. (2003): The Marketing of Luxury Goods: An Exploratory Study – Three Conceptual Dimensions, in: The Marketing Review, Vol. 3, No. 4, pp. 459-478.
[54] Williams, A.J. (2000): Consuming hospitality: learning from postmodernism, in Lashley, C., Morrison, A. (Eds), In Search of Hospitality, Butterworth-Heinemann: Oxford.
[55] Williams, A.J. (2006): Tourism and hospitality marketing: fantasy, feeling and fun, in: International Journal of Contemporary Hospitality Management, Vol. 18, No. 6, pp. 482-495.

22 Wine as Luxury Experience: A Taxonomy of Consumers Based on Best-Worst Scaling

Stefan Behrens, Klaus-Peter Wiedmann, Nadine Hennigs

22.1 Introduction

Due to the ever-growing internationalisation of traditional eating habits – including menus with more courses combined with premium drinks – wine has become an integral component of culture in many countries [7]. Particularly in widespread areas of Asia and in Middle and Eastern Europe, wine consumption has obtained socio-cultural significance and hedonistic value [43]. Hence, world consumption of wine slowly began to rise from 224 million hl in the early 1990s to an estimated 232 to 242 million hl in 2010 [49], [51].

Despite the increasing interest in wine, consumers seem to have a comparatively low level of knowledge with regard to wine; this lack of knowledge may be primarily explained by the existence of nearly 10,000 different grape varieties [56]. According to the terminology of the Nobel laureate George Akerlof, there is an asymmetric distribution of information regarding the wine market [1]. This situation prompts the important question of what alternative appraisal criteria are used by consumers to compensate for their lack of information. Potential answers vary and may include intrinsic cues (e.g., taste, aroma and alcohol content) or extrinsic cues (e.g., brand, price and country of origin) [13], [52]. Hence, this study will be of interest for researchers, managers and marketers because it seeks to measure and forecast the attributes with the greatest influence on consumers' wine choices.

According to this background, the purpose of the present study is twofold: First, to rate selective wine attributes, we apply the methodology of best-worst (BW) scaling (also known as maximum difference scaling or maxdiff), which assists in overcoming the limitations of classical scale-based surveys. The BW approach enables researchers to obtain a full ranking of product attributes based on the answers of respondents regarding their most important items ('most' or 'best') and their least important items ('least' or 'worst') on a paired preference test [16], [38]. Second, based on these results, we used the aggregated BW scores to develop a taxonomy of wine consumers according to their choice criteria for wine.

The paper is organised as follows: Section 2 briefly reviews consumer behaviour towards wine preferences before introducing the theoretical framework of the study in section 3. After providing an outline of the experimental design and the data collection with respect to the methodology of best-worst scaling in section 4, the main empirical results are discussed. In section 5, we conclude with the managerial implications and suggestions for further research in the field of wine marketing.

22.2 Theoretical Background

22.2.1 Effective Wine Marketing

Although the factors that influence consumer purchase behaviour are still not fully understood, the literature on wine perceptions provides a multitude of articles that emphasise that the implementation of effective wine marketing strategies assumes information about customers' characteristics and needs [42], [5]. Thus, given the increasingly competitive global market, the success of the wine industry primarily depends on the transition from a production orientation to a marketing orientation [39]. Pursuing a sophisticated market cultivation, Goldsmith and d'Hauteville (1998) [27] divided consumers according to their frequency of wine purchase. Their results show among other findings that heavy wine users are more enthusiastic about and interested in wine than light and non-users; thus, marketers who wish to capture the attention of heavy wine drinkers should focus their efforts on interest and involvement. Examining consumption frequency, Berni et al. (2005) [6] found that frequent drinkers, who are more knowledgeable about wines from the New World, consume red wine more frequently. In 2005, Cholette and Castaldi [14] conducted a segmentation of the US wine market by consumer types. Many Americans are *Non-Drinkers* (42.5%), whereas the dominant segment consists of *Marginal Drinkers* (46%), who consume wine primarily on special occasions. The remaining 11.5% of Americans are *Connoisseurs, Aspirants, Newcomers* and *Simple Wine Drinkers,* who belong to the *Core Wine Drinker* segment that consumes 90% of all wine. This segmentation based on the criteria of McKinna (1987) [47] emphasises the relatively low level of consumers' wine awareness [22], [40].

Therefore, many authors assume that consumers with little wine knowledge tend to rate wines primarily based on information such as expert opinions [5], [42]. Concerning this matter, Orth and Krška (2002) [52] ascertained that exhibition awards significantly influence consumers' wine purchases; this result is a sufficient reason to examine the influence of other wine attributes.

22.2.2 Wine Attributes and Consumer Choice

As noted in the introduction, wine is widely perceived as a bundle of attributes that renders the consumers' task of choosing wine as complex. The difficulty in processing this large amount of product information may lead to a high level of confusion in the (pre-)purchase phase, and this confusion might negatively affect consumers' decision-making abilities [11]. Hence, to avoid suboptimal choices, consumers attempt to reduce their confusion by concentrating on selective wine characteristics.

In this regard, researchers have determined that the attractiveness of front labels, varieties of grapes, brands and regions of origin seem to be important elements that influence the decision-making processes of wine customers [11]. Moreover, Tzimitra-Kalogianni (1999) [57], who investigated the wine preferences of Greek consumers, discovered that full taste, clarity, appellation of origin, aromas and attractive labels are the most characteristic attributes. Earlier, Keown and Casey (1995) [40] identified the countries of origin, brand names, grape varieties and regions of origin as the most important factors affecting customers' wine choices, and official quality indicators (i.e., classification) were perceived as least important.

Researchers from Australia have recently determined that the influence of wine attributes may differ considerably in each country. Thus, Goodman (2009) [29] found that taste, recommendations, grape varieties, and countries of origin tend to influence Australian, German, and French consumers, whereas Brazilian and Chinese wine drinkers rely on taste and brand names. Ultimately, Balestrini and Gamble (2006) [5] extended the geographical aspects from the regional level to the country level and discovered that the country of origin (COO) is the most influential choice criterion for Chinese wine consumers. Thus, the COO is expected to be more important than prices or brands. A similar result was presented in the study of Li et al. (2006) [43], who showed that the COO has a stronger influence when consumers evaluate wine for the purposes of gift giving and drinking in public than for consumers' own consumption.

Orth et al. (2005) [53] obtained responses from adult consumers in several states of the USA and discovered that Americans prefer wines from California, France, Italy and Australia. New World wines, which are produced outside of the traditional European countries, have been increasingly in the centre of public attention and thus have become an alternative to the Old World wines from the traditional European wine areas [10]. According to Alyward (2003) [2], who interviewed wine firms regarding their perception of the New World's innovative ability, New World wine industries are causing changes in the way wine is grown, produced and marketed.

Thus, the majority of the above-mentioned studies have focused on general purchase behaviour and have not specified the reasons for purchase or the places where consumers purchase their wine [46]. Nevertheless, Martínez-Carrasco Martínez et al. (2006) [46] found that the designations of origin and price are the most important attributes affecting wine purchases in restaurants, whereas wine purchases in shops are primarily influenced by the types of wine.

22.3 Conceptual Framework

The brief literature review is intended to create a basic understanding of the attributes that may be relevant to the wine selection process of consumers. However, not all consumers base their preferences upon all criteria [52]; thus, this paper focuses on 11 factors that were identified in previous studies as crucial for wine purchase (e.g., Goodman, 2009 [29]; Cohen et al., 2009 [17]; Drummond and Rule, 2005 [20]; Orth and Krška, 2002 [52]; Halstead, 2002 [34]; Hall and Lockshin, 2000 [32]). On the basis of this limitation, we conceptualise the framework of this paper as illustrated in **Figure 22.1**.

Figure 22.1 The Conceptual Model

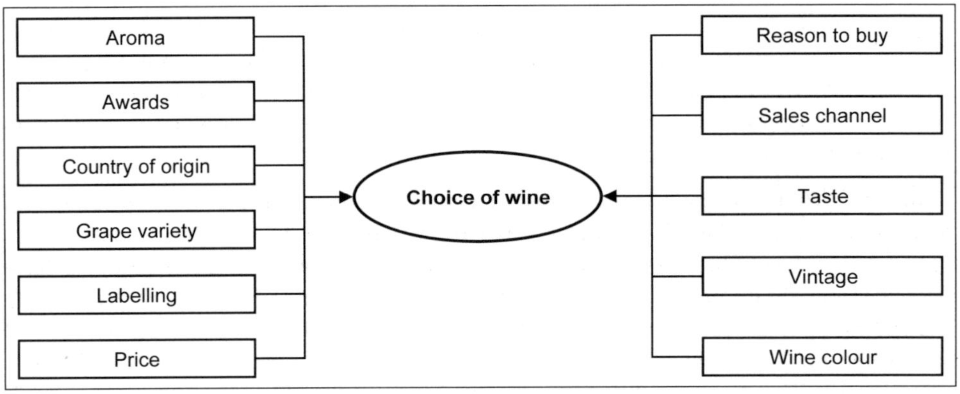

To achieve a common understanding which is of particular importance for the conducted survey, the elements of this framework can be described as follows:

Labelling. The label contains information that is regulated by law and optional information and thus acts as a communication medium between producers and consumers [55]. Most wines have front labels with basic information (e.g., alcohol content, type of wine, country of origin) and back labels with additional remarks (e.g., serving suggestions) [36], [55], [3].

Grape variety. According to botanical structuring, wines belong to the more than 60 million-year-old genus Vitis, which contains approximately 50 species [26], [36]. Each grape variety originated with a typical grape variation that influences the primary wine characteristics, such as colour, aroma and taste [26].

Taste. The taste includes all flavours perceived by the nose and the mouth and seems to be one of the most distinguishing features of wine [37], [15]. Whereas the tongue and palate only perceive the consistency of foods (e.g., viscous and thin or dry and creamy) and the four main flavours – sweet, sour, salty and bitter – the olfactory sense is able to identify fine flavour nuances [15], [54].

Aroma. The aroma structure is detected when particles of wine enter the nasal cavity and come in contact with the olfactory epithelium [26]. Thus, perceived aromas facilitate conclusions concerning producing regions, grape varieties and wine defects/olfactory defects [26], [28].

Wine colour. While wine colour is typically important for aesthetic reasons as well as for fundamental choices between red wine, white wine or rosé wine, the colour can be classified as a basic qualitative parameter on the occasion of wine degustation [37]. Thus, wine colour may provide information about the quality of vintage, grape varieties, production methods and vintage [37], [25].

Vintage. The style of wine varies each year depending on the vintage conditions and the different preparation methods of the winemaker [37], [15]. Climate particularly affects the ripeness of wine grapes, flavours and the fruit aromas in red and white wines [35].

Awards. "It is assumed that consumers rely to a great extent on easy recognizable labels or tags (i.e., medals) when choosing products (…)" (Orth and Krška, 2002, p. 387 [52]). Consequently, wine producers advertise important awards on the front or back labels to increase sales and enhance price acceptance [44].

Price. The price of bottled wine primarily depends on the major costs involved in wine production – the costs of grapes, machinery of production and packaging materials [21], [58]. Moreover, scarcity value, the quality of grape varieties and sales channels lead to higher prices [21].

Sales channel. Producers face two major decisions when choosing methods for distributing their goods: selecting a specific sales channel and deciding on the distribution intensity [8], [31]. Due to its qualitative attributes, wine can be sold via different sales channels (e.g., supermarkets, specialist shops, wineyards, etc.) to provide customers with choices regarding different purchase locations [9].

Reason to buy. "(…) several key studies have suggested that the main determinant that affects consumers when making wine purchasing decisions is the occasion at which the wine will be consumed." (Halstead, 2002, p. 10 [34]). In fact, the reason to buy, understood as a concrete purchase situation (e.g., wine as a present, wine for private consumption, etc.) seems to influence the willingness to pay and the selection of specific types of wine [33], [34].

Country of origin. Geographical declarations annotate an identity and guarantee the origin of wines that are produced in certain countries or regions [12], [4]. The indication of origin allows consumers to draw conclusions regarding the ambiance and location of wines and thus offers clues regarding the characteristics of wines [9], [41]. Given the importance of the COO as a wine choice criterion, a main focus of our empirical study is the preference of wines from the Old World versus the New World.

Drawing from prior findings and the conceptual framework, one main research question will be analysed using the best-worst method:

> RQ: Which of the presented criteria exerts the strongest influence upon consumers' wine choices, and what consumer segments can be distinguished according to these criteria?

22.4 Methodology

The instrument

For the purposes of our study focusing on consumer wine choices [30], following the suggestions of Cohen and Markowitz (2002) [19], Finn and Louviere (1992) [23], and Marley and Louviere (2005) [45], we decided to use the best-worst scaling (BWS) method. This approach assumes some underlying subjective interest and measures the degree of importance that respondents assign to selected (three or more) items. More specifically, the respondents are provided choice sets to choose one most important (BEST) and one least important (WORST) item in each set that is presented to them [19], [18]. As respondents are necessarily required to make trade-offs between items and benefits [16], BWS is a more discriminating method for measuring attribute importance compared with either rating scales or paired comparison methods [48], [18], [24].

For a detailed discussion of the method and its application in the wine sector, see Goodman, Lockshin and Cohen (2005) [30] and the study overview in **Table 22.1**.

Table 22.1 Existing Research Insights – Best-Worst Studies in Marketing

Author	Sales Channel	Country	n	Most Important	Least Important
Casini, L. et al. (2009)	Retail Stores	Italy	314	Tasted the wine previously Matching food Origin of the wine	An attractive front label Alcohol level below 13% Promotional display in-store
	Restaurants		308	I matched it to my food I have had the wine before, and I liked it I had read about it but never tasted it	Alcohol level below 13% Available in half bottle (375 ml) Promotion card on the table
Cohen, E. (2009)	Restaurants	Australia	283	I have had the wine before, and I liked it I matched it to my food Try something different	Varietal Available in half bottle (375 ml) Alcohol level below 13%
		France	147	I matched it to my food I have had the wine before, and I liked it I had read about it but never tasted it	Varietal Alcohol level below 13% Suggestion on the menu
		United Kingdom	304	I have had the wine before, and I liked it I matched it to my food Suggested by another person at the table	Promotion card on the table Available in half bottle (375 ml) Alcohol level below 13%

Author	Sales Channel	Country	n	Most Important	Least Important
Goodman, S. (2009)	Retail Stores	Australia	305	Tasted the wine previously Someone recommended it Grape variety	An attractive front label Promotional display in-store Alcohol level below 13%
		Austria	182	Grape variety Tasted the wine previously Origin of the wine	Information on the shelf Alcohol level below 13% Promotional display in-store
		Brazil	293	Brand name Tasted the wine previously Matching food	Promotional display in-store Medal/award An attractive front label
		China	197	Brand name Tasted the wine previously Origin of the wine	An attractive front label Matching food Alcohol level below 13%
		France	154	Matching food Origin of the wine Tasted the wine previously	An attractive front label Information on the shelf Alcohol level below 13%
		Germany	160	Tasted the wine previously Someone recommended it Matching food	Brand name Promotional display in-store Alcohol level below 13%
		Israel	184	Tasted the wine previously Matching food Someone recommended it	An attractive front label Information on the shelf Alcohol level below 13%
		Italy	314	Tasted the wine previously Matching food Origin of the wine	An attractive front label Alcohol level below 13% Promotional display in-store

Author	Sales Channel	Country	n	Most Important	Least Important
		New Zealand	364	Tasted the wine previously Grape variety Medal/award	An attractive front label Information on the shelf Alcohol level below 13%
		Taiwan	317	Tasted the wine previously Someone recommended it Origin of the wine	Information on the shelf An attractive front label Alcohol level below 13%
		United Kingdom	303	Tasted the wine previously Someone recommended it An attractive front label	Promotional display in-store Medal/award Alcohol level below 13%
		USA	196	Tasted the wine previously Someone recommended it Grape variety	Information on back label Promotional display in-store Alcohol level below 13%
Wiedmann, K.-P. et al. (2011)	Specialist Shops	Germany	108	Taste Country of origin Price	Vintage Labelling Sales channel
Yu, Y. et al. (2009)	Typical Consumers	China	197	Tasted the wine previously Origin of the wine Brand name	An attractive front label Matching food Alcohol level below 13%
	University Students		122	Brand name Origin of the wine Tasted the wine previously	An attractive front label Promotional display in-store Alcohol level below 13%

Based on the insights described above, **Figure 22.2** shows an exemplary choice set which we applied within the questionnaire to ask for an evaluation of the 11 above presented wine choice criteria (1=Labelling, 2=Grape variety, 3=Taste, 4=Aroma, 5=Wine colour, 6=Vintage, 7=Awards, 8=Price, 9=Sales channel, 10=Reason to buy, 11=Country of origin).

Figure 22.2 Example of Best-Worst Scaling (Wine choice factors)

Please choose the most and the least important Wine choice factors in an everyday choice situation by ticking 'Most/ Best' and 'Least/ Worst' per all choice set.

Choice set	**Least/ Worst**	Wine choice factor	**Most/ Best**
1	☐	Labelling	☐
	One answer → ☒	Country of origin	☐
	☐	Wine colour	☒ ← **One answer**
	☐	Awards	☐

The same design, presented in **Figure 22.3** was adapted in a second step to request the importance of 11 wine-producing countries [50], including the ten main producers and Austria, which is of critical interest within the German-speaking area (1=Italy, 2=France, 3=Spain, 4=USA, 5=Argentina, 6=Australia, 7=China, 8=Germany, 9=South Africa, 10=Chile, 11=Austria).

Figure 22.3 Example of Best-Worst Scaling (Country of origin)

Please choose the most and the least important Country of origin for wine in an everyday choice situation by ticking 'Most/ Best' and 'Least/ Worst' per all choice set.

Choice set	**Least/ Worst**	Country of origin	**Most/ Best**
1	☐	France	☐
	One answer → ☒	USA	☐
	☐	Spain	☒ ← **One answer**
	☐	Italy	☐

With regard to the 11 wine choice criteria and the 11 wine producing countries all respondents were asked to evaluate each 12 different choice sets with four, six or eight items; this method led to the design that is presented in **Figure 22.4**.

The choice set design was adopted from Finn and Louviere (1992) [23] to ensure that each item appears 6 times across all choice sets and can be evaluated as most important (BEST) or least important (WORST). The level of importance for each choice is determined by subtracting the negative evaluations from the positive evaluations and depends on the number of respondents and the frequency with which each item appears in the choice sets [30]. The total number of times that an attribute is mentioned as the most important (BEST) minus the total number of times that it is mentioned as the least important (WORST) results in the total best-worst (b-w) score [29]. The results are presented below.

Figure 22.4 Design for Choice Sets

	Choice set number												
Issues	**1**	**2**	**3**	**4**	**5**	**6**	**7**	**8**	**9**	**10**	**11**	**12**	**Appea-rences**
1	0	0	1	0	0	1	1	1	0	1	0	1	**6**
2	1	0	0	1	0	0	0	1	1	1	0	1	**6**
3	0	1	0	0	1	0	1	0	1	1	0	1	**6**
4	1	0	1	0	0	0	1	0	1	1	1	0	**6**
5	0	0	1	0	1	0	0	1	1	0	1	1	**6**
6	0	0	0	1	0	1	1	0	1	0	1	1	**6**
7	0	1	1	1	0	0	0	0	0	1	1	1	**6**
8	1	1	0	0	0	0	1	1	0	0	1	1	**6**
9	0	0	0	1	1	0	1	1	0	1	1	0	**6**
10	1	0	0	0	1	1	0	0	0	1	1	1	**6**
11	0	1	0	0	0	1	0	1	1	1	1	0	**6**
Issues in set	**4**	**4**	**4**	**4**	**4**	**4**	**6**	**6**	**6**	**8**	**8**	**8**	

Note: 1 indicates that an issue is present in a set; 0 indicates that the issue is not present in a set.

The sample

The people targeted in the survey were wine consumers that answered the questionnaire due to their personal interest in wine; the convenience sample consisted of 108 respondents who were recruited during an everyday purchase situation via personal interviews in German wine specialist shops in 2010. Personal interviews were perceived as adequately because they open the possibility to clarify comprehenison problems by explaining respondents the requested wine choice criteria in accordance with the underlying conceptual framework. Further information regarding the sample characteristics are provided in **Table 22.2**.

Table 22.2 Characteristics of the Sample

Variable	n	%
Gender		
Male	54	50.0%
Female	54	50.0%
Age		
< 35 years (male)	27	25.0%
< 35 years (female)	27	25.0%
≥ 35 years (male)	27	25.0%
≥ 35 years (female)	27	25.0%
Marital status		
Single	57	52.8%
Married	43	39.8%
Widowed	1	0.9%
Divorced	7	6.5%
Education		
Did not graduate from high school	1	0.9%
Lower secondary school	6	5.6%
Intermediate secondary school	24	22.2%
A-levels	46	42.6%
University degree	31	28.7%
Occupation		
Self-employed	4	3.7%
Employee	37	34.3%
Worker	4	3.7%
Civil servant	17	15.7%
Homemaker	6	5.6%
Retiree	9	8.3%
Pupil	2	1.9%
Student	27	25.0%
Formation	1	0.9%
Not employed	1	0.9%
Household income		
< 500 EUR	3	2.8%
500 EUR < 1000 EUR	17	15.7%
1000 EUR < 1500 EUR	4	3.7%

Variable	n	%
1500 EUR < 2000 EUR	10	9.3%
2000 EUR < 3000 EUR	23	21.3%
3000 EUR < 4000 EUR	14	13.0%
> 4000 EUR	13	12.0%
Not specified	24	22.2%
Wine consumption		
Regularly (several times a week)	19	17.6%
Frequently (once a week)	26	24.1%
Occasionally (twice a month)	38	35.2%
Rarely (once a month)	15	13.9%
Very rarely (less than once a month)	10	9.3%

In our sample, we attained the goal of a 50/50 gender distribution with an equal distribution across all age groups. With reference to wine consumption, 76.9% of the respondents consume wine several times a month.

22.5 Results and Discussion

Based on the given choice sets and the method described above, the results revealed the scores as presented in **Table 22.3**.

Table 22.3 Importance of Choice Factors to German Wine Consumers (n=108)

Choice factor	Total best	Total worst	B-W score
Taste	442	1	441
Country of origin	192	46	146
Price	114	42	72
Grape variety	148	76	72
Aroma	93	62	31
Awards	84	82	2
Wine colour	64	92	-28
Reason to buy	82	131	-49
Vintage	48	115	-67
Labelling	13	249	-236
Sales channel	16	400	-384

The taste of the wine and the country of origin were shown to be the most important choice criteria followed by price and grape variety. The respondents rated the sales channel and the label as the least important choice criteria. With reference to the existing research insights as summarised in **Table 22.1**, the results of our study show that wine consumers perceive intrinsic cues to be nearly as important as extrinsic cues.

Regarding the importance assigned to the country of origin, our results as shown in **Tabelle 22.4** revealed that France was the most important country followed by Italy, Germany and Spain. In contrast, Austria, the USA and China were perceived to be less important as wine-producing countries.

Tabelle 22.4 Importance of Countries of Origin to German Wine Consumers (n=108)

Country of Origin	Total best	Total worst	B-W score
France	349	7	342
Italy	272	9	263
Germany	243	18	225
Spain	215	9	206
South Africa	77	56	21
Australia	42	70	-28
Chile	39	109	-70
Argentina	29	123	-94
Austria	18	134	-116
USA	12	231	-219
China	0	530	-530

To develop our taxonomy of wine consumers according to their preferred choice criteria, we aggregated the mean scores of the 11 wine choice criteria and the 11 wine-producing countries to create the following aggregated set of variables encompassing four dimensions as the basis of our cluster analysis: *Extrinsic Cues* (labelling, price, country of origin, awards, sales channel, and vintage), *Intrinsic Cues* (taste, wine colour, grape variety, aroma, and reason to buy), *Old World* (France, Germany, Austria, Italy, and Spain), and *New World* (Argentina, Australia, Chile, China, South Africa, and the USA). Our classification into these categories relies on Cohen (2009) [16] who suggests that consumers' choice for wine is based on extrinsic cues such as price, brand, region, medal, variety and intrinsic cues such as taste, bouquet and other sensory characteristics of the wine that can only be assessed during consumption. The ratings referred to 12 different choice sets; thus, the values range from 0 to 12.

In our analysis, we used a combination of an initial hierarchical clustering procedure and non-hierarchical k-means clustering. Because it produces tight minimum variance clusters and is regarded as one of the best of the hierarchical clustering techniques, Ward's method of minimum variance was chosen to check the cluster differences in each stage of combinations and to maximize homogeneity within and heterogeneity between clusters. The results strongly suggested the presence of four clusters. Using non-hierarchical k-means clustering, this four-cluster solution as presented in **Table 22.5** produced the most interpretable and stable results. With regard to classification accuracy, a discriminant analysis validated the results of the cluster analysis; as presented in **Table 22.6**, the correct rate of the classification was 100.0%.

Table 22.5 Results of the cluster analysis

	Cluster 1 The Balanced Traditionalists	Cluster 2 The Traditional Tasters	Cluster 3 The Traditional Observers	Cluster 4 The Innovative Tasters	F	sig
Cluster characteristics						
Cluster size (% of the sample)	31.5%	34.3%	17.6%	16.7%		
Mean age (years)	40.0	40.3	39.2	33.2		
Gender: male	41.2%	56.8%	57.9%	44.4%		
Gender: female	58.8%	43.2%	42.1%	55.6%		
Old World	**11.21**	**11.30**	**11.32**	**4.61**	**120.212**	**.000**
France	**3.38**	**3.65**	**3.89**	1.39		
Germany	2.82	2.30	2.47	0.83		
Austria	0.12	0.27	0.16	0.06		
Italy	2.71	2.89	3.05	0.83		
Spain	2.18	2.19	1.74	**1.50**		
New World	**0.79**	**0.70**	**0.68**	**7.39**	**120.212**	**.000**
Argentina	0.21	0.08	0.11	0.94		
Australia	0.12	0.14	0.21	1.61		
Chile	0.09	0.11	0.05	1.72		
China	0.00	0.00	0.00	0.00		
South Africa	**0.32**	**0.32**	**0.32**	**2.67**		
USA	0.06	0.05	0.00	0.44		
Extrinsic Cues	**4.97**	**1.95**	**8.42**	**3.67**	**131.087**	**.000**
Labelling	0.09	0.03	0.37	0.11		
Price	0.85	0.49	**2.84**	0.72		
Country of origin	**2.28**	**1.05**	2.37	**1.50**		

	Cluster 1 The Balanced Traditionalists	Cluster 2 The Traditional Tasters	Cluster 3 The Traditional Observers	Cluster 4 The Innovative Tasters	F	sig
Awards	1.12	0.19	1.16	0.94		
Sales channel	0.15	0.03	0.53	0.00		
Vintage	0.38	0.16	1.16	0.39		
Intrinsic Cues	**7.03**	**10.05**	**3.58**	**8.33**	**131.087**	**.000**
Taste	**4.21**	**5.22**	**1.63**	**4.17**		
Wine colour	0.53	0.68	0.26	0.89		
Grape variety	1.41	1.43	1.00	1.56		
Aroma	0.62	1.51	0.11	0.78		
Reason to buy	0.26	1.22	0.58	0.94		

Table 22.6 Results of the discriminant analysis

Discriminant Function	Eigenvalue	Canonical Correlation	Wilk's Lambda	χ^2	Sig
1	20.996	.977	.013	447.971	0.000
2	2.093	.823	.284	129.611	0.000
3	.138	.348	.879	13.311	0.001

	Function 1	Function 2	Function 3
Centroids (group means)			
Cluster 1	-2.036	.859	-.462
Cluster 2	-2.010	.748	.438
Cluster 3	9.729	-.067	-.002
Cluster 4	-2.294	-3.089	-.025
Significant variable (structure matrix)			
Old World	.073	.990	.119
New World	-.045	-.614	-.074
Extrinsic Cues	.877	-.211	-.051
Intrinsic Cues	-.273	.010	.959

Classification matrix revealed that 100 % of the cases were classified correctly.

Based on consumer preferences as illustrated in **Figure 22.5**, the four clusters can be described as follows:

Figure 22.5 Taxonomy of Wine Consumers According to Their Choice Criteria

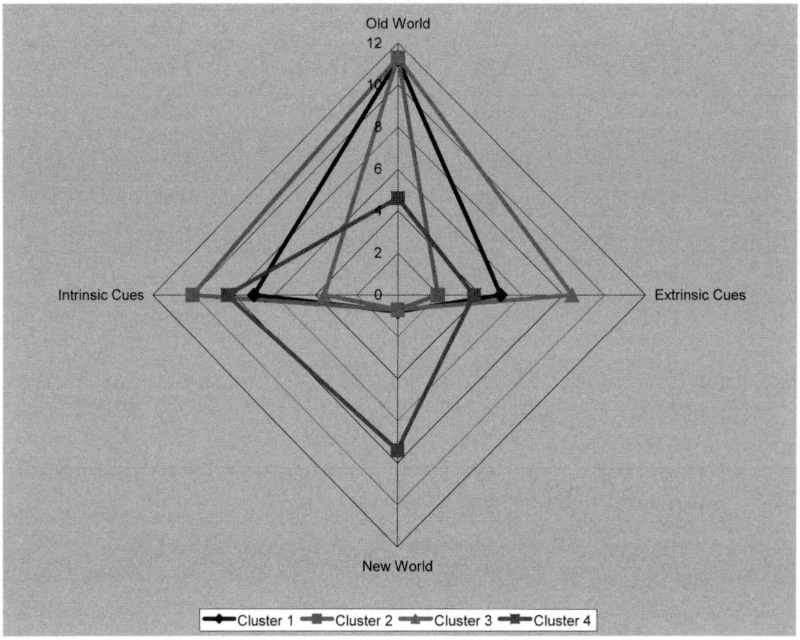

The Balanced Traditionalists (Cluster 1) form 31.5% of the sample with a mean age of 40.0 years; the respondents were 58.8% female and 41.2% male. Typical consumers in this cluster prefer to buy and consume wines from Old World countries (i.e., France, Germany, and Italy). Additionally, even if these consumers' wine choice decision schemes are more balanced compared with the decision schemes for other segments, consumers in this cluster perceive intrinsic cues (i.e., taste and grape variety) to be slightly more important than extrinsic cues like the country of origin and awards

The Traditional Tasters (Cluster 2) form 34.3% of the sample with a mean age of 40.3 years; the respondents were 43.2% female and 56.8% male. Similar to consumers in Cluster 1 and Cluster 3, consumers in Cluster 2 prefer wines from Old World countries (i.e., France, Italy, and Germany). Typical consumers in this cluster state that their wine choices are primarily based upon intrinsic cues (i.e., taste, aroma, and grape variety); this tendency is more prominent in this cluster compared with other clusters.

The Traditional Observers (Cluster 3) form 17.6% of the sample with a mean age of 39.2 years; the respondents were 42.1% female and 57.9% male. With regard to the country of origin, consumers from this cluster prefer wines from Old World countries (i.e., France, Italy, and Germany). In addition, they perceive extrinsic cues (i.e., price, country of origin, awards, and vintage) to be the most important attributes influencing their wine choices.

The Innovative Tasters (Cluster 4) form 16.7% of the sample with a mean age of 33.2 years; the respondents were 55.6% female and 44.4% male. This segment prefers New World wines (i.e., South Africa, Chile, and Australia) over those from the Old World. Besides, typical consumers in this segment perceive intrinsic cues (i.e., taste, grape variety, and wine colour) to be more important than extrinsic cues.

From a research and managerial perspective, the cluster analysis results revealing four groups of consumers based on their perception of wine choice criteria may help to gain a better understanding of wine consumers and to develop more effective segmentation and efficient positioning strategies aiming at targeting these diverse consumer segments. In this context, suggestions for further research and managerial practice are presented in the following section.

22.6 Managerial Implications and Further Research Steps

Given the variety of criteria that influence wine consumers' decision-making processes and in light of the taxonomy of wine consumers as described above, there are a number of implications for researchers and practitioners in the field of wine marketing. In an attempt to explore the fundamental motivation behind consumer choice, the main purpose of this study was to apply the methodology of BW scaling to gain insights in the attributes that influence the preference, perception, and behaviour of wine consumers. In an industry context, where consumers' product knowledge is comparatively low, a clear understanding of consumers' needs and behaviours and the way consumers combine intrinsic and extrinsic signals is an appropriate basis for developing effective strategies in order to address different segments of consumers appropriately.

Our study results based on BW scaling provide evidence that consumers use a bundle of product characteristics in their decision making. Apart from the ever-present extrinsic cues in wine marketing like country of origin and price, intrinsic cues like taste, grape variety, and aroma are significant aspects to address; for some consumers (the 'tasters'), the intrinsic cues are rated to be even more important than the extrinsic cues. In this context, consumers should be allowed to experience the physical properties of a wine – its smell, its colour and above all its taste. Wine tasting and the information regarding suitable dishes in retail stores or wine stores may increase consumers' wine-related interest and desire to buy.

Furthermore, the results from our German sample show that consumers have no specific preferences for domestic wines. Consequently, the reputations of local wine regions and the images of the wines produced by local viticulturists and winegrowers should be presented to a broader audience (e.g., with wine seminars, trips to wine regions and vineyards in collaboration with wine merchants and tourist offices). Nevertheless, wine tasting and other events require consumer-sided involvement; thus, less-involved consumer segments must be addressed explicitly by direct, personal and sensorial experiences in the context of wine consumption. Regarding other wine-producing countries, our results reveal that traditional countries of origin (e.g., France, Italy and Spain) are widely accepted, whereas the nations of the New World have only a subordinate role in the perception of consumers and appeal especially to younger consumer groups like cluster 4 (the innovative tasters). The negative assessment of the other countries (excluding South Africa) even suggest that conservative German wine drinkers may perceive the country of origin of New World wines to be a non-purchase rather than a pro-purchase criterion. Thus, a more comprehensive marketing strategy for imported wines and advertisements based on the superior reputations of well-established wine regions may transfer to the products from the New World countries. Considering linguistic-cultural analogies, we suggest a possible approach in which the products of Spanish, Chilean and Argentine winegrowers are bundled based on collective marketing efforts and event endorsements. In the German market, a combined offer of domestic and Austrian wines may be considered. An additional approach may be to identify the consumer groups who have the ability to influence other consumers in their wine-buying decisions via word-of-mouth communication. Consulting wine mavens and opinion leaders whose perceptions of important wine-growing nations are very similar to the perceptions of other consumers may be an important approach to address other consumer segments who are less interested in wine and related marketing communications. Consequently, marketers need to identify and target these wine mavens and opinion leaders to promote recommendation processes adequately. With regard to future research steps, as the BW method is independent from the cultural background of the respondents [30], it allows the extension of our approach and the resulting taxonomy to a cross-national study on the attributes that influence consumers' wine choices. Thus, by enlarging the size of the sample and collecting data in different countries, the results of this study could be compared on an international level to further explore the issues of product cues within the wine industry and to assess if our suggested taxonomy of wine consumers can be confirmed across national borders.

References

[1] Akerlof, G.A. (1979): The Market for 'Lemons': Quality Uncertainty and the Market Mechanism; in: The Quarterly Journal of Economics, Vol. 84, No. 3, pp. 488-500.

[2] Alyward, D.K. (2003): A Documentary of Innovation Support Among New World Wine Industries, http://ro.uow.edu.au/cgi/viewcontent.cgi?article=1008&context=commpapers&sei-redir=1#search="alyward+2003+new+world (Access on 10/04/2011).

[3] Ambrosi, H. (1976): Where the Great German Wines Grow, New York.

[4] Anderson, K. (2004): The World's Wine Markets: Globalization at Work, Cornwall.

[5] Balestrini, P./Gamble, P. (2006): Country-of-origin effects on Chinese wine consumers; in: British Food Journal, Vol. 108, No. 5, pp. 396-412.

[6] Berni, P./Begalli, D./Capitello, R. (2005): An Occasion-Based Segmentation Approach to the Wine Market in Denmark; in: Journal of International Food & Agribusiness Marketing, Vol. 17, No. 1, pp. 117-145.

[7] Bisson, L.F./Waterhouse, A.L./Ebeler, S.E./Walker, M.A./Lapsley, J.T. (2002): The present and future of the international wine industry, in: Nature, Vol. 418, 8 August, pp. 696-699.

[8] Boone, L.E./Kurtz, D.L. (2011): Contemporary Business: 2010 Update Edition, 13th Edition, New Jersey.

[9] Brostrom, G./Brostrom, J. (2009): The Business of Wine: An Encyclopedia, Westport.

[10] Campbell, G./Guibert, N. (2006): Old World strategies against New World competition in a globalising wine industry; in: British Food Journal, Vol. 108, No. 4, pp. 233-242.
[11] Casini, L./Corsi, A.M./Goodman, S. (2009): Consumer preferences of wine in Italy applying best-worst scaling; in: International Journal of Wine Business Research, Vol. 21, No. 1, pp. 64-78.
[12] Charters, S. (2006): Wine and Society: The Social and Cultural Context of a Drink, Oxford.
[13] Charters, S./Pettigrew, S. (2007): The dimensions of wine quality; in: Food Quality and Preferences, Vol. 18, No. 7, pp. 997-1007.
[14] Cholette, S./Castaldi, R. (2005): Analyzing the US Retail Wine Market Using Price and Consumer Segmentation Models (Refereed), http://academyofwinebusiness.com/wp-content/uploads/2010/05/AnalyzingUSRetail.pdf (Access on 10/04/2011).
[15] Clarke, R.J./Bakker, J. (2004): Wine Flavour Chemistry, Oxford.
[16] Cohen, E. (2009): Applying best-worst scaling to wine marketing; in: International Journal of Wine Business Research, Vol. 21, No. 1, pp. 8-23.
[17] Cohen, E./d'Hauteville, F./Sirieix, L. (2009): A cross-cultural comparison of choice criteria for wine in restaurants; in: International Journal of Wine Business Research, Vol. 21, No. 1, pp. 50-63.
[18] Cohen, S.H. (2003): Maximum Difference Scaling: Improved Measures of Importance and Preference for Segmentation, http://www.sawtoothsoftware.com/download/techpap/maxdiff.pdf (Access on 11/04/2011).
[19] Cohen, S.H./Markowitz, P. (2002) "Renewing Market Segmentation: Some new tools to correct old problems," ESOMAR 2002 Congress Proceedings, ESOMAR: Amsterdam, The Netherlands, pp. 595-612.
[20] Drummond, G./Rule, G. (2005): Consumer confusion in the UK wine industry; in: Journal of Wine Research, Vol. 16, No. 1, pp. 55-64.
[21] Edwards, F./Spawton, T. (1990): Pricing in the Australian Wine Industry; in: European Journal of Marketing, Vol. 24, No. 4, pp. 11-17.
[22] Felzensztein, C./Hibbert, S./Vong, G. (2004): Is the country of origin the fifth element in the marketing mix of imported wines? A critical review of the literature; in: Journal of Food Products Marketing, Vol. 10, No. 4, pp. 73-84.
[23] Finn, A./Louviere, J.J. (1992): Determining the Appropriate Response to Evidence of Public Concern: The Case of Food Safety, Vol. 11, No. 2, pp. 12-25.
[24] Finn, A./Louviere, J.J./Williams, M. (1993): Quantifying Public Priorities: Best-Worst Analysis Versus Conventional Ratings Data. Working Paper.
[25] Fischer, J.R. (2006): Wine and Food-101: A Comprehensive Guide to Wine and the Art of Matching Wine with Food, Bloomington.
[26] Gibson, M. (2010): The Sommelier Prep Course: An Introduction to the Wines, Beers, and Spirits of the World, New Jersey.
[27] Goldsmith, R.E./d`Hauteville, F. (1998): Heavy wine consumption: Empirical and theoretical perspectives; in: British Food Journal, Vol. 100, No. 4, pp. 184-194.
[28] Goode, J. (2005): The Science of Wine: From Vine to Glass, Berkeley and Los Angeles.
[29] Goodman, S. (2009): An international comparison of retail consumer wine choice; in: International Journal of Wine Business Research, Vol. 21, No.1, pp. 41-49.
[30] Goodman, S./Lockshin, L./Cohen, E. (2005): Best-Worst Scaling: A Simple Method to Determine Drinks and Wine Styles, http://www.winepreferences.com/resources/page28/files/page28_1.pdf (Access on 10/04/2011).
[31] Gorchels, L./West, C./Marien, E.J. (2004): The Manger's Guide to Distribution Channels, New York.
[32] Hall, J./Lockshin, L. (2000): Using Means-End Chains for Analysing Occasions – Not Buyers, http://smib.vuw.ac.nz:8081/www/ANZMAC1999/Site/H/Hall_Lockshin.pdf (Access on 10/04/2011).
[33] Hall, J./Winchester, M. (2001): Empirical Analysis of Spawton's (1991) Segmentation of the Australian Wine Market; in: Asia Pacific Advances in Consumer Research, Vol. 4, pp. 319-327.

[34] Halstead, L. (2002): How do Consumers Select Wine? Factors that Affect the Purchase Decision Making Process in the Wine Category, http://www.wineintelligence.com/pdfs/How_do_consumers_select_wine.pdf (Access on 22/10/2009).
[35] Harrington, R.J. (2008): Food and Wine Pairing: A Sensory Experience, New Jersey.
[36] Jackson, R.S. (2008): Wine Science: Principles and Applications, 3rd Edition, San Diego.
[37] Jackson, R.S. (2009): Wine Tasting: A Professional Handbook, 2nd Edition, San Diego.
[38] Jaeger, S.R./Cardeloo, A.V. (2009): Direct and indirect hedonic scaling methods: A comparison of the labelled affective magnitude (LAM) scale and best-worst scaling; in: Food Quality and Preferences, Vol. 20, No. 2, pp. 249-258.
[39] Johnson, R./Bruwer, J. (2007): Regional brand image and perceived wine quality: the consumer perspective; in: International Journal of Wine Business Research, Vol. 19, No. 4, pp. 276-297.
[40] Keown, C./Casey, M. (1995): Purchasing behaviour in the Northern Ireland wine market; in: British Food Journal, Vol. 97, No. 1, pp. 17-20.
[41] Kramer, M. (2003): Making Sense of Wine, Philadelphia.
[42] Lee, K./Zhao, J./Ko, J.-Y. (2005): Exploring the Korean Wine Market; in: Journal of Hospitality & Tourism Research, Vol. 29, No. 1, pp. 20-41.
[43] Li, L./Hu, X./Jun, Z. (2006): Do Chinese Consumer Prefer to Buy Imported Wine? – the Effect of Country-of-Origin, http://faculty.washington.edu/karyiu/confer/beijing06/papers/zhuo-li-hu.pdf (Access on 10/04/2011).
[44] Lockshin, L./Jarvis, W./d'Hauteville, F./Perrouty, J.-P. (2006): Using simulations from discrete choice experiments to measure consumer sensitivity to brand, region, price, and awards in wine choice; in: Food Quality and Preference, Vol. 17, No. 3-4, pp. 166-178.
[45] Marley, A.A.J./Louviere, J.J. (2005): Some probabilistic models of best, worst, and best-worst choices; in: Journal of Mathematical Psychology, Vol. 49, No. 6, pp. 464-480.
[46] Martínez-Carrasco Martínez, L./Brugarolas Mollá-Bauzá, M./del Campo Gomis, F.J./Martínez Poveda, A. (2006): Influence of purchase place and consumption frequency over quality wine preferences; in: Food Quality and Preferences, Vol. 17, No. 5, pp. 315-327.
[47] McKinna, D. (1987): Marketing Strategies for wine. Paper presented at the Victorian Wine Association.
[48] Mueller, S./Lockshin, L./Louviere, J./Hackman, D. (2005): Do Respondents use Extra Information Provided in Online Best-Worst Choice Experiments?, http://conferences.anzmac.org/ANZMAC2007/papers/Mueller_1.pdf (access on 10/04/2011).
[49] OIV (2010a): Global Vitiviniculture Economic Data, http://news.reseau-concept.net/images/oiv_uk/Client/Point_de_conjoncture_octobre2010_EN.pdf (Access on 10/04/2011).
[50] OIV (2010b): World Statistics: 8th General Assembly of the OIV Tibilisi 2010, http://news.reseau-concept.net/images/oiv_uk/Client/DIAPORAMA_STATISTIQUES_Tbilissi_2010_EN.ppt (Access on 10/04/2011).
[51] OIV (2011): State of the Vitiviniculture World Market, http://news.reseau-concept.net/images/oiv_uk/Client/2011_note_conj_mars_EN.pdf (Access on 10/04/2011).
[52] Orth, U.R./Krška, P. (2002): Quality signals in wine marketing: the role of exhibition awards; in: International Food and Agribusiness Management Review, Vol. 5, No. 4, pp. 385-397.
[53] Orth, U.R./McGarry Wolf, M./Dodd, T.H. (2005): Dimensions of wine region equity and their impact on consumer preferences; in: Journal of Product & Brand Management, Vol. 14, No. 2, pp. 88-97.
[54] Peynaud, E./Blouin, J. (1996): The Taste of Wine: The Art and Science of Wine Appreciation, Paris.
[55] Poyet, E. (2004): The Little Black Book of Wine: A Simple Guide to the World of Wine, New York.
[56] This, P./Lacombe, T./Thomas, M.R. (2006): Historical origins and genetic diversity of wine grapes; in: Trends in Genetics, Vol. 22, No. 9, pp. 512-519.
[57] Tzimitra-Kalogianni, I./Papadaki-Klavdianou, A./Alexaki, A./Tsakiridou, E. (1999): Wine routes in Northern Greece: consumer perceptions; in: British Food Journal, Vol. 101, No. 11, pp. 884-892.
[58] Wells, W.D./Prensky, D. (1996): Consumer Behavior, New York.
[59] Yu, Y./Sun, H./Goodman, S./Ma, H. (2009): Chinese choices: a survey of wine consumers in Beijing; in: International Journal of Wine Business, Vol. 21, No. 2, pp. 155-168.